STUDIES IN THEORETICAL PHILOSOPHY

Herausgegeben von Tobias Rosefeldt
und Benjamin Schnieder

in Zusammenarbeit mit

Elke Brendel (Bonn)
Tim Henning (Gießen)
Max Kölbel (Barcelona)
Hannes Leitgeb (München)
Martine Nida-Rümelin (Fribourg)
Christian Nimtz (Bielefeld)
Thomas Sattig (Tübingen)
Jason Stanley (New Brunswick)
Marcel Weber (Genf)
Barbara Vetter (Berlin)

vol. 2

VITTORIO KLOSTERMANN

NICK HAVERKAMP

Intuitionism vs. Classicism

A Mathematical Attack on Classical Logic

VITTORIO KLOSTERMANN

Bibliographische Information der Deutschen Nationalbibliothek
Die Deutsche Nationalbibliothek verzeichnet diese Publikation in der Deutschen
Nationalbibliographie; detaillierte bibliographische Daten sind im Internet über
http://dnb.dnb.de abrufbar.

Gedruckt auf Alster Werkdruck der Firma Geese, Hamburg,
alterungsbeständig ∞ ISO 9706 und PEFC-zertifiziert.
Druck: betz-druck GmbH, Darmstadt
Printed in Germany
ISSN 2199-5214
ISBN 978-3-465-03906-8

For Esther

Contents

Preface

This book is based on my dissertation *Intuitionistic Arguments Against Classical Logic*, which I wrote as a member of the Phlox research group between 2008 and 2012. For most of that time, I did not know I would write on this topic. Originally, I wanted to write about the semantics of explanation. My intent was to transfer techniques and results developed by relevant logicians to the semantical study of the sentential operator "because". However, this project failed. It took me quite a while to realise, but eventually I came to believe that implication and explanation are too dissimilar to use the achievements of relevant logicians for developing a semantics for "because". A longer period followed during which I had no clear idea which topic I wanted to take up instead, and, in 2009, I was happy to be distracted from philosophy by the birth of my first child and the ensuing chaos. When I started to do philosophy again, I realised that there was one aspect of the relevantists' project that puzzled me: They present a *coherent* criticism of *fundamental logical principles*, which I had always thought of as being too basic to allow for fruitful debate. I decided that on my second attempt I wanted to write about arguments against classical logic. Since I had (more or less) made up my mind about relevant logic already, I first turned to anti-classical arguments based on considerations about paradoxes of truth and then to the critique launched by mathematical intuitionists. The more I thought about intuitionism, the clearer it became to me that I wanted to focus my dissertation solely on this topic. Intuitionism is intimately related to fundamental questions about the nature of mathematics, epistemology, and semantics, and it stands out as one of a small number of enterprises whose adherents claim that they can rigorously establish the falsity of central classical logical principles. In my view, the conflict between classicists and intuitionists is exceptionally deep and far-reaching, and it is certainly one of the most radical disputes that can be pursued by rational means. Therefore, I am very happy to have been given the opportunity to develop my view on the nature of this argument.

Many friends and colleagues have made helpful comments on earlier drafts of my dissertation and this book. I am very grateful to all of them. In particular, I would like to thank: Sebastian Bünker, Miguel Hoeltje, Stephan Krämer, Raphael van Riel, Benjamin Schnieder, Moritz Schulz, Alexander Steinberg, and the members of Benjamin Schnieder's Research Colloquium in Hamburg, where I was able to discuss the second chapter of my dissertation.

A special thank you goes to the examiners of my dissertation: Benjamin Schnieder, Hannes Leitgeb, and Timothy Williamson. Benjamin has been a great supervisor. In particular, he helped me with much understanding and support when I realised that I had to abandon my plan to write about the semantics of "because". Hannes and Tim agreed, at very short notice, to examine my dissertation, for which I am infinitely grateful. The valuable comments provided by all three of them allowed me to write a much better book than I could have written without them.

I am indebted to the editors of this series: Tobias Rosefeldt and Benjamin Schnieder. It is a great privilege to have my book published in this series.

My greatest debt is to my family: my parents and siblings for all their help, my children Jannis and Kilian for the privilege of being able to spend so much time with them, and, most of all, my wife Esther, who has supported me much too much and who has prevented me from going crazy more than once during the last couple of years. While Leibniz was certainly wrong regarding the world as a whole, Esther, Jannis and Kilian are just as certainly a part of it that could not possibly be any better.

N.H.

Introduction

Logical reasoning is an especially secure means for obtaining knowledge. By logically inferring conclusions from known premises, we reliably come to know what was previously unknown to us, and, in various contexts, logical inferences are the *only* means by which we can extend our knowledge. *Mathematical truths* are particularly accessible items of human knowledge. While there is no guarantee that currently held empirical hypotheses remain to be found plausible in the future, one can be relatively confident that presently accepted mathematical statements will always be accepted by the mathematical community. The exceptional nature of logic and mathematics suggests that logical reasoning with purely mathematical concepts is one of the most reliable epistemic processes performed by humans. If someone does not accept a mathematical argument that is logically valid according to standard logical principles, the prospect for rational discussion may seem doubtful. According to Frege (1893: XVI), for example, such a person would simply be insane.

At the beginning of the previous century, Luitzen Egbertus Jan Brouwer declared standard mathematical practice and generally accepted forms of logical reasoning, referred to as *classical mathematics* and *classical logic*, to be seriously flawed. He propagated a new form of mathematics, called *intuitionism*, which incorporates principles that contradict basic assumptions of classical mathematics and even of classical logic. For example, Brouwer claimed that the following sentence is provable, although it is the negation of a classical logical truth that involves only logical and mathematical vocabulary:

Brouwer's Continuum Not every real number is or is not rational.

The nature and legitimacy of such mathematical and related philosophical claims are the topic of this book. My goal will be to reconstruct precise versions of intuitionistic arguments against classical logic, to analyse these arguments, and to evaluate them. I thereby hope to illustrate that it *is* possible to argue rationally even about the most fundamental principles of rational thought.

The first chapter focuses on logical theories and arguments against them. I will provide a general understanding of the notion of a logical theory and single out different types of arguments against a logical theory. In addition, I will take a look at two non-intuitionistic considerations which have been put forward against classical logic: considerations about relevance and considerations about empty singular terms. I will highlight those of their features which differentiate them from the anti-classical arguments that are brought up by intuitionists.

In the second chapter, I will discuss a famous claim about the nature of the dispute between classicists and intuitionists. According to Carnap, Quine, Dummett, and many others, classicists and intuitionists do not understand logical and mathematical vocabulary in the same way. It is claimed that logical and mathematical assertions of classicists and intuitionists are not contradictory. Their 'disagreement' is taken to be of a radically different nature than ordinary disagreements, in which one party affirms something that the other party denies. For example, when a classicist assertively utters the sentence "every real number is or is not rational", and an intuitionist assertively utters the sentence "not every real number is or is not rational", then, it is claimed, they do not contradict each other, because they do not attach the same meanings to the expressions "not", "or", "every", "real number", and "rational".

I will show that this claim is false. Drawing on results of Popper (1948) and Harris (1982), I will argue that classicists and intuitionists understand logical expressions in the same way. Furthermore, I will develop analogous results regarding mathematical terms and use them to show that classicists and intuitionists also understand these mathematical terms in the same way. The findings of Chapter 2 constitute an important precondition for the following chapters, in which intuitionistic arguments against classical logic are discussed. Since intuitionists attach the same senses to logical and mathematical expressions as classicists, their arguments, if cogent, would really show that classical logic and classical mathematics are incorrect.

In the first two sections of the third chapter, I will deal with two anti-classical arguments that are mathematical in character. Intuitionists develop their mathematics on the foundation of principles which contradict classical logical truths. I will introduce two such principles and discuss quasi-mathematical justifications for them that have recently been given by McCarty (2005) and de Swart (1992). The main objective will be to make the ideas of McCarty and de Swart sufficiently precise so that the most questionable premises of their reasoning can clearly be singled out.

In the third section of Chapter 3, I will introduce meta-mathematical arguments against classical logic. With those arguments, intuitionists try to show that certain basic classical principles are presently not known to be true by invoking assumptions about mathematical provability. The pertinent example will be an instance of the *principle of the excluded third*:

Goldbach's Disjunction Every or not every even number greater than 2 is the sum of two primes.

Intuitionists claim that Goldbach's Disjunction is presently not known to be true. Here, the crucial underlying thesis states that a mathematical disjunction is only provable if at least one of its disjuncts is provable, while, uncontroversially, neither of the disjuncts of Goldbach's Disjunction is known to be provable.

The question of how intuitionists might justify such a thesis leads to consideration of the notion of a mathematical proof and of the relation between mathematical activity and logical validity. This topic will be treated in the final section of the third chapter, in which I will discuss intuitionistic views about the relation between mathematics and logic. In particular, I will consider Brouwer's outlook on mathematical activity and Sundholm's (2004) views about legitimate inferential activity and the logical status of associated linguistic expressions. I will close Chapter 3 by presenting Prawitz' (2009, 2012a) anti-classical account of correct inferences and proofs. I will demonstrate that the fundamental principles of this account, as well as the other fundamental principles of the presented intuitionistic arguments, are in need of an independent justification if they are meant to undermine classical logic and mathematics. I will display that *semantic* considerations are the most promising candidates for such a justification.

The intuitionistic arguments in the third chapter are thus seen to lead to questions about the meanings of logical expressions and about the proper form of a semantic theory for a language that contains these expressions. In the fourth and fifth chapter, I will deal with intuitionistic arguments against classical logic which are explicitly meaning-theoretical in character. In recent philosophical discussions about the conflict between intuitionists and classicists, considerations about meanings and semantic theories play the most prominent role.

In the fourth chapter, I will discuss Dummett's famous *manifestation argument*. According to Dummett, the theory of classical logic depends on a truth conditional semantics that involves an epistemically unconstrained notion of truth. However, he takes the manifestation argument to show that a plausible truth conditional semantics must be based on an epistemically constrained notion of truth. The fundamental premises of the argument state that sentential understanding must be manifestable in linguistic behaviour, and that knowledge of truth conditions can only be manifested in such behaviour if the involved notion of truth is epistemically constrained.

The reasoning underlying Dummett's argument is rather complex, and it will be an important task to clarify its fundamental ideas and to rebut unconvincing refutations. In the final section of Chapter 4, I will show that the manifestation argument has to be rejected. It will be seen that it is possible to manifest one's knowledge of the truth condition of a sentence S by using the parts of S in various other sentences and that this behaviour does not require the ability to recognise the truth value of S in any possible situation. I will thus show that the manifestability of knowledge of truth conditions does not imply that the involved notion of truth is epistemically constrained.

In the fifth chapter, I will deal with *proof-theoretic arguments* against classical logic. These arguments combine semantic and epistemological considerations. Their proponents, e.g. Dummett (1991) and Tennant (1997), try to show that classical logic is false because it is not founded on a system of inferential rules that yields an account of the meanings of the logical operators. In addition to advancing the technical achievements of Read (2010), Humberstone (2011), and others, a major goal of Chapter 5 will be the development of a general understanding of proof-theoretic arguments and the precise reconstruction of the most important examples. It will be seen that disputes about the possible success of proof-theoretic arguments, such as the dispute between Dummett (1991) and Read (2000), have to be resolved by clarifying the general idea of a proof-theoretic argument. It will turn out that the plausibility of specific ex-

amples depends on subtle properties of systems of inferential rules that have
not received due attention.

The most persuasive proof-theoretic arguments are based on the idea that
the rules of a system that yields an account of the meanings of the logical
operators have to exhibit a certain kind of *harmony*: the rules for introducing
a logical operator must match the rules for eliminating it. Proponents of such
proof-theoretic arguments claim that classical logic has to be rejected because no
collections of harmonious rules of inference produce precisely those arguments
that are logically valid according to classical logic. I will single out what I
take to be the two most promising proof-theoretic arguments that are based
on the premise that a sound logical theory must correspond to a collection of
harmonious rules of inference. In the final section of Chapter 5, I will then show
that both of these arguments must be rejected by presenting classical systems
of harmonious rules of inference. Both proof-theoretic arguments will be seen
to rely on an unjustified assumption about the precise form of an inferential
account of the meanings the logical operators. Importantly, this assumption is
independent of the requirement that a sound logical theory must correspond to
a system of harmonious rules.

I will conclude the book with my view on the dispute between classicists
and intuitionists: the only plausible arguments that either side can put forward
against the other side are mathematical in character, but such arguments will
always beg the question by presupposing the falsity of a fundamental logical or
mathematical assumption of the view that is argued against; non-mathematical
arguments, such as semantic, epistemological, or ontological ones, cannot be
used to resolve the pertinent logical and mathematical disagreements. The dispute between classicists and intuitionists lacks a non-circular solution.

Chapter 1

Arguing about Classical Logic

The main subject of this book is the conflict between *classicists*, who endorse
the predominant canon of logical methods, and *intuitionists*, who favour alternative logical principles that derive from Brouwer's mathematical innovations.
Its principal aim consists in an evaluation of the intuitionistic challenges to classical logic. In preparation, I will discuss general features of arguments against
classical logic. Against this background it will then be possible to highlight the
peculiarities of the intuitionistic criticism.

In the first three sections, I will discuss the nature of arguments against
classical logic in general. To this end, I will introduce the notion of a logical
theory (1.1), present classical and intuitionistic first-order predicate logic (1.2),
and discuss four ways of arguing against a logical theory (1.3). In the remaining
two sections, I will present two non-intuitionistic considerations which have
been put forward against classical logic: considerations about relevance (1.4)
and considerations about empty singular terms (1.5). The discussion of these
anti-classical arguments will be taken only to such a level that it is possible to
compare them with those that are put forward by intuitionists.

1.1   Logical Theories

In one sense of the term, *logic* is one of the disciplines of pure mathematics,
comprising such sub-disciplines as model theory, proof theory, and recursion
theory. It is expressed in a mathematical language and pursued by mathematical methods. To argue against the standard execution of this discipline would
involve a critique of the assumptions made and the methods used. Very often,
however, arguments against classical logic are not directed against pure mathematics; they are based on a different conception of logic. In another sense of
the term, *logic* is the theory of logical validity and related notions like logical
truth and logical inconsistency. For what follows, this notion is the pertinent
one. I will try to elucidate it in the present section.

1.1.1   The Bearers of Logical Properties

Given that the fundamental notions of logic are notions like logical truth and logical validity, it is desirable to know to which kinds of entities these notions apply. There are four main possibilities: (i) linguistic objects like (declarative) sentences and arguments, (ii) speech acts in which tokens of these linguistic objects are produced, (iii) mental acts like judgements and inferences, and (iv) abstract objects like propositions and propositional arguments. I will choose the first of these options, and I would like to indicate briefly my reasons for doing so. For simplicity, I will concentrate on sentences and their counterparts: sentential utterances, judgements, and propositions.

The strategic reason for assuming that some *sentences* are logically true is that there are sufficiently developed and widely accepted theories about them. These theories have been suggested by Tarski (1931), and they have been developed by Quine (1940: ch. 7) and Martin (1958: ch. 3). Speech acts and mental episodes, by contrast, have not received a comparable mathematical treatment, and propositions gave rise to numerous incompatible theories. There is also a non-strategic reason for preferring sentences over propositions (and, consequently, over judgements) in this context. Notions like logical truth seem to be sensitive to differences which, on the face of it, exist only at a linguistic level. Compare the following two sentences:

(1) All furze is furze.   (2) All furze is gorse.

Although these two sentences are synonymous, only sentence (1) is a logical truth. If we make the reasonable assumption that they express the same proposition, then there is a logical difference which can only be accounted for at a linguistic level and not at a propositional level.[1]

For these reasons, I will assume that logic is primarily concerned with (declarative) sentences and other linguistic objects like arguments and theories. A fundamental assumption of the present approach is that for any natural language $\mathcal{L}$ there is the set of sentences of $\mathcal{L}$ which are logically true, the set of arguments of $\mathcal{L}$ which are logically valid, and the set of theories of $\mathcal{L}$ which are logically inconsistent. A logical theory is then put forward as a revealing description of these sets.

It has to be admitted that the choice of linguistic objects as the bearers of logical properties creates a number of problems. In particular, one has to mention the fact that expressions of natural languages may be structurally or

---

[1]   The claim that (1) and (2) express the same proposition is based on three assumptions:

(a) The general terms "furze" and "gorse" are synonymous.

(b) If "furze" and "gorse" are synonymous, then so are (1) and (2).

(c) If (1) and (2) are synonymous, they express the same proposition (in a fixed context).

It should be noted that there are theories of propositions according to which (c) fails: theories according to which synonymous sentences may express distinct propositions in the same context (see Künne 2003: 369-72). Furthermore, there are theories of meaning according to which (b) fails: theories according to which substitution of synonymous parts (in extensional contexts) does not always preserve the meaning of the whole (see Fine 2007: 37-42).

lexically ambiguous and the fact that many of them exhibit various kinds of context-sensitivity. Consider the following candidate for a logically true sentence:

(3) If you are a philosopher, then you are a philosopher.

This conditional could be used to say something false when the two tokens of the pronoun "you" are used to refer to different persons (of whom only the first is a philosopher). It could also be uttered without asserting anything at all (for example, if it is used as a sample sentence of English without fixing a referent for the two tokens of "you"). In the end, an adequate logical theory should be applicable to context-sensitive expressions and thus will not simply be about sentences and other linguistic objects.[2] For simplicity, however, these complications will not be taken into account here.

Given the assumption that linguistic entities are the bearers of logical properties, it is helpful to single out one kind of linguistic entity and a corresponding logical notion. I take *arguments* to be the basic bearers of logical properties, and, consequently, I take the notion of *logical validity* to be the basic logical notion. I will assume that an *argument* consists of a finite list of sentences, its *premises*, a single sentence, its *conclusion*, and an expression like the word "therefore" which separates the list of premises from the conclusion.

As is well known, the notion of logical validity can be used to define several further logical notions such as logical implication, logical truth, and logical inconsistency. Importantly, however, there are also kinds of linguistic entities that fall under logical notions for which it is less clear how they are related to the notion of logical validity for arguments. In what follows, three closely related kinds of linguistic entities are of special relevance: *argumentations*, *argumentative steps*, and *rules* for performing argumentative steps. I will return to these objects and their logical properties in 1.2.5 and in later chapters.

1.1.2   Total and Partial Logical Theories

The following subsection introduces a distinction which relates to the notion of logical validity. What has been said thus far suggests that a logical theory for some language $\mathcal{L}$ is a theory which yields a specification of all logically valid arguments of $\mathcal{L}$. However, not every logical theory is like this. Consider the theory of classical sentential logic, and regard the following argument:

(4) Not every philosopher is wise. Therefore, some philosopher is not wise.

A proponent of classical sentential logic does not have to deny that this argument is logically valid. It would be a mistake though to infer that a logical theory only makes *positive* claims of logical validity. Classical sentential logic, for example, includes a *negative* thesis about (4). It treats a number of expressions, which typically include "not", "and", "or", and "if", as *logical constants*,

---

[2]   Rumfitt (2007: 642), for example, proposes that such a theory is about what he calls *statements*, i.e. certain "ordered pairs whose first element is a meaningful, indeed disambiguated, declarative type-sentence and whose second element is a possible context of utterance".

and it involves the thesis that (4) is *not* validated by the logical properties of these constants.[3]

The moral to be drawn is that apart from what one might call a *total* logical theory for some language $\mathcal{L}$, which is put forward as an account of the totality of logically valid arguments of $\mathcal{L}$, there are also various *partial* logical theories for $\mathcal{L}$. Each of these theories may be put forward as an account of the class of arguments which are logically valid in $\mathcal{L}$ relative to a chosen set of logical constants. In fact, it suffices to consider partial logical theories since total logical theories can be treated as limiting cases of them: an argument is *logically valid* (*simpliciter*) iff it is logically valid with respect to the set of *all* logical constants.

1.1.3   The Shape of a Logical Theory

A logical theory specifies which arguments of the relevant language are logically valid in relation to the elements of the chosen set of logical constants. Let $\mathcal{L}$ be a natural language and suppose that $\mathcal{C}$ is a set of logical constants of $\mathcal{L}$. I will assume that a corresponding logical theory $\mathcal{T}$ is presented in a fragment of English. More precisely, the language of $\mathcal{T}$ is allowed to be an *extension* of a fragment of English; it may contain resources which are not part of English.

A logical theory $\mathcal{T}$ has to have a *syntactic part* $\mathcal{T}_{synt}$, a part for speaking about the syntactic properties of arguments of $\mathcal{L}$. This part should contain *canonical terms* for the expressions of $\mathcal{L}$, and it should contain syntactic predicates like "is an argument of $\mathcal{L}$". It will then consist of axioms which express basic principles about the syntactic properties of the expressions of $\mathcal{L}$. In addition to the syntactic part, the logical theory $\mathcal{T}$ needs to have a *main part*, a part for speaking about logical validity. I will assume that this part contains the unary predicate "is logically valid in $\mathcal{L}$ w.r.t. $\mathcal{C}$", which is meant to apply to those arguments of $\mathcal{L}$ that are logically valid in $\mathcal{L}$ with respect to the chosen set $\mathcal{C}$ of logical constants. The logical theory $\mathcal{T}$ is then the union of its syntactic part and its main part.

How could the syntactic part and the main part of a logical theory be stated? Given the syntactic complexity of natural languages this might seem extremely complicated. As a consequence, it might be thought that it is very difficult to compare competing logical theories. Fortunately, it is possible to deal at least with this second concern. For many pairs of competing logical theories it is possible to find surveyable simplifications of the pertinent natural languages such that the logical differences of the two theories can completely be stated with respect to these simplifications. The idea is to break up the main part of a logical theory into two parts: a *transformational part*, which reduces the syntactic complexity of the arguments through a formalisation process, and a *logical part*, which classifies the formalised versions of the original arguments (cp. Resnik 1985: 224 and Aberdein & Read 2009: 615-6).

---

[3]   The notion of a logical constant might lack a sharp boundary. Furthermore, it should be emphasised that the assumption that there is a (fuzzy) line between logical and non-logical expressions does not imply that this distinction is epistemologically significant. See Field 2009a: 342-3 for skepticism about whether "the demarcation between logic and nonlogic" is philosophically important.

To make this explicit, consider a formal language $\mathcal{L}^*$ and assume that $\mathcal{T}_{synt^*}$ is a collection of axioms which express basic syntactic principles about the expressions of $\mathcal{L}^*$. In addition, suppose that $\mathcal{T}$ contains a unary function symbol π which stands for a *formalisation function* that maps every argument of $\mathcal{L}$ to a formal counterpart of $\mathcal{L}^*$. Furthermore, suppose that $\mathcal{T}$ contains the predicate "is $\mathcal{L}^*$-valid" for the formal arguments of $\mathcal{L}^*$ that correspond to the arguments of $\mathcal{L}$ which are logically valid in $\mathcal{L}$ with respect to $\mathcal{C}$. The language $\mathcal{L}$ is called the *informal object language*, and $\mathcal{L}^*$ is called the *formal object language*. The main part of $\mathcal{T}$ is now split up: first, there is a set of transformational axioms $\mathcal{T}_{trans}$ for π, which state how an argument is to be formalised; second, there is a set of axioms $\mathcal{T}_{logic}$ for the predicate "is $\mathcal{L}^*$-valid"; third, there is the following bridge principle, which yields the desired axiomatisation of the predicate "is logically valid in $\mathcal{L}$ w.r.t. $\mathcal{C}$":

Bridge Principle (BP) For every α, α is logically valid in $\mathcal{L}$ w.r.t. $\mathcal{C}$ if and only if $\pi(\alpha)$ is $\mathcal{L}^*$-valid.

This principle characterises the logically valid arguments of the informal object language as those that are mapped to $\mathcal{L}^*$-valid arguments of the formal object language. In sum, a logical theory $\mathcal{T}$ of this type is then the union of five parts:

$$\mathcal{T} = \mathcal{T}_{synt} \cup \mathcal{T}_{trans} \cup \mathcal{T}_{synt^*} \cup \{\mathrm{BP}\} \cup \mathcal{T}_{logic}.$$

It has two syntactic parts, $\mathcal{T}_{synt}$ and $\mathcal{T}_{synt^*}$, which are linked via the transformational part $\mathcal{T}_{trans}$. In addition, it has a logical part, $\mathcal{T}_{logic}$, which is linked to the other parts by BP and which gives $\mathcal{T}$ its point.

I would like to emphasise that I do not consider the expressions of the formal object language $\mathcal{L}^*$ to be independently meaningful. They are only meaningful in a derivative sense according to which they inherit their meanings from their informal counterparts (see 2.1.2). Correspondingly, I consider logical validity to be fundamentally a property of arguments of the informal object language. The notion of $\mathcal{L}^*$-validity has only instrumental value, its job being to assist in axiomatising the fundamental notion of logical validity in $\mathcal{L}$ with respect to $\mathcal{C}$.

Dealing with logical theories of this type has one important advantage. In many cases competing logical theories can be taken to differ only in their logical parts. One can then restrict attention to the axioms for the predicate "is $\mathcal{L}^*$-valid", which are easier to survey than direct axioms for the predicate "is logically valid in $\mathcal{L}$ w.r.t. $\mathcal{C}$". In particular, the logical theories of classical and intuitionistic logic have the same syntactic parts and the same transformational part.[4] Thus, although the syntactic complexity of natural languages forces logical theories to be very complex, one does not have to take this complexity into account when one compares classical and intuitionistic logic.

It may be noted that the present conception of a logical theory is rather undemanding. In particular, its logical part is not required to involve axioms about entities like models or rules. This notion differs, therefore, from the notion

---

[4] The view that classical and intuitionistic logic have the same transformational part is partly based on the claim that classicists and intuitionists understand the relevant logical constants in the same way. This claim will be defended in Chapter 2.

of a logical theory put forward by Resnik (1985: 225) and the notion employed by
Aberdein & Read (2009: 618). My reasons for adopting such an undemanding
notion of a logical theory are purely pragmatic: it allows for rather simple
theories of classical and intuitionistic logic. This will facilitate the discussion in
the following chapters.

1.2   Classical and Intuitionistic Logic

I will now introduce the theories of classical and intuitionistic first-order predicate logic. These theories treat English as the informal object language; that
is, they describe the logical properties of English. This means that the informal
object language and the language of the logical theories coincide or, at least,
overlap. Most of what follows can easily be adapted to cases in which these
languages are disjoint. (For one exception, see the discussion of *basic criticism*
in 1.3.2.) The elements of the following set and their stylistic variants are the
logical constants of first-order predicate logic:

$$\mathcal{C} := \{ \text{"not"}, \text{"or"}, \text{"and"}, \text{"if"}, \text{"every"}, \text{"some"} \}.$$

They will be called *standard logical constants*. Throughout this book, biconditionals, i.e. sentences of the form "S_1 if and only if S_2", will be treated as
conjunctions of conditionals.

First, I will introduce the formal object language $\mathcal{L}^*$ and indicate the formal
syntactic part $\mathcal{T}_{synt^*}$, which is common to the logical theories discussed here
(1.2.1). I will then present the logical part of the *model-theoretic* logical theory
$\mathcal{T}^m$ (1.2.2), which is neutral between classical and intuitionistic logic, and the
logical parts of two *derivational* logical theories (1.2.3): the classical logical
theory $\mathcal{T}^C$ and the intuitionistic logical theory $\mathcal{T}^I$. Afterwards, I will discuss
some aspects of the transformational part $\mathcal{T}_{trans}$ and the informal syntactic part
$\mathcal{T}_{synt}$, which belong to the three introduced logical theories (1.2.4). Finally, I
will remark on the relation between logically valid arguments and logically valid
rules and their applications (1.2.5).

1.2.1   The Formal Syntactic Part

To begin with, the formal object language $\mathcal{L}^*$ will be specified. Its vocabulary
consists of logical symbols, punctuation symbols, and non-logical symbols:

$$\neg \quad \vee \quad \wedge \quad \rightarrow \quad \forall \quad \exists \quad (\quad) \quad , \quad \succ \quad x \quad a \quad c \quad f \quad R \quad \circ \quad *$$

The logical symbols $\neg$, $\vee$, $\wedge$, $\rightarrow$, $\forall$, $\exists$ correspond to the standard logical constants. The punctuation symbols are brackets, the comma for building finite
lists of sentences, and the symbol $\succ$ for building arguments from finite lists of
sentences and sentences. The non-logical symbols x, a, c, f, R, $\circ$, $*$ generate
the parameters, variables,[5] constants, function signs, and relation signs. The

---

[5]   I follow Prawitz (1965) in using parameters instead of allowing for free variables. The
use of different kinds of symbols for (bound) variables and parameters will simplify the
presentation of the logical calculi of natural deduction in 1.2.3.

symbol $^\circ$ is used to indicate the arities of function and relation signs, and $*$ is used to generate a sufficiently large supply of the different kinds of expressions. For example, the binary function signs are $f^{\circ\circ}$, $f_*^{\circ\circ}$, $f_{**}^{\circ\circ}$, etc. Whenever it is convenient, I will make informal use of further expressions as parameters, variables, constants, function signs, and relation signs; the context will make it clear to which category they belong.

The following standard syntactic notions are used in this text. Of fundamental importance is the notion of an expression of $\mathcal{L}^*$:

Expressions An *expression of $\mathcal{L}^*$* is a finite concatenation of symbols of $\mathcal{L}^*$.

I will use the symbols of $\mathcal{L}^*$ as names for themselves, and I will use concatenations of symbols of $\mathcal{L}^*$ as terms for expressions of $\mathcal{L}^*$. The other syntactic notions defined here apply only to expressions of $\mathcal{L}^*$. The notions of an open term and of an open sentence are defined recursively:[6]

Open Terms A parameter is an open term. A constant is an open term. If $t_1, \ldots, t_n$ are open terms and f is an n-ary function sign, then $ft_1 \ldots t_n$ is an open term.

Open Sentences A sequence of an n-ary relation sign followed by n open terms is an open sentence. If S_1 and S_2 are open sentences, then so are $\neg S_1$, $(S_1 \wedge S_2)$, $(S_1 \vee S_2)$, $(S_1 \rightarrow S_2)$. If S_1 is an open sentence which contains tokens of the parameter x, and S_2 results from S_1 by substituting tokens of a variable a for the tokens of x, then $\exists a\, S_2$ and $\forall a\, S_2$ are open sentences, provided that the newly introduced tokens of a are bound by the initial quantifier.

Note that in an open sentence every variable is bound by some quantifier and every quantifier binds some variable. A *term* is an open term without parameters, and a *sentence* is an open sentence without parameters. Finally, the notions of an open list and of an open argument are defined as follows:

Open Lists An *open list* is a finite and alternating concatenation of open sentences and commas which begins and ends with an open sentence.

Open Arguments An *open argument* is either a concatenation of an open list, the symbol $\succ$, and an open sentence or the concatenation of the symbol $\succ$ and an open sentence.

For example, the expression $R^\circ x, \forall a\, (R^\circ a \rightarrow \neg R^\circ x) \succ \neg R^\circ c$ is an open argument. A *list* is an open list without parameters, and an *argument* is an open argument without parameters.

The formal syntactic part $\mathcal{T}_{synt^*}$ contains the fundamental syntactic assumptions about $\mathcal{L}^*$. They are assumptions about the identity and existence of expressions (see the end of 1.2.3). Importantly, the axioms of $\mathcal{T}_{synt^*}$ can be accepted by classicists and intuitionists. They belong to each of the three logical theories introduced here.

---

[6]   In the following two definitions, f, x, and a are used as meta-variables.

1.2.2   A Model-Theoretic Logical Theory

In this and the following subsection, I will present the logical parts of three logical theories. First, I will introduce a *model-theoretic* approach originating with Tarski 1936. Then, I will present *derivational* approaches originating with Frege 1879 and Gentzen 1934. These or slightly different accounts can be found in any textbook on formal logic. I include them here for future reference.

The logical part of the model-theoretic logical theory $\mathcal{T}^m$ contains some purely set-theoretic axioms and a set MTA of model-theoretic axioms. I will indicate the axioms of MTA in an informal way here. To begin with, the notions of an M-term and of an M-sentence have to be introduced. Suppose that M is any set. Then, *M-terms* are certain finite sequences of symbols of $\mathcal{L}^*$ and elements of M: they are like open terms except that they contain elements of M instead of parameters.[7] One may think of M-terms of $\mathcal{L}^*$ as ordinary terms of an extension of $\mathcal{L}^*$: every element of M is added to $\mathcal{L}^*$ as a term for itself. *M-Sentences* are defined in analogy to M-terms: they are like open sentences except that they contain elements of M instead of parameters.

There is one axiom of MTA which contains the predicate "is $\mathcal{L}^*$-valid":

$\mathcal{L}^*$-Validity in Terms of Models

> For every argument α of $\mathcal{L}^*$, α is $\mathcal{L}^*$-valid iff the conclusion of α is true in every model in which each premise of α is true.

In addition to some syntactic notions from $\mathcal{T}_{synt^*}$, this axiom presupposes the notion of a model and the notion of a sentence being true in a model. These notions are dealt with in additional model-theoretic axioms.

A *model* is a pair $\mathcal{M} := (M, I)$ which consists of an inhabited set M, its *domain*, and an *interpretation function* I.[8] A set is referred to as *inhabited* iff it has at least one element. According to classical logic, a set is inhabited iff it is not empty. However, intuitionists doubt this biconditional and prefer the displayed notion of a model (cp. Dummett 1991: 27). An *interpretation function* I is a function that maps every constant c to some element $c^{\mathcal{M}}$ of M, every n-place function sign f to some n-place function $f^{\mathcal{M}}$ on M, and every n-place relation sign R to some set $R^{\mathcal{M}}$ of n-tuples of M.

To define the notion of being true in a model, one first recursively defines the notion of a denotation function $D^{\mathcal{M}}$ from the set of M-terms of $\mathcal{L}^*$ to M:[9]

$$D^{\mathcal{M}}(m) := m, \quad D^{\mathcal{M}}(c) := c^{\mathcal{M}}, \quad D^{\mathcal{M}}(ft_1 \ldots t_n) := f^{\mathcal{M}}(D^{\mathcal{M}}(t_1), \ldots, D^{\mathcal{M}}(t_n)).$$

That is, every element of M denotes itself, every constant denotes its interpretation, and every term in which a function sign is attached to some terms denotes that element of M to which the interpretation of the function sign maps the denotations of the attached terms.

---

[7]   More precisely: an element of M is an M-term, a constant is an M-term, and if $t_1, \ldots, t_n$ are M-terms and f is an n-ary function sign, then $ft_1 \ldots t_n$ is an M-term.

[8]   To be precise, it has to be required that M does not contain variables or terms of $\mathcal{L}^*$.

[9]   In the following definition, $\mathcal{M} = (M, I)$ is a model, m is an element in M, c is a constant, f is an n-ary function sign, and $t_1, \ldots, t_n$ are M-terms. If τ is a function and x is an argument of τ, then $\tau(x)$ is the image of x under τ.

Then, a recursive definition of the predicate "is true in $\mathcal{M}$" is given:[10]

(Tr$_R^{\mathcal{M}}$) $Rt_1 \ldots t_n$ is true in $\mathcal{M}$ iff $(D^{\mathcal{M}}(t_1), \ldots, D^{\mathcal{M}}(t_n))$ is an element of $R^{\mathcal{M}}$;

(Tr$_\neg^{\mathcal{M}}$) $\neg S_1$ is true in $\mathcal{M}$ iff S_1 is not true in $\mathcal{M}$;

(Tr$_\rightarrow^{\mathcal{M}}$) $(S_1 \rightarrow S_2)$ is true in $\mathcal{M}$ iff S_2 is true in $\mathcal{M}$ if S_1 is true in $\mathcal{M}$;

(Tr$_\vee^{\mathcal{M}}$) $(S_1 \vee S_2)$ is true in $\mathcal{M}$ iff S_1 is true in $\mathcal{M}$ or S_2 is true in $\mathcal{M}$;

(Tr$_\wedge^{\mathcal{M}}$) $(S_1 \wedge S_2)$ is true in $\mathcal{M}$ iff S_1 is true in $\mathcal{M}$ and S_2 is true in $\mathcal{M}$;

(Tr$_\forall^{\mathcal{M}}$) $\forall a S$ is true in $\mathcal{M}$ iff, for every $m \in M$, S_a^m is true in $\mathcal{M}$;

(Tr$_\exists^{\mathcal{M}}$) $\exists a S$ is true in $\mathcal{M}$ iff, for some $m \in M$, S_a^m is true in $\mathcal{M}$.

Since every sentence is an M-sentence (for every M), this yields as a limiting case the desired relation between models and sentences.

Now, the model-theoretic axioms of MTA have to be supplemented with set-theoretic axioms. There are various strong collections of set-theoretic assumptions accepted by many intuitionists and classicists. The two most prominent such collections are IZF, the so-called *intuitionistic Zermelo-Fraenkel set theory*, and CZF, the so-called *constructivist Zermelo-Fraenkel set theory*. In classical logic, both IZF and CZF are equivalent to the standard classical Zermelo-Fraenkel set theory; in intuitionistic logic, CZF is weaker than IZF by conforming to *predicativist* standards.[11] In addition to the model-theoretic axioms, one may choose the axioms of either of these collections as the axioms of the logical part of $\mathcal{T}^m$. For definiteness, I will assume that the logical part of $\mathcal{T}^m$ equals CZF $\cup$ MTA. Importantly, the resulting logical theory $\mathcal{T}^m$ is accepted by most classicists and most intuitionists.

Neither IZF nor CZF comprises choice principles. There are weak choice principles, e.g. the axiom of *countable* choice or the axiom of *dependent* choice (see Troelstra & van Dalen 1988: 189-91), which many intuitionists accept. These could be added to the logical part of $\mathcal{T}^m$ as well.

There are also set-theoretic assumptions which are only accepted by one of the two camps. Famously, the *Axiom of Choice* is a principle not accepted by intuitionists,[12] while the so-called *Uniformity Principle* (see 3.1.1) is not accepted by classicists. Thus, there are theories which extend $\mathcal{T}^m$, some of which are rejected by intuitionists and some of which are rejected by classicists. (Of course, not every mathematician who accepts classical logic accepts the

---

[10]   In the following definition, $\mathcal{M} = (M, I)$ is a model, m is an element of M, a is a variable, R is an n-place relation sign, $t_1, \ldots, t_n$ are M-terms, S_1 and S_2 are M-sentences, and S is like an M-sentence, except that it contains *unbound* tokens of the variable a. In the last two clauses, S_a^m results from S by replacing every unbound token of a with m. If, for example, S is $(Fa \rightarrow \exists a Ga)$, then S_a^m is $(Fm \rightarrow \exists a Ga)$.

[11]   Surveys of IZF are given in Beeson 1985: chs. 8 & 9 and in Ščedrov 1985. For CZF see Aczel 1978 and Aczel & Rathjen 2001.

[12]   The Axiom of Choice has been introduced by Zermelo in his proof of the Well-Ordering Theorem (see Kanamori 2004). Diaconescu (1975) has discovered that the Axiom of Choice implies the principle of the excluded third; see also Goodman & Myhill 1978.

Axiom of Choice, and not every mathematician who accepts intuitionistic logic accepts the Uniformity Principle.)

It might be asked how it is possible that intuitionists and classicists disagree about which arguments are logically valid given that they accept the same logical theory $\mathcal{T}^m$. The simple answer is that they disagree about which sentences are consequences of $\mathcal{T}^m$. For example, a classicist will affirm while an intuitionist will deny that the following sentence is a theorem of $\mathcal{T}^m$:

(5) $\neg\neg Fc \succ Fc$ is $\mathcal{L}^*$-valid.

The reason for this is the classicist belief that for every model $\mathcal{M}$ and for every sentence S of $\mathcal{L}^*$, if S is not not true in $\mathcal{M}$, then S is true in $\mathcal{M}$, something which an intuitionist does not believe. To deal with this situation, it is helpful to consider different kinds of logical theories for which the different views about their consequences have less dramatic effects.

1.2.3   Two Derivational Logical Theories

The main alternative to a model-theoretic approach to logical validity is a derivational one. Here the situation is to some extent reversed: classicists and intuitionists sharply disagree about what kind of logical part a derivational theory should have, but they do not quarrel about which arguments are logically valid according to such a theory (see 1.3). Concepts from the derivational approach will play an important role in Chapters 2 and 5.

The logical part of the intuitionistic derivational theory $\mathcal{T}^I$ and the logical part of the classical derivational theory $\mathcal{T}^C$ each consists of a set of broadly syntactic, proof-theoretic axioms referred to as PTA^I and PTA^C. I will indicate them in an informal way here. (At the end of this subsection, I will mention how they can be stated more precisely.) To begin with, one needs to define a couple of new concepts. A *transition* is a pair consisting of a finite set of open arguments of $\mathcal{L}^*$, the *initial arguments*, and another open argument of $\mathcal{L}^*$, the *final argument*. A *rule* is a set of transitions, and a *calculus* is a collection of rules. The elements of a rule are also called its *applications*.[13] A calculus $\mathcal{R}$ induces the following property of arguments:

$\mathcal{R}$-**Derivability** An open argument α is called $\mathcal{R}$-*derivable* iff there is a finite sequence $\alpha_0, \ldots, \alpha_n$ of open arguments such that $\alpha = \alpha_n$ and such that for every $i \leq n$ there is a subset J_i of $\{0, 1, \ldots, i-1\}$ such that $(\{\alpha_j : j \in J_i\}, \alpha_i)$ is an application of a rule in $\mathcal{R}$.

According to this definition, an open argument is $\mathcal{R}$-derivable iff it can be obtained by a finite number of applications of rules in $\mathcal{R}$. A logical theory corresponding to a calculus $\mathcal{R}$ contains the following axiom:

---

[13]   It may be noted that rules are tied to a single (formal) language. In 2.2.3 and 2.2.4, I will argue for the philosophical importance of a different conception of rules, according to which they are language-transcendent objects.

$\mathcal{L}^*$-Validity in Terms of $\mathcal{R}$-Derivations (V$\mathcal{R}$)

For every argument α of $\mathcal{L}^*$, α is $\mathcal{L}^*$-valid iff α is $\mathcal{R}$-derivable.

That is, $\mathcal{L}^*$-validity is identified with $\mathcal{R}$-derivability.

There are two pertinent calculi: the intuitionistic calculus $\mathcal{R}^I$ and the classical calculus $\mathcal{R}^C$, giving rise to the axiom V$\mathcal{R}^I$ of PTA^I and to the axiom V$\mathcal{R}^C$ of PTA^C. The rules of the calculi $\mathcal{R}^I$ and $\mathcal{R}^C$ are introduced by presenting schemata for their applications. An application is written as follows:

$$\frac{t_1 \ldots t_n}{t}$$

where $t_1, \ldots, t_n$ represent the initial arguments, and t represents the final argument. In the following presentation, Γ, Δ, Σ are schematic letters for lists of open sentences, φ, ψ, χ are schematic letters for open sentences, and a, x, t are schematic letters for variables, parameters, and terms.

The intuitionistic calculus $\mathcal{R}^I$ comprises the following rules. First, there are three so-called *structural* rules, *Assumption*, *Exchange*, and *Contraction*, which are represented by schemata in which no logical operator occurs:

$$\frac{}{\varphi \succ \varphi}\ (A) \qquad \frac{\Gamma, \varphi, \psi, \Delta \succ \chi}{\Gamma, \psi, \varphi, \Delta \succ \chi}\ (E) \qquad \frac{\Gamma, \varphi, \varphi \succ \psi}{\Gamma, \varphi \succ \psi}\ (C)$$

Second, there are so-called *introduction* and *elimination rules*. For every logical operator there are one or two introduction rules and one or two elimination rules in whose schemata the operator figures:[14]

$$\frac{\Gamma \succ \varphi \quad \Delta \succ \psi}{\Gamma, \Delta \succ \varphi \wedge \psi}\ (\wedge_I) \qquad \frac{\Gamma \succ \varphi \wedge \psi}{\Gamma \succ \varphi}\ (^1\wedge_E) \qquad \frac{\Gamma \succ \varphi \wedge \psi}{\Gamma \succ \psi}\ (^2\wedge_E)$$

$$\frac{\Gamma \succ \varphi}{\Gamma \succ \varphi \vee \psi}\ (^1\vee_I) \qquad \frac{\Gamma \succ \psi}{\Gamma \succ \varphi \vee \psi}\ (^2\vee_I) \qquad \frac{\Gamma \succ \varphi \vee \psi \quad \Delta, \varphi \succ \chi \quad \Sigma, \psi \succ \chi}{\Gamma, \Delta, \Sigma \succ \chi}\ (\vee_E)$$

$$\frac{\Gamma, \varphi \succ \psi}{\Gamma \succ \varphi \to \psi}\ (\to_I) \qquad \frac{\Gamma \succ \varphi \to \psi \quad \Delta \succ \varphi}{\Gamma, \Delta \succ \psi}\ (\to_E)$$

$$\frac{\Gamma, \varphi \succ \psi \quad \Delta, \varphi \succ \neg\psi}{\Gamma, \Delta \succ \neg\varphi}\ (\neg_I) \qquad \frac{\Gamma \succ \varphi \quad \Delta \succ \neg\varphi}{\Gamma, \Delta \succ \psi}\ (\neg_E)$$

$$\frac{\Gamma \succ \varphi}{\Gamma \succ \forall a\varphi_x^a}\ (\forall_I) \qquad \frac{\Gamma \succ \forall a\varphi}{\Gamma \succ \varphi_a^t}\ (\forall_E)$$

$$\frac{\Gamma \succ \varphi_a^t}{\Gamma \succ \exists a\varphi}\ (\exists_I) \qquad \frac{\Gamma \succ \exists a\varphi \quad \Delta, \varphi_a^x \succ \psi}{\Gamma, \Delta \succ \psi}\ (\exists_E)$$

The rules $(\forall_I)$ and $(\exists_E)$ are subject to restrictions: in $(\forall_I)$, the parameter x must not occur in Γ, and in $(\exists_E)$, the parameter x must not occur in $\Gamma, \Delta, \exists a\varphi, \psi$.

---

[14] The schemata for the quantifier rules are to be understood as follows: in $(\forall_I)$, φ_x^a results from φ by replacing every token of x by a token of a, where it is required that every such new token of a is bound by the newly introduced initial quantifier; in $(\forall_E)$ and $(\exists_I)$, φ_a^t results from φ by replacing every token of a which is bound by the initial quantifier by a token of t; in $(\exists_E)$, φ_a^x results from φ by replacing every token of a which is bound by the initial quantifier by a token of x.

The classical calculus $\mathcal{R}^C$ comprises the above rules and an additional rule that governs $\neg$:

$$\frac{\Gamma \succ \neg\neg\varphi}{\Gamma \succ \varphi} \quad (\neg_E)$$

This immediately implies that every argument derivable in the intuitionistic calculus $\mathcal{R}^I$ is also derivable in the classical calculus $\mathcal{R}^C$. It also turns out, though this result is less immediate, that not every argument derivable in the classical calculus is also derivable in the intuitionistic calculus.

I will close with some remarks about the possibility of making the formal syntactic part $\mathcal{T}_{synt*}$ and the logical parts of $\mathcal{T}^I$ and $\mathcal{T}^C$ more precise. The idea is to give a precise account of the language fragment in which the pertinent logical theory $\mathcal{T}$ is presented, and then to show how the axioms of $\mathcal{T}$ can be (re-)formulated in this fragment.

Consider first the formal syntactic part $\mathcal{T}_{synt*}$ which contains the fundamental syntactic assumptions about the formal object language $\mathcal{L}^*$. Apart from the standard logical constants of English, its underlying language contains names of the symbols of $\mathcal{L}^*$, a binary function symbol $^\wedge$ for building expressions of $\mathcal{L}^*$, the predicate "is an expression of $\mathcal{L}^*$", and the predicate "is identical to". Corresponding to these linguistic resources, $\mathcal{T}_{synt*}$ contains axioms which express basic principles about the expressions of $\mathcal{L}^*$. There are axioms that state what the expressions of $\mathcal{L}^*$ are, and there are axioms that state the conditions under which an expression e_1 is identical to an expressions e_2.[15] Each of these axioms is classically and intuitionistically acceptable. Finally, one defines the pertinent syntactic predicates like "is a sentence of $\mathcal{L}^*$" and "is an argument of $\mathcal{L}^*$" in terms of the mentioned linguistic resources (see Martin 1958: 76-86).

Second, consider the logical parts PTA^I and PTA^C of the derivational theories $\mathcal{T}^I$ and $\mathcal{T}^C$. Remarkably, the resources from the formal syntactic part $\mathcal{T}_{synt*}$ suffice to replace all predicates of $\mathcal{T}^I$ and $\mathcal{T}^C$ except "is $\mathcal{L}^*$-valid"; in particular, the predicates "is $\mathcal{R}^I$-derivable" and "is $\mathcal{R}^C$-derivable" are so replaceable.[16] Apart from V$\mathcal{R}^I$ and V$\mathcal{R}^C$, these two logical parts do not need axioms which are not already contained in $\mathcal{T}_{synt*}$.[17]

---

[15] For two slightly different proposals see Tarski 1931: 173-4 and Martin 1958: 75. The differences between Tarski and Martin concern the axiom which is meant to express that the expressions of $\mathcal{L}^*$ are precisely the finite concatenations of symbols of $\mathcal{L}^*$. Tarski employs additional set-theoretic vocabulary, while Martin uses a infinitary rule, i.e. a rule whose applications have infinitely many premises. I would prefer an axiom schema that is formulated in the original language and which corresponds to the induction schema that is known from first-order arithmetical theories.

[16] See Martin 1958: 86-91. Martin deals with Hilbert-style calculi. However, it poses no difficulty to transfer his definitions and results to the present setting.

[17] If one wished, one could also replace the predicate "is $\mathcal{L}^*$-valid" (either by the expression that already replaces "is $\mathcal{R}^I$-derivable" or by the one that replaces "is $\mathcal{R}^C$-derivable"). In that case, one would have to differentiate between two possible bridge principles (see 1.1.3). The intuitionistic theory $\mathcal{T}^I$ would contain a bridge principle BP^I that links logical validity with the expression that replaces "is $\mathcal{R}^I$-derivable". The classical theory $\mathcal{T}^C$ would contain a bridge principle BP^C that links logical validity with the expression that replaces "is $\mathcal{R}^C$-derivable". Nothing hangs on my choice of sticking with a single bridge principle and distinct collections of proof-theoretic axioms.

1.2.4   The Transformational Part

In this subsection, I will mention some aspects of the transformational part
of the pertinent logical theories. However, I will not indicate axioms for the
formalisation function π from English to $\mathcal{L}^*$ here. The reader is assumed to
have had some training in formalising arguments into the language of first-
order predicate logic, and this should suffice for the following discussion.

Apart from questions of detail, there are also various issues of a more general
nature. Consider the following argument:

(6) Peter hits Paul at midnight. Therefore, Peter hits Paul.

Some philosophers (e.g. Davidson 1967b) believe that an *underlying logical form*
of this argument makes it accessible to first-order predicate logic. They might
want to map this argument to a formal argument like the following, which is
logically valid according to classical and intuitionistic first-order predicate logic:

(7) $\exists a\,(R_1ac_1 \wedge R_2ac_2 \wedge R_3ac_3) \succ \exists a\,(R_1ac_1 \wedge R_2ac_2)$

Other philosophers (e.g. Strawson 1974) are skeptical about the idea that arguments may have logical forms that differ in such a radical way from their surface
structure. They would deny that (6) is logically valid relative to classical (or
intuitionistic) first-order predicate logic. In their view such an argument could
only be logically valid according to a logical theory which deals directly with
adverbial modification.

For reasons of space, I will set aside most issues concerning formalisation.
I will only make some remarks about *negation* and *conditionality* that concern
questions about formalisation which are relevant for some of what follows. (In
1.5, I will take a look at questions about formalisation which are connected with
the presence of empty terms in natural languages.)

As a preparation, I include a clarificatory remark about "that"-clauses and
singular terms. "That"-clauses are syntactic units of more complex expressions.
They occur in at least three different environments. They combine with certain (modified) verbs, with certain (modified) verbal nouns, and with certain
(modified) adjectives:

(8) Ann firmly <u>believes</u> *that Peter is tall.*

(9) The ridiculous <u>claim</u> *that Peter is small* was never made.

(10) It is absolutely <u>unbelievable</u> *that Peter is small.*

In all of these constructions, a "that"-clause forms the complement of a certain
expression. Now, as I use this phrase, something is a *singular term* only if it is a
singular determiner/noun phrase.[18] Thus, for syntactic reasons, "that"-clauses

---

[18]   Since Abney's dissertation (1987) many (but not all) linguists call phrases like "the pres-
ident" *determiner phrases* instead of *noun phrases*. Uncontroversially, however, "that"-
clauses are neither determiner phrases nor noun phrases. Note that there are certain
sentential fragments in which the syntactic difference between a "that"-clause on the one
hand and a sentence or a determiner/noun phrase on the other hand is invisible:

are not singular terms. The question of whether "that"-clauses resemble singular terms semantically, however, remains.

I will now turn to the first important question about formalisation: what is the negation of a declarative sentence? In English there are various rather different means for negating a sentence (cp. Napoli 2006: 239-45). Consider the following examples:

(11) Peter is tall. ⤳ Peter is not tall.

(12) Peter or Mary is tall. ⤳ Neither Peter nor Mary is tall.

(13) Some philosopher is tall. ⤳ No philosopher is tall.

Some philosophers think that there are also uniform methods in English for negating a sentence. Given any sentence S, the following sentences are often taught in logic classes to be negations of S:

(14) It is not the case that S;   (17) That S is not the case;

(15) It is not true that S;   (18) That S is not true;

(16) It does not hold that S;   (19) That S does not hold.

Note that in (14)-(19) the sentence S occurs as part of a "that"-clause. If one is willing to call sentences like (14)-(19) *negations* of S, then one will presumably accept corresponding equivalence principles. Such a principle states that a sentence S and one of the following sentences can be formalised in the same way:

(20) It is the case that S;   (23) That S is the case;

(21) It is true that S;   (24) That S is true;

(22) It does hold that S;   (25) That S does hold.

Given such an assumption, say the assumption that S and "it is true that S" are equivalent, it can then be said that there is a uniform method for negating a sentence because there is a uniform method for negating sentences of the form "it is true that S", and every sentence is equivalent to a sentence of such a form. According to this view, a sentence does not have to be an immediate constituent of its negation; it is allowed that it is only an immediate constituent of a "that"-clause which is, in turn, an immediate constituent of its negation.

The question of whether a sentence S and a corresponding sentence from (20)-(25) are to be formalised in the same way may be relevant for questions about logical validity. Consider a person who accepts the premise of the following argument but rejects its conclusion (she thinks that Peter is just a few inches

(!) Ann believes ((the proposition) that) Peter is tall.

  The two words "Ann believes" can be combined with a sentence, or a "that"-clause, or a determiner/noun phrase. Similarly, phrases like "is true" can not only be combined with a determiner/noun phrase but also with a "that"-clause. (However, this is not a general feature of singular predicative expressions:

(?) That Peter is tall exists (/is an object/is abstract).

  Each of the three displayed supplementations of "That Peter is tall" is ill-formed.)

taller than borderline cases of tallness but that he is not a definite example of a tall person):

(26) Peter is tall. Therefore, it is true that Peter is tall.

If this person wants to stick to the principle that every sentence logically implies itself, then she should not formalise a sentence S and the corresponding sentence "it is true that S" in the same way.

Those who believe that a logical theory should not distinguish between a sentence S and the corresponding sentences from (20)-(25) may use a formalisation function which treats the sentences in (14)-(19) as negations of S. They will then say that (26) is logically valid according to classical and intuitionistic first-order predicate logic. Those who believe that "that"-clauses make for a logical difference should use a formalisation function according to which the negation of a sentence S has to be a sentence in which S occurs as an immediate constituent. It will then be said that (26) is not logically valid according to classical and intuitionistic first-order predicate logic. (Though (26) could still be logically valid relative to a set which contains "true" and "that" if these are considered to be logical constants.)

Even if there is no uniform method for negating a sentence in English, for some of what follows it is useful to pretend otherwise. More precisely, I will consider a slight variant of English as the informal object language in which this possibility is realised. When it matters, I will use the following device: to form the negation of a sentence, one may prefix it with the expression "not-". (And if there is a risk of ambiguity, one may use brackets to indicate the scope of the negation operator.)

Compared to the logico-grammatical subtleties of *conditionality*, those of negation seem to be comparatively insignificant. I will only mention three of them which are especially important. The first one concerns the logico-grammatical category of "if". According to the orthodox view, "if" belongs to the same category as "or" and "and": all of them are binary connectives, i.e. operators which combine two sentences to form another sentence. This orthodoxy has been questioned by Lewis and Kratzer. Lewis (1975) claims that a token of "if" in a restrictive "if"-clause does not function as a sentential connective but rather as an adverb restrictor. Kratzer (1986, 1991) extends this treatment to other occurrences of "if". However, since all logical theories considered in this book are based on the assumption that "if" plays the role of a two-place connective, I will not consider the Lewis-Kratzer view here.

The second and the third point concerns the extent of the relevant kind of conditionality: which sentences of English should be mapped by the formalisation function to sentences which have $\rightarrow$ as their main operator? This can be split up into two questions:

(i) Should every sentence of the form "if S_1, then S_2" (and stylistic variants like "S_2 if S_1") be mapped to such a formal sentence?

(ii) Should some sentences which do not involve "if" be mapped to such formal sentences?

To answer the first question, consider the following argument:

(27) Socrates had not been a cook. Therefore, if he had been a cook, he would
     have lived in France.

Suppose someone argues to the conclusion that (27) is not logically valid; does
she thereby argue against a theorem of classical and intuitionistic logic? To resist
this conclusion, one should point out that in the above conclusion "if" has been
used to form a *counterfactual* or *subjunctive* conditional, and one should adopt
the view that only *indicative* "if"-conditionals are mapped to sentences whose
main operator is $\rightarrow$.[19]

Consider now the second question: should some sentences which do not
involve "if" be mapped to sentences in which $\rightarrow$ occurs? For what follows, there
is one construction which is of special importance. Some philosophers, especially
some of those who favour theories of relevant logic, believe that conditionality
can also be expressed by verbs like "imply" or "entail" (see 1.4). Consider the
following argument:

(28) Socrates was not a cook. Therefore, that Socrates was a cook implies that
     he lived in France.

It might be claimed that classical and intuitionistic logic fail if this argument
is not logically valid.

As in the case of negation, there are two options. Those who believe that
a logical theory should not distinguish between a sentence of the form "if S_1,
then S_2" and the corresponding sentence "that S_1 implies that S_2" may use a
formalisation function which treats implication sentences as conditionals. They
will then say that (28) is logically valid according to classical and intuitionistic
first-order predicate logic. Those who believe that implication and conditionality should be treated differently by a logical theory should deny that (28) is
logically valid according to classical and intuitionistic first-order predicate logic.

1.2.5   Logically Valid Rules

I would like to close this section with some remarks about the logical validity
of argumentative steps and rules for performing such steps. Recall that a rule is
here identified with a set of transitions, pairs of a finite set of open arguments
of $\mathcal{L}^*$, which are called the *initial arguments* of the transition, and another
open argument of $\mathcal{L}^*$, which is called the *final argument* of the transition. The
question at issue is whether a logical theory, as it is understood here, is suited
to give an account of the logical validity of such transitions and rules. For
simplicity, I will disregard the differences between the formal and the informal
object language in this subsection. In particular, I will not distinguish between
$\mathcal{L}^*$-validity and logical validity in $\mathcal{L}$ with respect to the set of standard logical
constants.

There is one important kind of difficulty which I will set aside, namely that
rules are applied to *open* arguments, i.e. expressions which may contain parameters. Rather, I will restrict attention to rules whose transitions are made up

---

[19]   See Edgington 1995 for more on these notions and for further references.

of (closed) arguments. In particular, I will consider the corresponding restrictions of the rules for the sentential operators of the calculi $\mathcal{R}^I$ and $\mathcal{R}^C$, and I will not consider the rules for the quantifiers.

In the calculi $\mathcal{R}^I$ and $\mathcal{R}^C$ from 1.2.3 there are two kinds of such rules: those which do and those which do not involve the discharge of assumptions. Consider first a rule of the former type, say the rule for introducing a conjunction:[20]

$$\frac{\Gamma \succ \varphi \quad \Delta \succ \psi}{\Gamma, \Delta \succ \varphi \wedge \psi} \ (\wedge_I)$$

where φ and ψ are schematic letters for sentences and where Γ is a schematic letter for finite lists of sentences of the formal object language. This rule permits the transition from two arguments $X \succ S_1$ and $Y \succ S_2$ to another argument $X, Y \succ S_1 \wedge S_2$. Put informally, this rule allows one to perform a simple inferential act: it allows one to infer a conjunction $S_1 \wedge S_2$ from its two conjuncts S_1 and S_2. The conjunction then depends on exactly those assumptions on which either S_1 or S_2 depends.

A logical theory, it might plausibly be said, should not only be concerned with the logical validity of arguments but also with the logical validity of rules. Furthermore, the above rule and its applications should be classified as logically valid according to both classical and intuitionistic logic. How can the type of logical theories considered here account for this? Is there a general recipe for translating claims about the logical validity of transitions and rules into claims about the logical validity of arguments? In the case of a rule which does not involve the discharge of assumptions, one can interpret the claim that it is logically valid as the claim that the argument from the conclusions of the initial arguments to the conclusion of the final argument is logically valid. In the case of $(\wedge_I)$ this would amount to the following sentence:

$(\wedge_I^V)$ For all sentences S_1 and S_2 of $\mathcal{L}^*$, $S_1, S_2 \succ S_1 \wedge S_2$ is $\mathcal{L}^*$-valid.

(It should be noted though that the sentence $(\wedge_I^V)$ might equally be treated as capturing the rule $\frac{\Gamma \succ \varphi \quad \Gamma \succ \psi}{\Gamma \succ \varphi \wedge \psi}$, where it is required that the initial arguments have the same list of premises.)

However, once one considers rules which involve the discharge of assumptions, it becomes apparent that the above recipe for translating claims about the logical validity of a rule into claims about the logical validity of arguments is not universally applicable. As an example, consider the rule of conditional proof:

$$\frac{\Gamma, \varphi \succ \psi}{\Gamma \succ \varphi \to \psi} \ (\to_I)$$

This rules permits the transition from an argument $X, S_1 \succ S_2$ to another argument $X \succ S_1 \to S_2$. Put informally, this rule allows one to perform a complex act in a situation in which one has established a sentence S_2 conditional on an assumption S_1: it allows one (i) to withdraw one's temporary commitment to

---

[20]   Due to the restriction to sentences, this is not the rule $(\wedge_I)$ of the calculi of natural deduction from 1.2.3. I will nevertheless use the same name.

S_1, and, at the same time, (ii) to put forward the conditional $S_1 \rightarrow S_2$. The conditional then depends on exactly those assumptions on which S_2 depended apart from the discharged assumption S_1. Again, this rule should be classified as logically valid according to both classical and intuitionistic logic.

It would be obviously inadequate to interpret the claim that this rule is logically valid as the claim that for all sentences S_1 and S_2 the argument from S_2 to $S_1 \rightarrow S_2$ is logically valid. What is needed is a sentence which takes into account that in moving to $S_1 \rightarrow S_2$ one may discharge the assumption S_1.

The best candidate for a sentence of the language of $\mathcal{T}$ which expresses the claim that the rule $(\rightarrow_I)$ is logically valid, seems to be the following universally quantified conditional:

$(\rightarrow_I^V)$ For every finite list X of sentences of $\mathcal{L}^*$ and for all sentences S_1 and S_2 of $\mathcal{L}^*$, if $X, S_1 \succ S_2$ is $\mathcal{L}^*$-valid, then so is $X \succ S_1 \rightarrow S_2$.

However, although this is indeed a theorem of classical and intuitionistic logic, it does not capture the full force of the claim that $(\rightarrow_I)$ is logically valid. To see this, consider the following rule:

$$\frac{\succ Fc}{\succ Fd} \ (?)$$

where F is a schematic letter for unary relation symbols and where c and d are schematic letters for constants. In my view, neither a classicist nor an intuitionist should classify $(?)$ as a logically valid rule of first-order predicate logic. Classical or intuitionistic logic does not give permission to make the transition from an argument $\succ Fc$ to an argument $\succ Fd$. Put informally, neither classical nor intuitionistic logic allows one to infer a sentence Fd from a sentence Fc which depends on no assumptions.[21]

Suppose it is granted that $(?)$ is not logically valid by either classical or intuitionistic standards. If one now uses the above recipe, one obtains the following sentence:

$(?^V)$ For every unary relation symbol F of $\mathcal{L}^*$ and for all constants c and d of $\mathcal{L}^*$, if $\succ Fc$ is $\mathcal{L}^*$-valid, then so is $\succ Fd$.

This sentence, however, *is* a theorem of the logical theories of intuitionistic and classical first-order predicate logic. The reason is simply that no atomic sentence is logically true according to these theories.

In my view this suggests that a conditional of the form

(29) if α is $\mathcal{L}^*$-valid, then so is β

does not ensure the logical validity of the transition from α to β and that a universal quantification which corresponds to a set of such conditionals does not ensure the logical validity of the corresponding set of transitions, i.e. of the corresponding rule. What the above recipe offers is a statement to the effect that

---

[21] Perhaps not everyone will agree with this verdict. A justification requires a closer inspection of the significance of natural deduction rules (see Chapter 5).

the rule (?) is *admissible* in the following sense: if one adds (?) to the calculi $\mathcal{R}^C$ or $\mathcal{R}^I$ of classical or intuitionistic first-order logic, no new arguments become derivable. Consequently, if my claim that (?) is not logically valid is correct, then admissibility falls short of guaranteeing logical validity.

According to the proposed conception of a logical theory, a such theory is concerned with the logical validity of arguments, and the notion of logical validity of an argument does not yield a notion of logical validity of a rule from sets of arguments to arguments. If one is interested in the logical validity of such rules, new machinery has to be set up.

1.3 Arguing Against a Logical Theory

What does it take to argue against a logical theory? In one sense, this is a rather theoretical undertaking in which logical theories are explicitly mentioned: arguments to the conclusion that this or that logical theory is not sound. (A theory is called *sound* iff all its axioms are true.) In another sense, one can argue against a logical theory without mentioning this theory at all. Consider someone who argues to the conclusion that the following argument is not logically valid according to the set of standard logical constants $\mathcal{C}$:

(30) Socrates was not a cook. Therefore, if he was a cook, he lived in France.

As I want to use this phrase, such a person argues against the logical theories $\mathcal{T}^C$ and $\mathcal{T}^I$ no matter whether they play any role in her argumentation. In general, there are two kinds of critique of a logical theory $\mathcal{T}$ in this sense. First, one might argue that one of the theorems of $\mathcal{T}$ should not be one. Second, one might argue that a sentence of the language of $\mathcal{T}$ which is not a theorem of $\mathcal{T}$ should be one. In this section, the main focus will be on the first kind of critique.

As was said in 1.2.2, there is an immediate difficulty. There is no general agreement about which sentences are theorems of a logical theory. To take the most important example: a classical logician and an intuitionistic logician might accept the same logical theory, e.g. the model-theoretic theory $\mathcal{T}^m$, and, nevertheless, the classical logician will affirm while the intuitionistic logician will deny that the following argument is logically valid:

(31) If Socrates did not live in France, he was not a cook. Therefore, if he was cook, he lived in France.

One way out of this difficulty is to concentrate on derivational logical theories. In this section, I will follow this path. With respect to a derivational logical theory $\mathcal{T}$ there is general agreement about the question of whether certain especially important sentences (see below) are theorems of $\mathcal{T}$.

1.3.1 Types of Sentences of the Language of $\mathcal{T}$

Suppose you are concerned with arguments against classical logic. No matter which kind of theory $\mathcal{T}$ is at issue, a model-theoretic one or a derivational one,

it is certainly counter-intuitive to treat every argument against a theorem of
$\mathcal{T}$ as an argument against classical logic. The axioms of the formal syntactic
part $\mathcal{T}_{synt}$ imply that every expression e is contained in a larger expression (e.g.
in the concatenation of e with itself). Now, consider some finitist who believes
that there are only a finite number of objects. Such a person rejects a theorem
that is common to all the logical theories considered here. Intuitively, however,
he does not thereby reject classical logic. His mathematics is unorthodox, his
logic might be absolutely standard.

This points to the fact that only certain sentences of the language of $\mathcal{T}$
concern the logical properties of the object language. Furthermore, it seems that
those sentences concern the logical properties of the object language in different
ways: some sentences deal with practically important logical properties, while
other sentences are of a more theoretical and meta-logical character. I will
now give some illustrating examples of such different types of sentences of the
language of $\mathcal{T}$.

To begin with, I will introduce sentences which are of special importance
for the logical properties of the object language. Suppose that t is a canonical
term for an argument α of $\mathcal{L}$. Then, the following sentence is called a *canonical
logicality sentence for α*:

CLS(α) t is logically valid in $\mathcal{L}$ w.r.t. $\mathcal{C}$.

A canonical logicality sentence states of some canonically specified argument
that it is logically valid in $\mathcal{L}$ with respect to $\mathcal{C}$. If such a sentence is a theorem
of some logical theory $\mathcal{T}$, then it is called a *canonical logicality theorem* of
$\mathcal{T}$. I assume that all canonical logicality sentences are equally important for
a characterisation of the logical properties of the object language. This may
be doubted. It might be held, for example, that canonical logicality sentences
about simple arguments are more important than those about complex ones.
(One possible measure of simplicity would be the number of (tokens of) logical
constants which are present in the argument.) I will not try to undermine such
proposals here. As will be seen, the pertinent sentences always concern fairly
simple arguments.

Canonical logicality sentences concern especially fundamental logical properties of the object language. However, there are some further interesting classes
of sentences. First, one might consider sentences which are built up from canonical logicality sentences by means of the usual sentential connectives. Second,
and more interestingly, there are sentences which state general claims about
logical validity. Consider the following two examples:

(32) For some argument α of $\mathcal{L}$, α is logically valid in $\mathcal{L}$ w.r.t. $\mathcal{C}$.

(33) For every sentence S of $\mathcal{L}$, an argument in which S is the only premise
and the conclusion is logically valid in $\mathcal{L}$ w.r.t. $\mathcal{C}$.

The first one states that some argument is logically valid, and the second one
states that every sentence logically implies itself. These and similar sentences are
clearly concerned with the logical properties of the object language. However,

the second one in particular is of a more theoretical and meta-logical character than canonical logicality sentences.

Third, there are sentences about the calculus underlying the pertinent logical theory. Consider the following example:

(34) For every argument α, α is logically valid in $\mathcal{L}$ w.r.t. $\mathcal{C}$ iff $\pi(\alpha)$ is $\mathcal{R}^C$-derivable.

This sentence states that an argument is logically valid iff its formalisation is derivable by means of the rules of the classical calculus $\mathcal{R}^C$. In some sense, this is surely a very important statement about the logical properties of $\mathcal{L}$. On the other hand, it is conceptually quite demanding and should not be treated on a par with sentences which just state of some canonically specified argument that it is logically valid.

Finally, there are also sentences which do not concern the logical properties of the object language at all. Consider the following example:

(35) For every expression e of $\mathcal{L}$, the concatenation of e with itself is also an expression of $\mathcal{L}$.

Such sentences have to be part of the language of a logical theory, but they do not deal with logical features of the object language.

For someone who is interested in arguments against a certain logical theory, say classical logic, the moral to be drawn from these examples is that not all arguments against theorems of classical logic are equally important. Some theorems - like (35) - are, so to speak, only present by accident; one can reject them from an orthodox classical logician's viewpoint. Other theorems state ingredients of the classical enterprise which are especially important from a more theoretical perspective; if someone accepts all finite arguments which are logically valid according to classical logic while she refuses to accept general statements like (34), then she might not count as a classical logician but nevertheless as a person who accepts classical logic. For present purposes, however, most important are canonical logicality theorems. These I will now discuss at greater length.

1.3.2   Four Ways of Arguing Against a Logical Theory

An especially important way of arguing against a logical theory $\mathcal{T}$ is to argue against some canonical logicality theorem of $\mathcal{T}$. Suppose that α^* is the argument "$S_1 \ldots S_n$ Therefore, S" of $\mathcal{L}$. Furthermore, suppose that t^* is a canonical term for α^*, and suppose that the canonical logicality sentence for α^* is a theorem of $\mathcal{T}$:

CLS(α^*) t^* is logically valid in $\mathcal{L}$ w.r.t. $\mathcal{C}$.

I will now introduce four kinds of arguments which could be offered against (consequences of) CLS(α^*).

To begin with, there is the possibility of arguing in favour of the negation of CLS(α^*):

LV-Criticism t^* is not logically valid in $\mathcal{L}$ w.r.t. $\mathcal{C}$.

As an example of LV-Criticism, consider an intuitionistic argument to the conclusion that the argument (31) (from the introduction to 1.3) is not logically valid in $\mathcal{L}$ with respect to $\mathcal{C}$.[22]

In addition to this way of arguing against a canonical logicality theorem, there is another one that is no less important. Consider someone who does not master the concept of logical validity and who argues to the following conclusion:

(36) The argument "Socrates was not a cook. Therefore, if Socrates was a cook, then he lived in France." is not valid.

Such a person does not thereby argue against a canonical logicality theorem of the theory $\mathcal{T}^C$ of classical logic (or against any other of its theorems). Nevertheless, she argues for a non-classical conclusion in a very strong sense. Another notion of questioning arguments declared to be logically valid by some logical theory is therefore useful. A second way to argue against $\mathcal{T}$ is to argue for the following conclusion:

V-Criticism t^* is not valid.

The idea is that *validity* is a weaker property than logical validity - weaker in the sense that every logically valid argument is valid but not *vice versa* - and the attack amounts to the assertion that an argument that is claimed to be logically valid is not even valid. What does validity amount to? The standard account is a modal one: an argument is valid iff it is necessary that if all its premises are true, then so is its conclusion.[23] However, I will not commit myself to this or any other explication of this notion.[24] Since "valid" is not a technical invention but a pre-theoretic term, one can go some way along with it without subscribing to any analysis of it.

There is a third way of questioning arguments that are declared to be logically valid by some logical theory. I will say that an argument α is *truth-preserving* iff the following holds: if the premises of α are true, then the conclusion of α is true. Then, a third way to argue against $\mathcal{T}$ is to argue for the following conclusion:

TP-Criticism t^* is not truth-preserving.

[22] Another important way of arguing against a logical theory $\mathcal{T}$ is to argue in favour of a canonical logicality sentence which is not a theorem of $\mathcal{T}$. If someone argues, for example, that the argument (31) *is* logically valid in $\mathcal{L}$ w.r.t. $\mathcal{C}$, then he argues against the intuitionistic theory $\mathcal{T}^I$ in this sense.

[23] It has to be presupposed that one keeps the meanings of the premises and the conclusion constant when one evaluates the argument with respect to counterfactual situations: the argument "This is furze. Therefore, this is gorse." is valid according to the modal explication although there are counterfactual situations in which "furze" and "gorse" have such meanings that something satisfies the former but not the latter.

[24] Furthermore, in contrast to Read (1988: 19-20) and Mares (2004: 3), I will not take the modal explication of validity to be an essential ingredient of classical logic.

That is, one argues against $\mathcal{T}$ in this sense if one argues in favour of the negation of the sentence "if S_1 is true, and $\ldots$, and S_n is true, then S is true". Here it is assumed that if an argument is valid, then it is truth-preserving.[25]

The special case of arguments without premises is worth noting. Suppose that S is logically true according to $\mathcal{T}$. Then, the above way of arguing against $\mathcal{T}$ amounts to an argument in favour of the conclusion that S is not true. For example, intuitionists claim that the sentence "every real number is or is not rational" is not true.

Note that each of the mentioned forms of criticism is meta-linguistic in the sense that it explicitly speaks about a certain argument. Closely related to TP-Criticism, there is also an indirect way of arguing against a canonical logicality theorem which is not meta-linguistic. Suppose that the informal object language is English. Then, a fourth way to argue against $\mathcal{T}$ is to argue for the following conclusion:

B-Criticism Not-(if S_1 and $\ldots$ and S_n, then S).

Arguing for such a conclusion will be referred to as *basic criticism*. For example, intuitionists believe that not every real number is rational or irrational, and they also believe that there is no real number which is neither rational nor irrational. Consequently, they accept the following sentence:

(37) Not-(if not every real number is rational or irrational, then some real number is neither rational nor irrational).

When intuitionists argue in favour of (37), then they argue against the classical theory $\mathcal{T}^C$ in this basic sense.

I take basic criticism of a logical theory to be of special importance. Consider a person who is prepared to use all arguments which are logically valid according to the classical theory $\mathcal{T}^C$, but who rejects all sentences which contain terms for arguments of $\mathcal{L}$ because their truth presupposes the existence of abstract objects (namely the existence of arguments of $\mathcal{L}$). It should be possible to describe this person as someone who accepts classical logic: she accepts the arguments which are logically valid according to $\mathcal{T}^C$ although she rejects the sentences which state of these arguments that they are logically valid, or valid, or truth-preserving. This person does not argue against $\mathcal{T}^C$ in the sense of levelling basic criticism against it. And this separates her from intuitionists who do question $\mathcal{T}^C$ in this sense. Intuitionism stands out as one of a small number of enterprises that put forward basic criticism of classical logic.

It should be noted that arguments against a logical theory $\mathcal{T}$ for some object language $\mathcal{L}$ in the present sense have conclusions which are sentences of the object language $\mathcal{L}$. Thus, if one puts forward basic criticism of a logical theory, then the object language has to be English. To obtain a meaningful expression from the sentential frame "Not-(if $\cdot$ and $\ldots$ and $\cdot$, then $\cdot$)", one has to insert declarative sentences of *English*. (One has to use *translations* of

---

[25]   This assumption has been questioned by Field (2009c) for reasons that relate to the semantic paradoxes. I will not go into this here.

this sentential frame if one deals with basic criticism of logical theories for other object languages.)

In sum, there are four ways of arguing against (consequences of) the canonical logicality theorem for α^*:

LV-Criticism   "$S_1 \ldots S_n$ Therefore, S" is not logically valid in $\mathcal{L}$ w.r.t. $\mathcal{C}$.

V-Criticism   "$S_1 \ldots S_n$ Therefore, S" is not valid.

TP-Criticism "$S_1 \ldots S_n$ Therefore, S" is not truth-preserving.

B-Criticism   Not-(if S_1 and $\ldots$ and S_n, then not-S).

In the first case, one argues directly in favour of the negation of some theorem of $\mathcal{T}$. In the other cases, one argues against important consequences of some theorem of $\mathcal{T}$.

1.4   Relevance

In this and the following section, I will discuss two types of considerations which have been used to argue against the theory of classical logic: considerations about relevance and considerations about empty singular terms. The main goal will be to highlight those aspects of these anti-classical arguments which differentiate them from intuitionistic arguments against classical logic.

In this section, I will discuss arguments against the theory of classical logic which concern the notion of *relevance* in relation to the notions of *conditionality*, *implication*, and *validity*.[26] The proponents of these arguments claim that the theory of classical logic is not sound since it does not satisfy *relevance constraints* like the following:

R₁ A conditional is true only if the truth of its antecedent is relevant for the truth of its consequent.

R₂ An implication is true only if the truth of its antecedent is relevant for the truth of its consequent.[27]

R₃ An argument is valid only if the truth of its premises is relevant for the truth of its conclusion.

These constraints are not meant to imply that every true conditional and implication has a true antecedent and a true consequent and that every valid argument has true premises and a true conclusion. The point is rather that the information about whether an antecedent or a premise is true is relevant for the question of whether the corresponding consequent or conclusion is true.

Adherents of these principles are called *relevantists*, and their preferred logical theories are called *theories of relevant logic*. I will give a short introduction to the concerns of relevantists and elucidate and defend the following two theses:

---

[26]   These arguments are equally directed against the theory of intuitionistic logic. For ease of exposition, I will consider them only in relation to classical logic.

[27]   Here an implication is a sentence of the form "that S_1 implies that S_2", and the sentences S_1 and S_2 are its antecedent and its consequent.

T$_1$-Rel Relevantists criticise classical logic on the basis of the claim that there are sentences/arguments which are intuitively classified as untrue/invalid, although they are logically true/valid according to classical logic.

T$_2$-Rel Relevantists do not criticise the argumentations of classical mathematics.

Intuitionists differ from relevantists in both respects. As will be seen, intuitionists do not refer to widely shared intuitions about the classification of certain sentences and arguments, and they *do* criticise standard mathematical practice.

1.4.1   Conditionality, Implication, and Validity

I will now illustrate the concerns of relevantists as regards the classical treatment of conditionality, implication, and validity.[28] To begin with, I will focus on conditionality. Consider the following two indicative conditionals:

(38) If Brouwer was and was not born in Amsterdam, then 2 equals 3.

(39) If Brouwer was not born in Amsterdam, then 2 equals 3 if Brouwer was born in Amsterdam.

Both of these sentences are logically true according to classical logic.

Relevantists object to these classifications. According to theories of relevant logic, these sentences are not logically true, which seems to speak in favour of these theories because the sentences (38) and (39) are intuitively rejected. Consider the second sentence. It has a true antecedent (Brouwer was born in Overschie, now part of Rotterdam) but a seemingly untrue consequent: it seems incorrect to say that 2 equals 3 if Brouwer was born in Amsterdam. It therefore has to be acknowledged, I think, that the theory of classical logic declares some sentences to be logically true which are intuitively rejected.

The motivation to adopt a theory of relevant logic is not fully disclosed by mentioning the fact that sentences like (38) and (39) seem to be unacceptable. A characteristic of relevantists is their diagnosis of the unacceptability of such sentences. They claim that a conditional is true only if the proposition which is expressed by the antecedent and the proposition which is expressed by the consequent are suitably related: the truth of the former has to be relevant for question of whether the latter is true. Now, the contradictory supposition that Brouwer was and was not born in Amsterdam is irrelevant for the question of whether 2 equals 3; therefore, relevantists deny that (38) is true. Similarly, the supposition that Brouwer was born in Amsterdam is irrelevant for the question of whether 2 equals 3; therefore, relevantists deny that (39) has a true consequent. And since (39) has a true antecedent, they infer that it is not true.

---

[28]   For further examples of sentences and arguments whose classical treatment relevantists object to, see Anderson & Belnap 1975: 17-8, Routley et al. 1982: 5-8, and Read 1988: 23-4.

Relevantists are not only concerned with the logical behaviour of the sentential operator "if"; they also apply their reasoning to transitive verbs like "imply" or "entail".[29] Focussing on implications, one can adapt the above example:

(40) That Brouwer was and was not born in Amsterdam implies that 2 equals 3.

As in the case of the conditional (38), this sentence is intuitively rejected.

It might now be asked whether conditionals and implications have to be treated differently or whether they are to be treated in the same way (see 1.2.4). Anderson and Belnap opt for the latter alternative:

> [I]t is philosophically respectable to 'confuse' implication or entailment with the conditional, and indeed philosophically suspect to harp on the dangers of such a 'confusion' (Anderson & Belnap 1975: 473).

They believe that there is a single logical concept expressed in different linguistic guises. I am skeptical about this claim. Suppose that Peter thinks about throwing a fair coin and consider the following two sentences:

(41) If Peter throws the coin, it comes up heads.

(42) That Peter throws the coin implies that it comes up heads.

Then, suppose that Peter throws the coin and that it comes up heads. In this case I am inclined to say that (41) is true while (42) is false. As regards (41), suppose that Mary had assertively uttered this sentence before Peter had thrown the coin and that Paul bet against Mary's assertion. Then, since Peter threw the coin and it came up heads, Mary would have won. And this suggests, I think, that Mary would have uttered a true sentence. As regards (42), I am inclined to think that this implication is false because it was equally likely that the coin would come up tails.[30] However, although it does not seem to be philosophically respectable to confuse implication and conditionality, it still holds (see below) that implications which are intuitively rejected can be used in arguments against classical logic.

Up to this point one might think that theories of relevant logic differ only locally from theories of classical logic. The idea would be that relevantists and classicists treat the logical constants "not", "or", "and", "every", and "some" in the same way and that they have only conflicting beliefs about the logical properties of "if". This is not true, however. Relevantists do not only argue in favour of a non-classical treatment of conditionality. Certain arguments that are logically valid according to classical logic and do not contain the conditional operator are rejected as invalid. Here is one example:

---

[29] As a background to this, it is worth recalling that not every logician takes the conditional operator "if" to be a basic logical constant (like "not", "or", "and", "some", and "every"). For example, Russell & Whitehead (1910: 9) use the verb "imply" in this context, and so does Kleene (1952: 69).

[30] This does not mean that a sentence of the form "that S_1 implies that S_2" implies the corresponding sentence "necessarily, if S_1, then S_2" (cp. Dunn & Restall 2002: 5).

(43) Brouwer was born in Amsterdam. Brouwer was not born in Amsterdam. Therefore, 2 equals 3.

This argument is logically valid according to classical logic, but relevantists reject it as invalid. Furthermore, they claim that the invalidity of this argument results from a lack of relevance: the truth-values of the propositions expressed by the premises are irrelevant for the truth-value of the conclusion. Thus, logical theories of relevant logic do not only differ locally from classical logic.

These considerations about the validity of arguments are connected to the previously mentioned considerations about the truth of implications. This connection can be expressed by the following thesis (cp. Anderson & Belnap 1975: 480-9):

Equivalence Thesis An argument of the form "$S_1 \ldots S_n$ Therefore, S" is valid iff the sentence "that S_1 and $\ldots$ and S_n implies that S" is true.

On the assumption that the Equivalence Thesis holds, it can be seen why untrue implications can form the basis for an argument against classical logic; they do so whenever the corresponding argument is logically valid according to classical logic.

My present aim is not to evaluate these relevantistic arguments. What I wanted to illustrate is that relevantistic criticism of classical logic is based on the existence of certain intuitively unacceptable sentences and seemingly invalid arguments which are logically true and logically valid according to classical logic. Relevantists try to put forward logical theories and corresponding semantic theories which fit to intuitive judgements of competent speakers of the pertinent language. For example, Mares (2004: 15) claims that the theory of relevant logic is superior to the theory of classical logic because the consequences of the former fit better to intuitive judgements of native speakers than the consequences of the latter. As will be seen in the following chapters, this makes relevantistic criticism very different from intuitionistic criticism. Intuitionists do not try to capture the intuitive judgements of native speakers about the status of various sentences and arguments. Intuitionistic arguments against classical logic turn out to be less accessible than relevantistic ones.

1.4.2   Relevantism and Classical Mathematics

I will now turn to the thesis T$_2$-Rel which states that relevantists do not object to classical mathematics. Its truth can not only be inferred from the absence of critical remarks about proofs and results of classical mathematics in the writings of relevantists. It characteristically shows itself when relevantists present proofs of their own. Unlike intuitionists, who do mathematics in a distinctively non-classical way, relevantists present classical mathematical proofs and results. Following John Burgess, I take the best explanation for this fact to be the thesis that relevantists put forward a kind of descriptive criticism rather than a prescriptive one:

> There is a difference between *prescriptive* criticism, alleging that classical logic correctly describes unacceptable practices in the com-

munity of orthodox mathematicians, and *descriptive* criticism, alleging that classical logic endorses incorrect principles that are *not* operative in the practice of that community (Burgess 2005: 729).

Intuitionists criticise classical mathematicians for relying on anti-intuitionistic principles, but relevantists do not criticise classical mathematicians at all. They rather reject the claim that anti-relevantistic principles are really relied on in classical mathematics.

This observation fits very well with the motivation for adopting a theory of relevant logic as reported in the previous subsection. Relevantists advertise a logical theory that fits intuitive verdicts about certain sentences and arguments. It would be astonishing if they combined such a view with the claim that the alleged untrue sentences and invalid arguments are routinely relied on in ordinary mathematical contexts. And indeed, relevantists rather mention hypothetical anti-relevantistic arguments in mathematical contexts to illustrate that such arguments are *not* employed in standard mathematics.[31]

What might be the idea behind the claim that anti-relevantistic principles are not really relied on in mathematical contexts? It would be naive to claim that all arguments put forward in ordinary mathematical proofs are logically valid according to theories of relevant logic. In particular, in many ordinary mathematical proofs a sentence S_2 is inferred from two sentences "S_1 or S_2" and "not-S_1", and this argument, corresponding to the rule *disjunctive syllogism* (see 2.2.4), is not logically valid according to theories of relevant logic. How does this cohere with the claim that anti-relevantistic principles are not relied on in mathematical contexts? Relevantists acknowledge that this is a problem for their preferred logical theory:

> The fact that disjunctive syllogism cannot be added to relevant logic is a problem. The original point of introducing relevant logic was to provide an intuitive characterisation of deductive inference. But we use disjunctive syllogism all the time (Mares 2004: 176).

If the distinctively classical rule of disjunctive syllogism is used in mathematical proofs, how is it possible that standard mathematics is independent of anti-relevantistic principles?

According to what I take to be the most plausible strategy for dealing with this problem, one abandons the idea that mathematical argumentations are always to be interpreted as purely logical argumentations (from certain mathematical principles). No one claims that all valid arguments are logically valid (cp. "This is furze. Therefore, this is gorse."). Similarly, no one should claim that all valid arguments used in mathematical argumentations are logically

---

[31]   Compare Anderson & Belnap 1975: 17-8 and Mares 2004: 3. Mares here criticises classical logic for declaring certain mathematical arguments with true premises to be valid which are, for good reasons, never used in mathematical proofs. In my view, this criticism is not convincing. According to nearly all theories of relevant logic, including the one that Mares favours, there are arguments with only true premises which are declared to be logically valid although they are, for good reasons, never used in mathematical proofs: arguments of the form "S Therefore, S".

valid. An argument by disjunctive syllogism in a mathematical proof might be valid without being logically valid. It is thus not incoherent to combine the descriptive criticism of classical logic with an affirmative attitude towards classical mathematics (cp. Mares 2004: 186-8).

To sum up, it has been argued that relevantistic criticism of classical logic is based on the claim that there are classical logical truths and classical logical validities intuitively classified as untrue and invalid. As will be seen, intuitionistic criticism of classical logic is not based on similar claims about the untruth and invalidity of certain classical logical truths and classical logical validities. Furthermore, it has been seen that relevantists are best characterised as endorsing a descriptive criticism of classical logic. In particular with respect to mathematical argumentations, it is not claimed that unacceptable principles and rules are at work so that standard mathematics is in need of a reform. Rather, it is claimed that standard mathematics does not depend on principles and rules which are part of classical logic but not of relevant logic. Again, this sharply contrasts with intuitionistic criticism that aims at a reform of standard mathematics, which is seen to be based on unacceptable principles and rules.

1.5   Empty Singular Terms

In this final section, I will deal with arguments against the theory of classical logic which concern the phenomenon of empty singular terms. The proponents of these arguments claim that the theory of classical logic is not sound since it does not account for the following principles:

F_1 A universal quantification does not imply those of its instances which are obtained by means of an empty singular term.

F_2 An existential quantification is not implied by those of its instances which are obtained by means of an empty singular term.

The theory of classical logic is claimed to be inadequate because it cannot deal with the peculiarities of arguments containing singular terms that do not denote anything.

I will call those who believe that the theory of classical logic fails because it does not account for the principles F_1 and F_2 *freedom fighters*, and I will call the logical theories they advocate *theories of free logic*. I will now introduce some ideas of freedom fighters and argue in favour of the following three theses:

T_1-Fr Freedom fighters criticise classical logic on the basis of the claim that there are sentences/arguments which are intuitively classified as untrue/ invalid, although they are logically true/valid according to classical logic.

T_2-Fr Freedom fighters do not criticise the argumentations of classical mathematics.

T_3-Fr Theories of free logic should be taken to differ from theories of classical logic with respect to their transformational parts.

As I have illustrated in 1.4, the theory of intuitionistic logic does not have one of the first two features. As was claimed in 1.2 and as will be defended in Chapter 2, it also lacks the third feature.

1.5.1   Theories of Free Logic

What kind of critique do freedom fighters put forward?[32] As in the previous section, my goal will only be to illustrate those aspects that make their criticism different from the criticism put forward by intuitionists. For simplicity, I will restrict attention to a proper fragment of English as the informal object language: the fragment of what I will call *simple sentences*. Here, a sentence is called simple iff it is built up from simple proper names and simple predicates by means of the standard logical constants.[33] In particular, simple sentences do not contain complex singular terms.

In the following discussion, I will concentrate on one empty singular term: the empty proper name "Vulcan".[34] Consider the following two arguments:

(44) Everything is an object. Therefore, Vulcan is an object.

(45) Vulcan is not an object. Therefore, something is not an object.

These arguments are invalid. Consider the first one. Its premise is true: everything is an object.[35] Its conclusion is not true, because "Vulcan" does not denote anything. Consequently, this argument is not even truth-preserving. On the face of it, however, the arguments (44) and (45) are logically valid according to the theory of classical logic. They seem to correspond to the formal arguments $\forall a\, Fa \succ Fc$ and $\neg Fc \succ \exists a\, \neg Fa$, which are $\mathcal{L}^*$-valid according to $\mathcal{T}^C$.

The claim that arguments like (44) and (45) are logically valid according to classical logic is based on the assumption that classical logic is a standard logical theory. Here, a logical theory $\mathcal{T}$ is called *standard* iff the formalisation function π of its transformational part $\mathcal{T}_{trans}$ (see 1.2.4) treats empty and nonempty singular terms in the same way: whether an argument is logically valid according to $\mathcal{T}$ is independent of whether the contained singular terms are empty or not. For example, if there are two arguments α_1 and α_2 such that α_1 contains some tokens of "Vulcan" but no tokens of "Mercury", and α_2 can be

---

[32]   For a number of different theories of free logic see the overviews of Lambert (2001), Bencivenga (2002), Lehmann (2002), and Posy (2007).

[33]   A proper name is called *simple* iff it does not contain a meaningful proper part that contributes to its meaning. Thus, for example, "London" and "Socrates" are simple while "Lake Michigan" and "Gottlob Frege" are not. A predicate is called *simple* iff it is sufficiently similar to predicates like "is larger than", "is a sister of", or "sleeps". Simple predicates are formed by means of an adjective, a noun, or a verb and additional 'purely grammatical' material.

[34]   In the 19th century, it was suggested that a small planet between Mercury and the Sun is responsible for certain peculiarities in Mercury's orbit, and it was given the name "Vulcan". As it turned out, there is no planet between Mercury and the Sun, and the name "Vulcan" disappeared from astronomical discussions.

[35]   I consider this to be an absolute triviality (see Haverkamp 2011). If someone doubts that the adjective "object" applies to everything, she might replace "an object" with "self-identical".

obtained from α_1 by substituting tokens of "Mercury" for the tokens of "Vulcan", then α_1 is logically valid according to $\mathcal{T}$ iff α_2 is logically valid according to $\mathcal{T}$. Now, theories of free logic are standard logical theories: they treat empty singular terms as they treat non-empty singular terms. The important question is whether classical logic should also be conceived of as such a standard logical theory.[36]

To begin with, I would like to note that the theses T_1-Fr and T_2-Fr are rather uncontroversial. According to the first one, theories of free logic are motivated by the existence of classical logical truths and validities intuitively classified as untrue and invalid. This thesis is clearly true: freedom fighters criticise classical logic for declaring invalid arguments like (44) and (45) to be logically valid and untrue sentences such as "if everything is an object, then Vulcan is an object" to be logically true. According to T_2-Fr, theories of free logic are not meant to be incompatible with classical mathematics. As in the case of relevantists, freedom fighters put forward a descriptive rather than a prescriptive criticism of classical logic (see 1.4.2). Now, in contrast to the difficulties for relevantists to show how their rejection of classical logic is compatible with an acceptance of classical mathematics, it is unproblematic for freedom fighters to substantiate the claim that their criticism does not pertain to classical mathematics: since classical mathematicians avoid the use of empty singular terms in the presentation of their proofs, one can criticise the classical treatment of arguments that contain such terms without criticising classical mathematics.

1.5.2   Free Logic and Non-Standard Classical Logic

Up to this point, I have only considered standard logical theories which treat empty and non-empty singular terms in the same way. However, this is not mandatory. Some philosophers reject sentences like "Vulcan is not a planet" (although "Vulcan" does not denote a planet), and this might speak in favour of a non-standard logical theory. I will now make some remarks about the motivation to adopt a logical theory which treats empty and non-empty proper names differently.

As a preparation, I have to include some clarifying remarks about the notion of a proposition.[37] Propositions are the contents of mental episodes and states. Someone who, for example, judges, recognises, grasps, believes, knows, or hopes something, performs a mental act or is in a mental state which has a proposition as its content. These mental episodes and states will be called *propositional attitudes* henceforth. Propositional attitudes and their contents differ in crucial respects. A propositional attitude is a *concrete object* (i.e. located in

---

[36]   I will assume that the formalisation function π of the transformational part $\mathcal{T}_{trans}$ of a non-standard logical theory maps every argument whose premises and conclusion are simple sentences, at least one of which contains an empty proper name, to a formal argument which is not logically valid, say $\succ \forall a\, Fa$. Since I have assumed that the formalisation function is total, there have to be images for such arguments. A more elegant approach would allow for *partial* formalisation functions. Then, one could say that the relevant arguments have no formal counterparts.

[37]   For a more detailed elucidation of the notion of a proposition, see Künne 2003: 249-63.

space and time), and it is *subjective*, i.e., no two persons can have the same propositional attitude. A proposition, by contrast, is an *abstract object*, and it is *inter-subjective*, i.e., different subjects can have propositional attitudes every one of which has the same proposition as its content. Moreover, propositional attitudes of different kinds can have the same proposition as their content: Peter's hope may have the same content as Paul's fear and Mary's belief. (It is contentious whether *every* proposition can be the content of propositional attitudes of different persons; cp. Künne 1997.) There are many intricate questions for which a theory of propositions would have to supply answers. One would like to know, for example, under which conditions two propositional attitudes have the same content (see Künne 2003: 42-52) and, more importantly here (see below), under which conditions a mental episode or state has a (propositional) content.

Now, should a logical theory be a standard one, or should it treat empty and non-empty singular terms differently? An early adherent of a non-standard logical theory was Frege. According to Frege (1881-3a: 67, 1881-3b: 190), logical rules presuppose that proper names are non-empty because sentences with empty proper names do not express judgements or thoughts.[38] Later adherents of non-standard logical theories are Evans (1982) and Williamson (2007b). It should be noted that Evans is only in some cases prepared to treat empty and non-empty proper names differently; in other cases he (1982: 31) favours a standard free logical approach. Williamson, by contrast, adopts a more radical view:

> In assessing validity, our concern is with truth-preservation only when the relevant formulas are fully interpreted. For a sentence to be fully interpreted, [...] it must [...] express a proposition as used in the relevant context. [...] Truth-preservation is required only once any singular terms in the argument have been assigned a reference (Williamson 2007b: 382).

Williamson suggests that a sentence with an empty singular term does not express a proposition and that arguments which contain such sentences are outside the scope of logical theories.

The motivation to adopt a non-standard logical theory stems from the claim that simple sentences which contain empty proper names do not express propositions. It is said that there cannot be propositional attitudes whose contents are expressed by such sentences. Now, on the face of it, the claim that simple sentences with empty names do not express propositions is rather implausible (cp. Sainsbury 2005: 86-90). Suppose that a person in the 19th century sincerely uttered the sentence "Vulcan is a planet" with the intention of asserting something. It then seems implausible to deny that she has asserted something. Similarly, it seems implausible to deny that some people had a belief whose content is expressed by this sentence.

---

[38]   It is controversial as to what extent Frege's later views retreat from those that are referred to here. See Evans 1982: 28-30 for the view that Frege retreated from them only partially; see Sainsbury 2002: 9-14 for a critique of this interpretation.

Part of the implausibility of the claim that these sentences do not express propositions derives from the fact that it presupposes a certain *intransparency thesis* about one's own mental states:

Intransparent Beliefs It may happen that someone falsely believes that she satisfies the predicate "believes that Vulcan is a planet", while she is not in a position to find out that she does not satisfy this predicate.

A person can be in a mental state which is introspectively indistinguishable from a belief, but which is not a state of belief because the relevant content that could be believed is not there. Similarly, a person can perform a mental act which is introspectively indistinguishable from a judgement, but which is not an act of judgement because the relevant content for such judgment is not there.

On closer inspection, however, it seems to turn out that the mentioned consequences have to be accepted anyway, and that the claim that simple sentences with empty proper names do not express propositions has less dramatic effects than it seems to have. Suppose that someone invents a new simple proper name, say "Lulcan", without even trying to fix a referent for it. Suppose, for example, that an astronomer invents the name "Lulcan" so that she can use it if she ever discovers a new planet. With respect to such a name it is, I think, quite plausible to say that simple sentences which contain it do not express propositions. Suppose that her husband finds the word "Lulcan" written on a list that is headed "Planets" and that he then sincerely utters the sentence "Lulcan is a planet" with the intention of asserting something. Then, I think, he would not have succeeded in asserting something. Although he would believe that he satisfies the predicate "believes that Lulcan is a planet", he would not really satisfy it, because the relevant proposition that could be the content of his belief is not there.

Now, I suspect that some freedom fighters would not accept my description of simple sentences that contain "Lulcan", and even if this description is correct, it does not follow that the same holds true of simple sentences that involve "Vulcan". To resolve the dispute between freedom fighters and their opponents, one would have to deal with theories about mental acts and states and theories about the relation between sentences and such acts and states. This task will not be pursued here. I only wanted to give some indication of how the conflict between freedom fighters and their opponents is to be understood. As will be seen, the conflict between intuitionists and their opponents cannot be understood in a similar way.

It has been argued that the criticism of relevantists and freedom fighters is based on the claim that there are classical logical truths and validities which are intuitively classified as untrue and invalid. The criticism of intuitionists, by contrast, is based on no such claim. Furthermore, it has been illustrated that relevantists and freedom fighters endorse a kind of descriptive criticism and that they do not object to standard mathematical practice. Intuitionists, by contrast, do not claim that classical mathematics is independent of classical logic; rather, they reject classical mathematics partly because they reject classical logic. Finally, I have tried to indicate how the dispute between freedom fighters and their opponents is to be understood and what must be done to resolve it. It is a dispute about how informal arguments are to be formalised; in particular, it concerns the question of whether empty and non-empty singular terms are to be treated in the same way. It has to be resolved by answering questions about the contents of mental acts and states and their relation to sentences. Here too, it will be seen that the dispute between intuitionists and their opponents is to be understood differently and cannot be resolved by similar means.

In Chapters 3 to 5, I will discuss intuitionistic arguments against classical logic. In the following chapter, I will deal with a central presupposition of these disputes, namely the thesis that the participants understand the involved logical and mathematical expressions in the same way.

Chapter 2

A Purely Verbal Dispute?

Is mathematical intuitionism a topic for philosophical inquiry? It is debatable to which extent intuitionistic mathematics should be conceived of as a philosophical position; perhaps it is better understood as a distinctive way of *doing mathematics* (cp. McCarty 2008b: 38). But unlike other possibilities of doing mathematics in different ways, for example, doing mathematics in solitude versus doing mathematics in work groups, the ways of doing mathematics classically and doing mathematics intuitionistically deliver incompatible results. Classical mathematicians claim to know mathematical truths which contradict basic intuitionistic assumptions, whereas intuitionistic mathematicians claim to know mathematical truths which contradict basic classical assumptions. Intuitionists even believe that they have refuted sentences which are logically true according to classical logic. An intelligible enterprise which seems to produce results that contradict basic principles of classical mathematics and classical logic deserves philosophical attention.

Mathematical intuitionism is a kind of mathematical constructivism.[1] An important commonality of all those who do mathematics constructively lies in their desiderata on existential proofs. Constructivists are dissatisfied with certain argumentative moves that are used by classical mathematicians to establish the existence of some mathematical object, say, a natural number, which satisfies a given condition. Suppose that it has been shown that not every number is not so-and-so ($\neg \forall n \, \neg \varphi$). A classical mathematician is then willing to conclude that there is a number which is so-and-so ($\exists n \, \varphi$). A constructivist rejects this kind of transition. According to her, to be able to prove that some number satisfies φ, it does not suffice to be able to refute that every number fails to satisfy φ; to be able to prove that some number satisfies φ, one has to be able to 'present' a number which satisfies φ. The meaning and plausibility of such constructivist claims will be a central topic in what follows.

Intuitionism is the oldest fully developed version of constructivism.[2] It was initiated by the mathematician Luitzen Egbertus Jan Brouwer at the beginning

---

[1]   Two important other types of mathematical constructivism are those initiated by Markov and by Bishop. See Troelstra & van Dalen 1988: 25-9 and the references given there.

[2]   Paul du Bois-Reymond (see McCarty 2009: 316-9) and Émile Borel (see Troelstra & van Dalen 1988: 19-20) can be mentioned as important ancestors of constructivism.

of the last century.[3] Its most characteristic aspects, those which differentiate
it from other forms of constructivism, concern mathematical domains which
are more complex than the domain of the natural numbers, e.g. the domain
of the real numbers. According to intuitionists, fundamental principles of such
domains imply the falsity of sentences which are logically true according to classical logic. A second goal in what follows will be the exploration of anti-classical
arguments based on such distinctively intuitionistic mathematical principles.

In this chapter, I will deal with a central presupposition of my discussion
of the dispute between classicists and intuitionists: the assumption that it is
not a purely verbal dispute, in which the participants talk past each other. I
will argue that intuitionists and classicists understand logical and mathematical
expressions in the same way. First, I will clarify my understanding of the notion
of a mathematical language (2.1). Then, I will defend the claim that intuitionists
and classicists understand logical vocabulary in the same way (2.2) and the
claim that they also understand mathematical terms in the same way (2.3).
Finally, I will substantiate the defended claims by criticising an argument that
is put forward against them (2.4).

2.1   Mathematical Languages

Logical theories concern the notion of logical validity, a property which applies
to certain arguments. Now, arguments are made up of sentences, and these sentences belong to some natural language (see 1.1.1). How is it then possible, it
might be asked, that mathematics is relevant for the evaluation of logical theories? Is not the use of sentences of a natural language rather different from the
use of mathematical formulae? In this section, I will present my understanding of the phrase "mathematical language" and explain the contrast between
mathematical languages and formal languages. It will then become apparent
why mathematics can be relevant for the evaluation of logical theories.

2.1.1   Natural Mathematical Languages

A mathematical language is a language which mathematicians normally *use*
when they communicate their mathematics. It is the language in which they
give their definitions, state their results, and present their proofs. How is such
a language related to a natural language? Hannes Leitgeb characterises this
relation as follows:

> What mathematicians do is to extend natural language by some
> (maybe new) distinguished symbols and to use this language in much
> the same way as lawyers extend natural language by legal terms for
> their own purposes. The language(s) of mathematics are thus best
> viewed as "organic" entities which grow and change in time, both

---

[3]   Brouwer's dissertation *Over de Grondslagen der Wiskunde* from 1907 was his first contribution to intuitionistic mathematics. For its English translation see Heyting 1975: 11-101.
Biographical information about Brouwer can be found in van Dalen 1999, 2005.

with the natural languages they extend and with respect to their
specific mathematical resources (Leitgeb 2009: 268).

I will follow Leitgeb's idea that languages are 'organic' entities which come
into existence, change in time, and may eventually cease to exist. However, the
presented account is compatible with two slightly different conceptions of the
relation between natural languages and mathematical resources.

According to the first conception, there is, on the one hand, a natural language like English and, on the other hand, one or more mathematical extensions of it. This may raise the question of whether there is an all-encompassing
language which contains the resources from English and all mathematical extensions, or whether it is important to employ different (partly overlapping)
languages in different contexts. According to the second conception, extensions
by mathematical terms should not be understood as processes of creating new
mathematical languages which then accompany a non-mathematical kernel language. In this view, which I take to be the most plausible one, mathematical
terms are terms of natural languages. New expressions are introduced into natural languages for all sorts of reasons. And, according to the view adopted
here, some of these reasons are mathematical. Sometimes a new mathematical
expression is introduced by means of an explicit stipulation. Here is an example:

Graph A *graph* is a pair of sets such that the elements of the second set are
subsets of the first set with exactly two elements.

By means of this definition, a new count noun is introduced into English: the
mathematical term "graph". In other cases, new expressions come into use in a
more implicit way, e.g. variables. Consider the following two sentences:

(1) For every number n there is a number m such that m is larger than n.

(2) For every number there is a number such that the second one is larger
than the first one.

These sentences are very close in meaning, but only the first one contains variables. Furthermore, the use of variables has become part of English without
having been introduced into the language by means of an explicit definition.[4]

In the present view, a mathematical language is nothing but a natural language containing mathematical vocabulary. There is thus no reason to claim
that a mathematician stops speaking her mother tongue when presenting a
mathematical proof or when putting forward a mathematical conjecture. This
view is strengthened by the observation that basic mathematical expressions
like "two", "number", or "set" are frequently used in non-mathematical contexts. The claim that such expressions could not be part of the natural language
English because they are mathematical simply does not seem warranted.

---

[4] Timothy Williamson has reminded me that some mathematical expressions, such as two-
dimensional matrices, do not belong to a natural language like English. Fortunately, such
expressions do not figure in the intuitionistic arguments against classical logic and can
therefore be set aside here.

Given that mathematical languages are natural languages, it becomes clear how mathematics can be relevant for logic: mathematicians present arguments and theories which consist of sentences of natural languages and which are, therefore, potential bearers of logical properties. Indeed, in the chapters to come, the logical properties of certain mathematical arguments will play an essential role. As a preparation, I will argue that intuitionists and classicists understand the mathematical language they use in the same way. However, before I can come to that, I would like to say something about certain mathematical structures which are closely related to (fragments of) natural languages but which should not be identified with them, namely *formal languages*.

2.1.2 Formal Mathematical Languages

Natural languages have to be distinguished from *ordinary formal languages*. The sentences of natural languages are meaningful; they are used to perform linguistic acts like making an assertion, asking a question, and giving a command. The expressions of ordinary formal languages are not meaningful in the same way; if no special conventions have been set up, these *formal sentences* cannot be used to perform linguistic acts. Formal languages are certain mathematical objects that are studied in relation to other mathematical objects in disciplines like model theory or proof theory. One can talk *about* formal sentences but, normally, one cannot *use* formal sentences to talk.

The philosophical significance of formal languages does not primarily consist in their (intrinsic or relational) mathematical properties. Rather, their philosophical significance derives from their relation to non-mathematical objects, especially from their relation to fragments of natural languages. Of special importance for this book is the use of formal languages as simplified substitutes of certain fragments of English. For such use there will be some translation function in the background, often only partly specified, that maps the sentences of a fragment of English to the sentences of a formal language (cp. 1.1.3). The semantic properties of a formal sentence are then taken to be derived from those of its translation. In a derivative sense, it can be said that the formal sentences are meaningful; they inherit their meaning from their natural language ancestors.

The advantage of speaking about formal languages rather than the corresponding fragments of English is mainly one of clarity and simplicity. The syntactic structure of a formal sentence is not ambiguous and can be apprehended more easily than the syntactic structure of its natural language counterpart. Moreover, there are fundamental syntactic differences among the expressions of a natural language which are irrelevant for some purposes and can be fruitfully neglected if one speaks about a corresponding formal language. For example, the difference between nouns and verbs, which is fundamental to the syntactic structure of the sentences of English, is irrelevant in the following and does not have to be reflected in corresponding formal languages.

Apart from ordinary formal languages, there are also what I will call *special formal languages*: formal languages whose sentences were invented to be usable to perform linguistic acts (cp. Sundholm 1997: 191-4). I have in mind special formal languages like those of Frege's *Grundgesetze der Arithmetik* or Russell and Whitehead's *Principia Mathematica*. The inventors of these formal languages wanted to create something that can be used to express mathematical propositions and to present mathematical proofs in a way that cannot be achieved by the use of a natural language. However, special formal languages do not have to be considered here. Formal languages are only used as syntactic simplifications of fragments of English.

2.2 Logical Vocabulary

Intuitionists and classicists seem to make incompatible assertions. As an example, consider the following two sentences:

Cl₁ Every real number is rational or not rational.

In₁ Not every real number is rational or not rational.

When a classical mathematician utters the first one assertively and an intuitionist utters the second one assertively, they seem to disagree. To all appearances, there is something which is affirmed by the one and denied by the other.

However, sometimes appearances are deceptive, and there is an influential tradition whose adherents claim that intuitionists and classicists do not really make incompatible assertions. According to their view, classicists and intuitionists do not understand and use mathematical and logical vocabulary in the same way. Prominent adherents of this view are Quine and Dummett:

> [W]hoever denies the law of excluded middle changes the subject. [...] In repudiating 'p or $\neg p$' he is indeed giving up classical negation, or perhaps alternation, or both [...] (Quine 1970: 83).

> The intuitionist should not be viewed as controverting us as to the true laws of certain fixed logical operations, namely, negation and alternation. He should be viewed rather as opposing our negation and alternation as unscientific ideas, and propounding certain other ideas, somewhat analogous, of his own (Quine 1970: 87).

> Someone who rejects the law of excluded middle, for example, cannot mean the same by 'or' and 'not' as one who accepts it [...] (Dummett 1991: 54).

This view has been combined with different attitudes towards the respective enterprises of classical and intuitionistic mathematics. Following Carnap, another prominent adherent of this view, some philosophers opt for a *Principle of Tolerance* (Carnap 1934: 51-2, 1963: 55) which allows for intuitionism and classical mathematics as equally legitimate enterprises:

> [L]et any postulates and any rules of inference be chosen arbitrarily; then this choice, whatever it may be, will determine what meaning is to be assigned to the fundamental logical symbols. By this method,

also, the conflict between the divergent points of view on the problem of the foundations of mathematics disappears (Carnap 1934:
XV).

According to this view, there is an intuitionistic understanding of logical and
mathematical vocabulary, which is active when intuitionists do mathematics,
and there is a different classical understanding of logical and mathematical
vocabulary, which is active when classicists do mathematics. I will call this
view *logical/mathematical dualism*. Others adopt an exclusive attitude which
involves the claim that one or the other way of doing mathematics is illegitimate.
According to their view, either intuitionists or classicists do not understand the
logical and mathematical vocabulary they use. I will call this view *exclusive
logical/mathematical monism.*[5]

I think theories of dualism and exclusive monism are mistaken. I will try to
defend the claim that classicists and intuitionists are both competent users of
logical and mathematical expressions and that they understand these expressions in the same way. This view will be called *inclusive logical/mathematical
monism.*

Before I present the defence of inclusive monism, I would like to make a
remark about equivalent ways of formulating it. On the one hand, I take the
mentioned positions to be definable in terms of the notions of *understanding
an expression* and *understanding an expression in the same way*. (I will use the
notions of *understanding an expression* and of *being a competent user of an
expression* interchangeably.) According to inclusive logical monism, intuitionists and classicists understand logical vocabulary, and they understand it in
the same way. According to logical dualism, intuitionists and classicists understand logical vocabulary, but they understand it in different ways. According
to exclusive logical monism, either intuitionists or classicists do not understand
logical vocabulary.

On the other hand, these views can also be characterised in terms of the
concept of *meaning*, more precisely, in terms of the concept of *attaching a meaning to some expression* and the concept of *attaching the same meaning to some
expression*. This way of speaking presupposes certain theses about meaning.
In particular, it has to be assumed that there *are* meanings which meaningful expressions of a language *have* and which are *attached* to these expressions
by individual language users. More specifically, the following equivalences are
assumed to hold:

Understanding₁ Some person x understands an expression e of a language $\mathcal{L}$
iff the expression e has some meaning m in $\mathcal{L}$ and x attaches m to e as
an expression of $\mathcal{L}$.

---

[5]   Dummett (1991: 17) endorses a view at the border between dualism and exclusive monism. Either it could be classified as a dualist position because Dummett says that intuitionists and classicists attach different meanings to mathematical terms, or it could
be classified as a variant of exclusive monism because Dummett denies that classical
mathematicians attach *coherent* meanings to mathematical terms. For what follows, the
correct classification of Dummett's position can be left unresolved.

Understanding₂ Some persons X understand an expression e in the same way iff there is a meaning m such that each of X attaches m to e.

Using this terminology, inclusive logical monism is the view that a logical expression has a single meaning which intuitionists and classicists attach to it. Logical dualism is the view that for each logical expression there are two distinct meanings, the first of which intuitionists attach to it and the second of which classicists attach to it. And exclusive logical monism is the view that either intuitionists or classicists do not attach to a logical expression (any of) its meaning(s).[6] For what follows, it is often convenient to speak about *attaching meanings to expressions* and not only about *understanding* expressions.

2.2.1 Inclusive Monism and the Burden of Proof

Do intuitionists and classicists understand the logical and mathematical expressions they use? And if so, do they understand them in the same way? To make an initial case in favour of affirmative answers to these questions, let me quote an illustrative paragraph, which McCarty has used for the same purpose:

> 'Weak' mathematical induction is a true principle. Consequently, for any natural number $n > 3$, $2^n < n!$. The inequality holds of 4, since $2^4 = 16 < 24 = 4!$. Assuming that it holds of k, we see easily that it also holds of $k + 1$, because
>
> $$2^{k+1} = 2 \times 2^k < (k+1) \times k! = (k+1)!$$
>
> for $k > 3$. However, 'least number' induction is false. Let S be any mathematical statement whatsoever. Consider the set A^S that definitely contains the number 1 but contains 0 if and only if S is true. In familiar notation, A^S is
>
> $$\{n|n = 0 \wedge S\} \cup \{1\}.$$
>
> A^S has a member, for 1 belongs to it no matter what. For the sake of argument, suppose that it has a least member. As you can see from the definition of A^S, the putative least member has to be either 0 or 1, as required by the meaning of the symbol '$\cup$'. On the one hand, if the least member is 0, then 0 belongs to A^S and S has to be true. On the other hand, if the least member of A^S is 1, 0 must fail to belong to A^S. There is only one way that can happen: S is false. So, if A^S has a least member, then S is either true or false. This result is generalizable: if every nonempty set of numbers had a least member, then, for any statement S, either S is true or S is false. The last clause, an expression of the classical law of bivalence

---

[6] Of course, there is the possibility of adopting different positions with respect to different logical expressions. For example, one might claim that intuitionists and classicists attach the same meaning to "and" but different meanings to "not". Since I will defend inclusive monism with respect to all standard logical constants, I will mostly not bother to make such distinctions.

or *tertium non datur*, is false. Therefore, the least number principle
is false as well (McCarty 2008a: 56-57).

Someone who endorses classical arithmetic will not believe what McCarty is
claiming here. However, the essential point is that acquaintance with intuitionism is not necessary for *understanding* this paragraph. McCarty claims that
weak mathematical induction is true, and he infers a simple arithmetical statement. He then demonstrates that least number induction implies the principle
of bivalence, expresses his disbelief in the latter, and infers the falsity of the
former. He uses English sentences, including some common symbols of arithmetic and set theory, and there is no reason to doubt that he intends to use
the sentences with their common meaning in English. The default assumption
should be that McCarty understands the expressions he uses, that he uses them
competently with their meaning in English, and that someone without any intuitionistic background can grasp what McCarty is saying here because she
understands these expressions in the same way as he does.

Now, there are also simple mathematical texts written by classical mathematicians in which they use the logical and mathematical terms that occur in
the above paragraph. In ordinary circumstances, there is no reason to doubt that
the relevant authors intend to use the sentences with their common meaning
in English. Again, the default assumption should be that they understand the
expressions they use, that they use them competently with their meaning in
English, and that, for example, an intuitionist can grasp what they are saying
because she understands these expressions in the same way as they do.

On the face of it, it then seems that intuitionists and classicists understand
the expressions they use, and that they understand them in the same way.
Initially, it seems more plausible to assume that intuitionists and classicists
disagree about whether every non-empty set of natural numbers has a least
member than to assume that they simply speak past each other by attaching
different senses to the sentence "every non-empty set of natural numbers has
a least member" or to assume that either of them does not understand this
sentence at all.

In the rest of this chapter, I will try to argue that this view is not only
initially plausible but actually true. I will do so in two steps. First, I will deal
with standard logical vocabulary. The question is whether classicists and intuitionists understand the logical constants "not", "or", "and", "if", "some",
and "every" in the same way. In this section, I will try to show that they do.
Second, I will deal with mathematical terms, in particular, with arithmetical,
set-theoretic, and function-theoretic vocabulary. My defence of inclusive mathematical monism will be presented in 2.3.

2.2.2 Meta-Linguistic Uniqueness Results

Is it plausible to deny that intuitionists and classicists understand words like
"not", "and", "or", "if", "some", and "every" in the same way? Initially, this
denial seems implausible. English speaking children become competent users
of these words long before they come into contact with mathematics. They

acquire an understanding of these words independently of mathematical argumentations, and they do not lose this understanding when they become classical mathematicians, intuitionists, or something else. I will now defend inclusive logical monism, i.e. the assumption that intuitionists and classicists understand logical vocabulary in the same way. I will do so mainly by criticising logical dualism, which I take to be more plausible than exclusive logical monism. (I will return to this point at the end of 2.2.4.) According to logical dualism, intuitionists and classicists attach different meanings to logical vocabulary. They claim that words like "not" and "if" are ambiguous and that people who do mathematics intuitionistically attach 'intuitionistic meanings' to the logical operators, while people who do mathematics classically attach different 'classical meanings' to them. In what follows, I will concentrate on the negation operator "not". Analogous considerations apply to the other standard logical operators.

The basic problem with logical dualism is that there are simple provable results, where the proofs are classically and intuitionistically acceptable, which seem to show that it is impossible that there are these two kinds of meanings: the intuitionistic ones and the classical ones. I will now introduce these results and explain their philosophical significance.[7] As is usual, instead of speaking directly about English and the word "not", I will speak about formal languages $\mathcal{L}$ which contain the symbol $\neg$ corresponding to "not". The philosophical significance of the following considerations depends upon this correspondence between "not" and $\neg$ (see 2.1.2). For simplicity, I will apply the notion of logical validity to arguments of $\mathcal{L}$, where, strictly speaking, I should make use of the formal counterpart of logical validity (see 1.1.3). Furthermore, I will often identify a formal language $\mathcal{L}$ with the set of sentences of $\mathcal{L}$ and write $X \subseteq \mathcal{L}$ to state that X is a list of sentences of $\mathcal{L}$.

I will mention four versions of the announced result, three of them in this subsection and another one in the following subsection.[8] The first one is formulated in terms of logical validity. Consider the following conditions on formal languages $\mathcal{L}$ which contain a unary sentential operator $\sharp$:

RAA$_\sharp^\mathcal{L}$ For all $X, Y \subseteq \mathcal{L}$ and for all $S_1, S_2 \in \mathcal{L}$:
if $X, S_1 \succ S_2$ and $Y, S_1 \succ \sharp S_2$ are logically valid, then so is $X, Y \succ \sharp S_1$.

EFQ$_\sharp^\mathcal{L}$ For all $X, Y \subseteq \mathcal{L}$ and for all $S_1, S_2 \in \mathcal{L}$:
if $X \succ S_1$ and $Y \succ \sharp S_1$ are logically valid, then so is $X, Y \succ S_2$.

DNE$_\sharp^\mathcal{L}$ For every $X \subseteq \mathcal{L}$ and for every $S \in \mathcal{L}$:
if $X \succ \sharp\sharp S$ is logically valid, then so is $X \succ S$.

These sentences correspond in an obvious way to the rules $(\neg_I)$, $(\neg_E)$, $(\bar{\neg}_E)$

---

[7]   The result which concerns the negation operator was discovered by Popper (1948: 113). Here Popper retreats from the dualist view he had put forward in 1947: 219-20. Apparently without awareness of Popper's article, this result and analogous results for the other logical operators were (re)discovered by Harris (1982). See also Humberstone 2011: 578-630.
[8]   The results presented by Popper and Harris (see n. 7) correspond to the fourth version presented in the next subsection.

from the calculus of natural deduction $\mathcal{R}^C$ (see 1.2.3).[9] They express general claims to the effect that if one or two arguments of a certain form are logically valid, then an argument of a related form is also logically valid.

Suppose that $\mathcal{L}$ is a formal language which contains the operator $\neg$ corresponding to "not". On the one hand, the sentences $\mathrm{RAA}_{\neg}^{\mathcal{L}}$, $\mathrm{EFQ}_{\neg}^{\mathcal{L}}$, and $\mathrm{DNE}_{\neg}^{\mathcal{L}}$ are all theorems of the theory $\mathcal{T}^C$ of classical logic. On the other hand, only the sentences $\mathrm{RAA}_{\neg}^{\mathcal{L}}$ and $\mathrm{EFQ}_{\neg}^{\mathcal{L}}$ are theorems of the theory $\mathcal{T}^I$ of intuitionistic logic. Now, in accordance with logical dualism, one might propose that there are two possible concepts of negation, i.e. two possible meanings for "not", such that an operator which expresses the first concept behaves classically, while an operator which expresses the second concept behaves intuitionistically. To avoid purely verbal disputes, one might then introduce two signs to express both of these concepts, say $\neg_1$ and $\neg_2$. The idea would then be that $\mathrm{RAA}_{\neg_1}^{\mathcal{L}}$, $\mathrm{EFQ}_{\neg_1}^{\mathcal{L}}$, $\mathrm{DNE}_{\neg_1}^{\mathcal{L}}$ are true, while $\mathrm{RAA}_{\neg_2}^{\mathcal{L}}$, $\mathrm{EFQ}_{\neg_2}^{\mathcal{L}}$ are true and $\mathrm{DNE}_{\neg_2}^{\mathcal{L}}$ is false. This claim constitutes one possibility for giving more substance to the thesis of logical dualism.

To present the obstacle to this view, the following notion has to be introduced:

Logically Equivalent Operators Two unary sentential operators $\sharp_1$ and $\sharp_2$ of $\mathcal{L}$ are called *logically equivalent* in $\mathcal{L}$ iff the following holds for every sentence S of $\mathcal{L}$: the arguments $\sharp_1 S \succ \sharp_2 S$ and $\sharp_2 S \succ \sharp_1 S$ are logically valid.

That is, two operators are logically equivalent iff they always produce logically equivalent sentences. Now, the following results are easily seen to hold:

Theorem $\mathrm{T}_1^{\neg}$ For every language $\mathcal{L}$ which contains two unary sentential operators $\neg_1$ and $\neg_2$ such that $\mathrm{RAA}_{\neg_1}^{\mathcal{L}}$, $\mathrm{EFQ}_{\neg_1}^{\mathcal{L}}$, $\mathrm{RAA}_{\neg_2}^{\mathcal{L}}$, $\mathrm{EFQ}_{\neg_2}^{\mathcal{L}}$ hold: $\neg_1$ and $\neg_2$ are logically equivalent in $\mathcal{L}$.

Corollary $\mathrm{C}_1^{\neg}$ For every language $\mathcal{L}$ which contains two unary sentential operators $\neg_1$ and $\neg_2$ such that $\mathrm{RAA}_{\neg_1}^{\mathcal{L}}$, $\mathrm{EFQ}_{\neg_1}^{\mathcal{L}}$, $\mathrm{RAA}_{\neg_2}^{\mathcal{L}}$, $\mathrm{EFQ}_{\neg_2}^{\mathcal{L}}$ hold: $\mathrm{DNE}_{\neg_1}^{\mathcal{L}}$ is true iff $\mathrm{DNE}_{\neg_2}^{\mathcal{L}}$ is true.

Proof. To establish $\mathrm{T}_1^{\neg}$, let S be any sentence of $\mathcal{L}$. The arguments $\neg_1 S \succ \neg_1 S$ and $S \succ S$ are logically valid. By $\mathrm{EFQ}_{\neg_1}^{\mathcal{L}}$, the argument $\neg_1 S, S \succ \neg_2 S$ is logically valid. Thus, the arguments $S \succ S$ and $\neg_1 S, S \succ \neg_2 S$ are logically valid. But then $\mathrm{RAA}_{\neg_2}^{\mathcal{L}}$ implies that the argument $\neg_1 S \succ \neg_2 S$ is logically valid. The second claim is proven analogously.

To establish $\mathrm{C}_1^{\neg}$, one first shows that for every sentence S of $\mathcal{L}$, $\neg_1 \neg_1 S$ and $\neg_2 \neg_2 S$ are logically equivalent. To this end, let S be any sentence of $\mathcal{L}$. By $\mathrm{T}_1^{\neg}$, the argument $\neg_2 S \succ \neg_1 S$ is logically valid. Furthermore, the argument $\neg_1 \neg_1 S \succ \neg_1 \neg_1 S$ is logically valid. By $\mathrm{EFQ}_{\neg_1}^{\mathcal{L}}$, the argument $\neg_1 \neg_1 S, \neg_2 S \succ \neg_2 \neg_2 S$ is logically valid. Thus, the arguments $\neg_2 S \succ \neg_2 S$ and $\neg_1 \neg_1 S, \neg_2 S \succ \neg_2 \neg_2 S$ are logically valid. But then $\mathrm{RAA}_{\neg_2}^{\mathcal{L}}$ implies that the argument $\neg_1 \neg_1 S \succ \neg_2 \neg_2 S$ is logically valid. The claim $\neg_2 \neg_2 S \succ \neg_1 \neg_1 S$ is proven analogously.

---

[9] The acronyms stand for *Reductio Ad Absurdum*, *Ex Falso Quodlibet*, and *Double Negation Elimination*.

Now, assume that $\text{DNE}^{\mathcal{L}}_{\neg_1}$ is true. Let $X \subseteq \mathcal{L}$ and $S \in \mathcal{L}$ be such that $X \succ \neg_2\neg_2 S$ is logically valid. By the forgoing observation it follows that the argument $X \succ \neg_1\neg_1 S$ is logically valid.[10] But then $\text{DNE}^{\mathcal{L}}_{\neg_1}$ implies that the argument $X \succ S$ is logically valid. The other conditional is proven analogously.

□

The corollary $\text{C}^{\neg}_1$ directly shows the mentioned logical dualists to be mistaken: if the sentences $\text{RAA}^{\mathcal{L}}_{\neg_1}$, $\text{EFQ}^{\mathcal{L}}_{\neg_1}$, $\text{DNE}^{\mathcal{L}}_{\neg_1}$, $\text{RAA}^{\mathcal{L}}_{\neg_2}$, $\text{EFQ}^{\mathcal{L}}_{\neg_2}$ are true, then the sentence $\text{DNE}^{\mathcal{L}}_{\neg_2}$ is also true.

Now, the logical dualist view does not have to be understood as implying the above mentioned thesis about logical validity, which is directly refuted by $\text{C}^{\neg}_1$. According to a natural alternative, it incorporates certain theses about what is true and what is truth-preserving in virtue of the meaning of "not".[11]

To spell out the new version of logical dualism, one needs to consider the following conditions on formal languages $\mathcal{L}$ which contain a unary sentential operator $\sharp$:

*RAA$^{\mathcal{L}}_{\sharp}$ For all $X, Y \subseteq \mathcal{L}$ and for all $S_1, S_2 \in \mathcal{L}$:
     if $X, S_1 \succ S_2$ and $Y, S_1 \succ \sharp S_2$ are truth-preserving, then so is $X, Y \succ \sharp S_1$.

*EFQ$^{\mathcal{L}}_{\sharp}$ For all $X, Y \subseteq \mathcal{L}$ and for all $S_1, S_2 \in \mathcal{L}$:
     if $X \succ S_1$ and $Y \succ \sharp S_1$ are truth-preserving, then so is $X, Y \succ S_2$.

*DNE$^{\mathcal{L}}_{\sharp}$ For every $X \subseteq \mathcal{L}$ and for every $S \in \mathcal{L}$:
     if $X \succ \sharp\sharp S$ is truth-preserving, then so is $X \succ S$.

These conditions are obtained from those above by replacing every token of the phrase "logically valid" by a token of "truth-preserving". The new dualist idea might then be spelled out as follows: there are two possible unary sentential operators $\neg_1$ and $\neg_2$ such that the following holds: the meaning of $\neg_1$ guarantees that $\text{*RAA}^{\mathcal{L}}_{\neg_1}$, $\text{*EFQ}^{\mathcal{L}}_{\neg_1}$, and $\text{*DNE}^{\mathcal{L}}_{\neg_1}$ are true, while the meaning of $\neg_2$ guarantees that $\text{*RAA}^{\mathcal{L}}_{\neg_2}$, and $\text{*EFQ}^{\mathcal{L}}_{\neg_2}$ are true but does not guarantee that $\text{*DNE}^{\mathcal{L}}_{\neg_2}$ is true.[12]

To see that this version of logical dualism does not fare better than the previous one, consider the following notion corresponding to the notion of logical equivalence for operators:

Materially Equivalent Operators Two unary sentential operators $\sharp_1$ and $\sharp_2$ of $\mathcal{L}$ are called *materially equivalent* in $\mathcal{L}$ iff the following holds for every sentence S of $\mathcal{L}$: the arguments $\sharp_1 S \succ \sharp_2 S$ and $\sharp_2 S \succ \sharp_1 S$ are truth-preserving.

---

[10]   It is assumed here that for all $X, Y \subseteq \mathcal{L}$ and for all $S_1, S_2 \in \mathcal{L}$, if the arguments $X, S_1 \succ S_2$ and $Y \succ S_1$ are logically valid, then so is the argument $X, Y \succ S_2$.

[11]   Recall that an argument $X \succ S$ is called *truth-preserving* iff the following holds: if every member of X is true, then S is true (see 1.3.2).

[12]   Note that it is not claimed that the meaning of $\neg_2$ guarantees that $\text{*DNE}^{\mathcal{L}}_{\neg_2}$ is false. Intuitionists do believe that $\text{*DNE}^{\mathcal{L}}_{\neg_2}$ is false, but it is doubtful whether they believe that the meaning of $\neg_2$ is responsible for the falsity of $\text{*DNE}^{\mathcal{L}}_{\neg_2}$.

That is, two operators are materially equivalent iff they always produce materially equivalent sentences. Corresponding to T_1^- and C_1^-, there are the following results:

Theorem T_2^- For every language $\mathcal{L}$ which contains two unary sentential operators $\neg_1$ and $\neg_2$ such that $^*RAA^{\mathcal{L}}_{\neg_1}$, $^*EFQ^{\mathcal{L}}_{\neg_1}$, $^*RAA^{\mathcal{L}}_{\neg_2}$, $^*EFQ^{\mathcal{L}}_{\neg_2}$ hold: $\neg_1$ and $\neg_2$ are materially equivalent in $\mathcal{L}$.

Theorem C_2^- For every language $\mathcal{L}$ which contains two unary sentential operators $\neg_1$ and $\neg_2$ such that $^*RAA^{\mathcal{L}}_{\neg_1}$, $^*EFQ^{\mathcal{L}}_{\neg_1}$, $^*RAA^{\mathcal{L}}_{\neg_2}$, $^*EFQ^{\mathcal{L}}_{\neg_2}$ hold: $^*DNE^{\mathcal{L}}_{\neg_1}$ is true iff $^*DNE^{\mathcal{L}}_{\neg_2}$ is true.

The proofs can be obtained from those that are given above by replacing each token of "logically valid" with a token of "truth-preserving".

The corollary C_2^- deprives the logical dualist view mentioned above of all its possible attractions. Suppose that the meaning of $\neg_1$ guarantees that $^*RAA^{\mathcal{L}}_{\neg_1}$, $^*EFQ^{\mathcal{L}}_{\neg_1}$, and $^*DNE^{\mathcal{L}}_{\neg_1}$ are true, and that the meaning of $\neg_2$ guarantees that $^*RAA^{\mathcal{L}}_{\neg_2}$ and $^*EFQ^{\mathcal{L}}_{\neg_2}$ are true. Then, by the simple proof of C_2^-, the sentence $^*DNE^{\mathcal{L}}_{\neg_2}$ is also guaranteed to be true. It might be replied that this does not show that the meaning of $\neg_2$ guarantees that $^*DNE^{\mathcal{L}}_{\neg_2}$ is true. But although this possibility cannot be rejected offhand, it cannot be of much value for the logical dualist. Logical dualism is motivated by the desire to offer a meaning-theoretical ground for the claim that two mathematical practices which seem to be incompatible are not really incompatible. The idea is that classical mathematicians are allowed to reason as they do because the negation concept they employ functions one way, while intuitionists are allowed to reason differently because they employ a different negation concept. Now, even if it is not directly guaranteed by the meaning of the intuitionistic negation operator $\neg_2$ that $^*DNE^{\mathcal{L}}_{\neg_2}$ is true, but only in a slightly more complicated way, it would still follow that classical mathematicians and intuitionists are allowed to reason with their respective negation concepts in exactly the same way. I take this to deprive logical dualism of all its motivation.

As was said in 1.3.2, sandwiched between the two notions of logical validity and truth-preservation is a further notion relevant for the evaluation of logical theories: the notion of validity. Now, it is easily possible to present a third version of logical dualism by means of this notion. However, this version of logical dualism is easily shown to be mistaken by a result T_3^- corresponding to T_1^- and T_2^-. I will not spell out the obvious modifications of the previous versions. I will rather turn to a fourth version of logical dualism which has a somewhat different character.

2.2.3 Rule-Based Uniqueness Results

The introduced versions of logical dualism all involved certain meta-linguistic elements. They were based on the observation that classicists accept sentences like $DNE^{\mathcal{L}}_{\neg}$ and $^*DNE^{\mathcal{L}}_{\neg}$ while intuitionists do not. The sentences in dispute are sentences about arguments and properties like *logical validity*, *validity*, and *truth-preservation*. It is natural to think, however, that the different attitudes towards these sentences are based on different attitudes which are not meta-linguistic.

Intuitionists and classicists are not only committed to the meta-linguistic sentence *RAA$^{\mathcal{L}}_{\neg}$. They are also committed to an inferential rule corresponding to this sentence. This rule might be captured by a statement about commitments to accept sentences conditional on certain assumptions:

raa$^*_{\neg}$ If you have accepted the sentence S_2 conditional on the set of assumptions $X \cup \{S_1\}$ and accepted the sentence $\neg S_2$ conditional on the set of assumptions $Y \cup \{S_1\}$, then you are committed to accepting the sentence $\neg S_1$ conditional on the set of assumptions $X \cup Y$.

The rule captured by this statement justifies certain argumentative moves. In an argumentation in which you have established the sentences S_2 and $\neg S_2$ relative to $X \cup \{S_1\}$ and $Y \cup \{S_1\}$, you are justified in performing a certain complex act: you are justified to infer the sentence $\neg S_1$ while discharging the assumption S_1. You then have brought about an argumentative situation in which the sentence $\neg S_1$ has been established relative to $X \cup Y$. If raa$^*_{\neg}$ is true, then, by arguing in accordance with the rule captured by this statement, you do not incur new commitments.

The inferential rules corresponding to the sentences *EFQ$^{\mathcal{L}}_{\neg}$ and *DNE$^{\mathcal{L}}_{\neg}$ are simpler. The first one can be captured by the following statement:

efq$^*_{\neg}$ If you have accepted the sentence S_1 conditional on the set of assumptions X and accepted the sentence $\neg S_1$ conditional on the set of assumptions Y, then you are committed to accepting any sentence S_2 conditional on the set of assumptions $X \cup Y$.

The rule captured by this statement justifies argumentative moves of a simpler kind. In an argumentation in which you have established the sentences S_1 and $\neg S_1$ relative to X and Y, you are justified to infer any sentence S_2. You then have brought about an argumentative situation in which the sentence S_2 has been established relative to $X \cup Y$.

Finally, the inferential rule corresponding to *DNE$^{\mathcal{L}}_{\neg}$ can be captured by the following statement:

dne$^*_{\neg}$ If you have accepted the sentence $\neg\neg S$ conditional on the set of assumptions X, then you are committed to accepting the sentence S conditional on the set of assumptions X.

The rule captured by this statement justifies argumentative moves of another simple kind. In an argumentation in which you have established the sentence $\neg\neg S$ relative to X, you are justified to infer the sentence S. You then have brought about an argumentative situation in which the sentence S has been established relative to X.

By arguing in accordance with the rules captured by efq$^*_{\neg}$ and dne$^*_{\neg}$ one makes simple inferences. In the former case one infers some sentence from another sentence and its negation. In the latter case, one infers a sentence from its double negation. By arguing in accordance with the rule captured by raa$^*_{\neg}$

one performs a complex act: one infers the negation of some sentence S and, simultaneously, one discharges one's commitment to S.

Now, another variant of logical dualism can be formulated in terms of these rules. Briefly put, it says that there is a classical version $\neg_1$ of the negation operator whose meaning allows one to argue in accordance with the rules captured by $\text{raa}^*_{\neg_1}$, $\text{efq}^*_{\neg_1}$, and $\text{dne}^*_{\neg_1}$, and an intuitionistic version $\neg_2$ of the negation operator whose meaning allows one to argue in accordance with the rules captured by $\text{raa}^*_{\neg_2}$ and $\text{efq}^*_{\neg_2}$, but whose meaning does not (in general) allow one to argue in accordance with the rule captured by $\text{dne}^*_{\neg_2}$.[13] This form of logical dualism does not aim to reconcile intuitionistic and classical attitudes towards certain meta-linguistic sentences but to reconcile intuitionistic and classical attitudes towards certain rules. The idea is that classicists are committed to the rules captured by $\text{raa}^*_{\neg_1}$, $\text{efq}^*_{\neg_1}$, and $\text{dne}^*_{\neg_1}$, while intuitionists are only committed to the rules captured by $\text{raa}^*_{\neg_2}$ and $\text{efq}^*_{\neg_2}$, and that the different meanings of $\neg_1$ and $\neg_2$ account for the diverging commitments.

It has been emphasised that it is helpful to consider formal languages if one is interested in certain properties of fragments of natural languages. Similarly, it is also helpful to deal with formal rules if one is interested in certain properties of ordinary rules. I will assume here that calculi of natural deduction are adequate models of collections of such ordinary rules.[14] There is one crucial aspect, however, in which the conception of rules (as presented in 1.2.3) has to be revised: rules should no longer be conceived of as tied to a single language. The new conception of rules has been put forward by Hodes:

> [I]t is important to conceive of a deductive rule, and with it of a logical concept, as a language-transcendent object (Hodes 2004: 145).

The language-transcendent conception of rules is also defended by Humberstone (2011: 588-9). In 1.2.3, a rule was identified with a set of transitions in some language. For example, relative to some fixed formal language $\mathcal{L}$, the rule $(\neg_I)$ was identified with the set of pairs

$$(\{(X, S_1 \succ S_2), (Y, S_1 \succ \neg S_2)\}, (X, Y \succ \neg S_1))$$

such that $X, Y \subseteq \mathcal{L}$ and $S_1, S_2 \in \mathcal{L}$. In the rest of this chapter, I will call such entities *specialised rules*. A (general) *rule*, by contrast, will be identified with a function from languages to specialised rules (see Humberstone ibid.). A specialised rule is obtained by applying a rule to some language.

Important kinds of rules are denoted by special expressions, namely schemata for their application. The rules that are most important here are denoted by schemata of the following form, in which $\sharp$ is some unary sentential operator:

$$\frac{\Gamma, \varphi \succ \psi \quad \Delta, \varphi \succ \sharp\psi}{\Gamma, \Delta \succ \sharp\varphi} \; \text{raa}_\sharp \qquad \frac{\Gamma \succ \varphi \quad \Delta \succ \sharp\varphi}{\Gamma, \Delta \succ \psi} \; \text{efq}_\sharp \qquad \frac{\Gamma \succ \sharp\sharp\varphi}{\Gamma \succ \varphi} \; \text{dne}_\sharp$$

---

[13]   That is, the meaning of $\neg_2$ does not *support* reasoning in accordance with the rule captured by $\text{dne}^*_{\neg_2}$. Of course, intuitionists do not object to *every* application of this rule. For example, they accept an inference from $\neg_2\neg_2\neg_2 S$ to $\neg_2 S$ for every sentence S.

[14]   The philosophical significance of calculi will be the central topic of Chapter 5.

These rules yield specialised rules whenever they are applied to a formal language containing the operator $\sharp$. In contrast to the understanding of such schemata in 1.2.3, I understand them here as denoting (general) rules. When applied to some language $\mathcal{L}$, they yield specialised rules for which I will write $\text{raa}_\sharp(\mathcal{L})$, $\text{efq}_\sharp(\mathcal{L})$, and $\text{dne}_\sharp(\mathcal{L})$. More generally, if $\mathcal{R}$ is a set of rules, then I will write $\mathcal{R}(\mathcal{L})$ for the corresponding set of specialised rules. The idea is now that the ordinary rules captured by $\text{raa}^*_\neg$, $\text{efq}^*_\neg$, and $\text{dne}^*_\neg$ can be modelled by the formal rules $\text{raa}_\neg$, $\text{efq}_\neg$, and $\text{dne}_\neg$.

Note that $\sharp$ is the only operator which occurs in the above schemata. I will express this by saying that these rules *govern* the operator $\sharp$. For the following result one also has to mention so-called *structural* rules represented by schemata in which no logical operator occurs:

$$\frac{}{\varphi \succ \varphi}\ (\text{A}) \qquad \frac{\Gamma, \varphi, \psi, \Delta \succ \chi}{\Gamma, \psi, \varphi, \Delta \succ \chi}\ (\text{E}) \qquad \frac{\Gamma, \varphi \succ \psi \quad \Delta \succ \varphi}{\Gamma, \Delta \succ \psi}\ (\text{T})$$

These correspond to assumptions about logical validity which were used in the proofs of T^-_1 and C^-_1.[15] Intuitionists and classicists are committed to the ordinary rules corresponding to these formal rules.

The new dualist position can now be spelled out as follows: there are two possible concepts of negation, which can be expressed by two unary sentential operators $\neg_1$ and $\neg_2$, such that the following holds. On the one hand, the meaning of $\neg_1$ allows one to reason in accordance with $\text{raa}_{\neg_1}$, $\text{efq}_{\neg_1}$, $\text{dne}_{\neg_1}$; it justifies the commitment to these rules. On the other hand, the meaning of $\neg_2$ allows one to reason in accordance with $\text{raa}_{\neg_2}$ and $\text{efq}_{\neg_2}$, but, in general, it does not allow one to reason in accordance with $\text{dne}_{\neg_2}$; it justifies the commitment to the rules $\text{raa}_{\neg_2}$ and $\text{efq}_{\neg_2}$ but not to the rule $\text{dne}_{\neg_2}$.

It is not difficult to see that this version of logical dualism is also to be rejected. Consider the following notions corresponding to the notions of logical and material equivalence of operators:

$\mathcal{R}(\mathcal{L})$-**Equivalent Operators** Two unary sentential operators $\sharp_1$ and $\sharp_2$ are called $\mathcal{R}(\mathcal{L})$-*equivalent* iff the following holds for every sentence S of $\mathcal{L}$: the arguments $\sharp_1 S \succ \sharp_2 S$ and $\sharp_2 S \succ \sharp_1 S$ are $\mathcal{R}(\mathcal{L})$-derivable.

$\mathcal{R}$-**Equivalent Operators** Two unary sentential operators $\sharp_1$ and $\sharp_2$ are called $\mathcal{R}$-*equivalent* iff they are $\mathcal{R}(\mathcal{L})$-equivalent for every language $\mathcal{L}$ which contains them.

If two $\mathcal{R}$-equivalent operators are applied to any sentence of $\mathcal{L}$, then the resulting sentences are '$\mathcal{R}(\mathcal{L})$-equivalent'. Now, there are the following results:

---

[15] The rule (A) corresponds to the assumption that for every sentence S the argument $S \succ S$ is logically valid. The rule (E) corresponds to the assumption that logical validity is preserved if two premises switch places. The rule (T) corresponds to the assumption that for all lists of sentences X, Y and for all sentences S_1, S_2 the following holds: if the arguments $X, S_1 \succ S_2$ and $Y \succ S_1$ are logically valid, then so is the argument $X, Y \succ S_2$ (see n. 10).

Theorem $T_4^\neg$ For every collection of rules $\mathcal{R}$ containing (A), (E), raa$_{\neg_1}$, efq$_{\neg_1}$, raa$_{\neg_2}$, efq$_{\neg_2}$: $\neg_1$ and $\neg_2$ are $\mathcal{R}$-equivalent.

Corollary $C_4^\neg$ For every collection of rules $\mathcal{R}$ containing (A), (E), (T), raa$_{\neg_1}$, efq$_{\neg_1}$, raa$_{\neg_2}$, efq$_{\neg_2}$: dne$_{\neg_1}$ and dne$_{\neg_2}$ are interderivable in $\mathcal{R}$.[16]

The result $T_4^\neg$ is also expressed by saying that the rules (A), (E), raa$_\neg$, efq$_\neg$ *uniquely characterise* the operator $\neg$ (see Humberstone 2011: 578-80).

Proof. I present the derivations in the suggestive tree-notation (see 5.1). Here is one of the two required derivations for $T_4^\neg$ (the other derivation can be obtained by switching the indices):

$$
\cfrac{\cfrac{\cfrac{\cfrac{S \succ S}{} \text{ (A)} \quad \cfrac{\neg_1 S \succ \neg_1 S}{} \text{ (A)}}{S, \neg_1 S \succ \neg_2 S} \text{ efq}_{\neg_1}}{\neg_1 S, S \succ \neg_2 S} \text{ (E)} \qquad \cfrac{S \succ S}{} \text{ (A)}}{\neg_1 S \succ \neg_2 S} \text{ raa}_{\neg_2}
$$

If Π is written for the above derivation, then let Σ be the following derivation:

$$
\cfrac{\cfrac{\cfrac{\cfrac{\Pi}{\neg_1 S \succ \neg_2 S} \quad \cfrac{\neg_2 \neg_2 S \succ \neg_2 \neg_2 S}{} \text{ (A)}}{\neg_1 S, \neg_2 \neg_2 S \succ \neg_1 \neg_1 S} \text{ efq}_{\neg_2}}{\neg_2 \neg_2 S, \neg_1 S \succ \neg_1 \neg_1 S} \text{ (E)} \qquad \cfrac{\neg_1 S \succ \neg_1 S}{} \text{ (A)}}{\neg_2 \neg_2 S \succ \neg_1 \neg_1 S} \text{ raa}_{\neg_1}
$$

Then, this is one of the two required derivations for $C_4^\neg$:

$$
\cfrac{\cfrac{X \succ \neg_2 \neg_2 S \quad \cfrac{\Sigma}{\neg_2 \neg_2 S \succ \neg_1 \neg_1 S}}{X \succ \neg_1 \neg_1 S} \text{ (T)}}{X \succ S} \text{ dne}_{\neg_1}
$$

$\square$

Now, the corollary $C_4^\neg$ shows the present form of logical dualism to be incorrect. If someone is committed to the rules (A), (E), (T), raa$_{\neg_1}$, efq$_{\neg_1}$, dne$_{\neg_1}$, raa$_{\neg_2}$, and efq$_{\neg_2}$, then she is also committed to the rule dne$_{\neg_2}$. If the meaning of $\neg_1$ and $\neg_2$ is such that the commitment to the former rules is justified, then a simple argumentation (corresponding to the derivation for $C_4^\neg$ displayed above) shows that the commitment to the rule dne$_{\neg_2}$ is also justified.

This completes the presentation of uniqueness results for the negation operator $\neg$. It should be noted that the presented results about two possible negation operators $\neg_1$ and $\neg_2$ do not by themselves imply that $\neg_1$ and $\neg_2$ are *synonymous*. For example, they are compatible with the possibility that for every sentence S, the sentence $\neg_2 S$ is synonymous with the sentence $\neg_1 \neg_1 \neg_1 S$ rather than with

---

[16] This means that for every language $\mathcal{L}$ containing $\neg_1$ and $\neg_2$ the following two claims hold: (i) the specialised rule dne$_{\neg_1}(\mathcal{L})$ is derivable in $\mathcal{R}(\mathcal{L}) \cup \{dne_{\neg_2}(\mathcal{L})\}$, and (ii) the specialised rule dne$_{\neg_2}(\mathcal{L})$ is derivable in $\mathcal{R}(\mathcal{L}) \cup \{dne_{\neg_1}(\mathcal{L})\}$.

the sentence $\neg_1 S$. However, the presented results show that the motivation for logical dualism is ill-founded. There can be no meaning-theoretical justification for the claim that intuitionistic and classical mathematics are equally legitimate enterprises. The involved negation concepts obey the same logical rules. Therefore, there is no reason to postulate two logically equivalent but non-synonymous negation operators. Consequently, one should stick to the simplest assumption according to which intuitionists and classicists understand "not" in the same way.

It is not difficult to see that there are analogous results for the other logical constants: $\vee, \wedge, \rightarrow, \exists, \forall$. More precisely, for each of these constants the rules of the intuitionistic calculus $\mathcal{R}^I$ that govern it, together with the structural rules of $\mathcal{R}^I$, uniquely characterise it (cp. Harris 1982). Logical dualism is equally implausible no matter which of the mentioned logical constants is under discussion. In the following subsection, I will discuss and reject a possible objection from a logical dualist.

2.2.4   Philosophical Consequences of Uniqueness

In my view, the result $T_4^{\neg}$ shows that logical dualism is false. However, there is a consideration that has to be dealt with if one wants to defend this verdict. The issue concerns the nature of the commitments of intuitionists and classicists concerning the negation operator. Consider the following objection to the claim that logical dualism has been refuted by $T_4^{\neg}$:

> There is a language $\mathcal{L}_1$ with a classical negation operator $\neg_1$: the meaning of $\neg_1$ justifies commitment to the specialised rules $\text{raa}_{\neg_1}(\mathcal{L}_1)$, $\text{efq}_{\neg_1}(\mathcal{L}_1)$, and $\text{dne}_{\neg_1}(\mathcal{L}_1)$. Furthermore, there is a language $\mathcal{L}_2$ with an intuitionistic negation operator $\neg_2$: the meaning of $\neg_2$ justifies commitment to the specialised rules $\text{raa}_{\neg_2}(\mathcal{L}_2)$ and $\text{efq}_{\neg_2}(\mathcal{L}_2)$ but not to the specialised rule $\text{dne}_{\neg_2}(\mathcal{L}_2)$. If someone uses a language $\mathcal{L}$ containing the combined resources of $\mathcal{L}_1$ and $\mathcal{L}_2$, then she is justified to reason in accordance with all mentioned specialised rules. But she is not justified to reason in accordance with the new specialised rules $\text{raa}_{\neg_1}(\mathcal{L})$, $\text{efq}_{\neg_1}(\mathcal{L})$, $\text{dne}_{\neg_1}(\mathcal{L})$, $\text{raa}_{\neg_2}(\mathcal{L})$, and $\text{efq}_{\neg_2}(\mathcal{L})$. Consequently, the result $T_4^{\neg}$ does not apply.

According to this objection, the relevant commitments are commitments to *specialised* rules; they are sensitive to an underlying language. The idea is that if a language is enriched by some new word, it does not simply follow that someone's commitments also pertain to argumentative moves involving the new word. In particular, it is claimed that classicists are only committed to specialised rules which do not pertain to arguments involving the intuitionistic negation operator and that intuitionists are only committed to specialised rules which do not pertain to arguments involving the classical negation operator.

The crucial question is whether intuitionists and classicists are committed to the relevant (general) rules, or whether they are only committed to certain specialisations of them. In my view, the mentioned objection underestimates the commitments incurred by intuitionists and classicists. I think that someone's commitments to natural classes of argumentative moves, representable by inference schemata, are her commitments to rules which also apply to extensions of her current language and which are thus language-transcendent (cp. Humberstone 2011: 592-3).

The view that the commitment to an inference rule is in this sense *open-ended*, pertaining to any possible extension of the language under consideration, is also defended by Williamson:

> The sort of open-ended commitment just described is typical of our commitment to rules of inference. For example, my commitment to reasoning by disjunctive syllogism is not exhausted by my commitment to its instances in my current language; when I learn a new word, I am not faced with an open question concerning whether to apply disjunctive syllogism to sentences in which it occurs. Indeed, open-ended commitment may well be the default sort of commitment: one's commitment is open-ended unless one does something special to restrict it (Williamson 2007b: 377).

To take up Williamson's example, consider the rule of *disjunctive syllogism*, a rule that is accepted by classicists and intuitionists alike:

$$\frac{\Gamma \succ \varphi \vee \psi \quad \Delta \succ \neg\varphi}{\Gamma, \Delta \succ \psi} \; ds_{\vee,\neg}$$

The informal rule corresponding to $ds_{\vee,\neg}$ allows one to infer a sentence S_2 from two sentences "S_1 or S_2" and "not-S_1". More explicitly, in an argumentation in which one has established the sentence "S_1 or S_2" relative to the set of assumptions X and in which one also has established the sentence "not-S_1" relative to the set of assumptions Y, one is justified to infer the sentence S_2. One then has brought about an argumentative situation in which the sentence S_2 has been established relative to the set of assumptions $X \cup Y$.

Now, imagine the following situation. Some chemist wants you to extend your language with the addition of the expression "pook". Syntactically, the word "pook" is meant to function like an ordinary adjective. Furthermore, he claims that "pook" is a very useful word since it allows you to describe certain chemical substances better than any expression in your current language. However, he says, you must be careful with the inference rule of disjunctive syllogism afterwards: you are not allowed to apply this rule if some involved sentence contains the new word. To illustrate the difficulty, he then gives the following example: "The substance in front of you is either an acid or pook, and it is not an acid. But, in this case, the substance in front of you is not pook either. You cannot apply the rule of disjunctive syllogism here since the involved sentences contain the new word 'pook'."

As far as I can see, it does not matter how much the chemist praises the potential of the expression "pook" to describe certain phenomena. I cannot follow his advice if I stick to my commitments, because my commitment to applications of $ds_{\vee,\neg}$ does not depend on the meanings of the involved words apart from "or" and "not". I am committed to a rule which allows me to infer a sentence S_2 from sentences "S_1 or S_2" and "not-S_1" for *any* sentences S_1 and S_2.

The problem is not that it is incoherent to assume that the rule of disjunctive syllogism is invalid. Relevant logicians doubt the legitimacy of this rule, and their position is not incoherent (it seems to me). The problem is rather the assumption that someone's commitments to rules of inference are restricted to the vocabulary of her language at a certain time. If our commitment to a rule of inference were like that, it would be extremely fragile; it could be undermined by unnoticed intrusions by new words. This seems to be a misconception of our commitments to rules of inference.

Consequently, the objection that is made by the logical dualist has to be rejected. If someone uses a classical negation operator $\neg_1$ and an intuitionistic negation operator $\neg_2$, then he is committed to all applications of the rules raa$_{\neg_1}$, efq$_{\neg_1}$, raa$_{\neg_2}$, and efq$_{\neg_2}$. Therefore, the corollary D$_4^\neg$ shows that he is committed to the rule dne$_{\neg_1}$ if and only if he is committed to the rule dne$_{\neg_2}$.

I would like to note that there are interesting technical results on 'combining logics', which show in detail the possibility of setting up frameworks in which classical and intuitionistic logical treatments of sentences and arguments can live side-by-side (see Schechter 2011 and the references given there). However, the crucial observation not undermined by these results is that the open-endedness of the commitments to the rules raa$_\neg$ and efq$_\neg$, which is shared by classicists and intuitionists, precludes the possibility of combing all classical and intuitionistic logical commitments while preserving their differences (cp. Schechter 2011: 603-4). If these open-ended commitments are taken seriously, then there is no room for a complete combination of classical and intuitionistic logic.

There is also a slightly different objection to the claim that logical dualism has been refuted by T$_4^\neg$:[17]

> There is a language $\mathcal{L}_1$ with a classical negation operator $\neg_1$: the meaning of $\neg_1$ justifies commitment to the specialised rules raa$_{\neg_1}(\mathcal{L}_1)$, efq$_{\neg_1}(\mathcal{L}_1)$, and dne$_{\neg_1}(\mathcal{L}_1)$. Furthermore, there is a language $\mathcal{L}_2$ with an intuitionistic negation operator $\neg_2$: the meaning of $\neg_2$ justifies commitment to the specialised rules raa$_{\neg_2}(\mathcal{L}_2)$ and efq$_{\neg_2}(\mathcal{L}_2)$ but not to the specialised rule dne$_{\neg_2}(\mathcal{L}_2)$. But there is no language $\mathcal{L}$ containing a unary sentential operator with the meaning of $\neg_1$ in $\mathcal{L}_1$ and another unary sentential operator with the meaning of $\neg_2$ in $\mathcal{L}_2$. Consequently, the result T$_4^\neg$ does not apply.

According to this objection, there is a language with a classical negation operator and a language with an intuitionistic negation operator, but there cannot be a language with both a classical and an intuitionistic negation operator.

Imagine the following situation. Some biologist wants you to use a new expression: "schmook". Syntactically, the word "schmook" is meant to function like an ordinary count noun. Furthermore, she claims that "schmook" is a very useful word since it allows you to describe certain organisms better than

---

[17]   Thanks to Miguel Hoeltje for clarifying discussion of this point.

any expression in your current language. However, she says, you can only use "schmook" if you no longer use the word "virus" (with its present meaning). Each of the words "schmook" and "virus" are meaningful count nouns in a language which contains all the (other) words of English, but, the biologist claims, they cannot both belong to such a language with their original meanings because their meanings are incompatible.

The idea of the present objection is that the imagined situation about the two count nouns "virus" and "schmook" is analogous to a situation about a classical unary sentential operator "not_1" and an intuitionistic unary sentential operator "not_2": each of these two operators is meaningful in a language which contains, say, all the non-logical and non-mathematical words of English, but there is no such language which contains both of them with their original meanings.

I find the idea of incompatible meanings quite obscure. Correspondingly, I find the following principle rather plausible:

Compatibility of Meanings If there is a language $\mathcal{L}$ and two expressions e_1 and e_2 such that e_1 has a meaning m_1 in an extension of $\mathcal{L}$ and e_2 has a meaning m_2 in an extension of $\mathcal{L}$, then there is an extension of $\mathcal{L}$ in which e_1 has the meaning m_1 and e_2 has the meaning m_2.

Meanings, according to this principle, are never incompatible; one can always enlarge a language with a new meaningful word without incurring the risk that some other word which was meaningful before no longer is.

Suppose that the objection is correct: there is a 'classical English language' with a classical negation operator "not_1", there is an 'intuitionistic English language' with an intuitionistic negation operator "not_2", but there cannot be an extension of either language containing both negation operators. Now, imagine a child who learns English from her parents: from her mother who speaks (only) classical English and from her father who speaks (only) intuitionistic English. I take it to be possible that the child can understand both of her parents in such a situation; she then has learned two different languages, classical and intuitionistic English, and she uses classical English when she talks to her mother and intuitionistic English when she talks to her father. Assume now that the child wants to complain to herself about what her parents have banned her from doing. (Her mother has said: "Do not_1 eat so much chocolate!", and her father has said: "Do not_2 eat so much ice cream!".) Then, she can meaningfully utter the following sentences:

(3) That's unfair! I'm not_1 allowed to eat chocolate. And I'm not_2 allowed to eat ice cream.

However, it is impossible for her to combine the resources of her languages in such a way that she can put her complaint in a single conjunctive sentence:

(4) That's unfair! I'm not_1 allowed to eat chocolate, and I'm not_2 allowed to eat ice cream.

This seems to be strikingly counter-intuitive.

In my view, the objection leads to unacceptable theses about understanding and competent use. It seems to be far more plausible to assume that any two expressions which are separately meaningful and which belong to syntactic categories of some language $\mathcal{L}$ can also be jointly meaningful in an extension of $\mathcal{L}$. Therefore, the dualist objection has to be rejected.

Logical dualism is thus seen to be false. If there were two negation operators, a classical one $\neg_1$ and an intuitionistic one $\neg_2$, then there would be an extension of English containing both of them (with their previous meanings). Someone who spoke this extension of English would then be committed to all applications of the rules $raa_{\neg_1}$, $efq_{\neg_1}$, $raa_{\neg_2}$, and $efq_{\neg_2}$. And, therefore, the corollary D_4^- shows that she would be committed to the rule $dne_{\neg_1}$ iff she were committed to the rule $dne_{\neg_2}$.

Before I turn to exclusive logical monism, I would like to mention an example of a logical dualist view which is refuted by the presented results. It concerns a view that Field attributes to Kripke:

> Kripke has argued in lectures for a view [...] according to which intuitionism is best viewed as a view that takes mathematical sentences to assert the existence of mental constructions; the classical $\neg$ and $\rightarrow$ are *applicable* to such sentences, but result in sentences of no mathematical interest ("I haven't performed a construction of ..."); to get a mathematically interesting claim about mental constructions one must give $\neg$ and $\rightarrow$ special readings, which turn out to obey intuitionist laws (Field 2009a: 344).

According to Field's Kripke, one must give a *special reading* to the negation operator used by intuitionists, a reading which obeys *the intuitionistic laws*, but, I take it, a reading which does not also obey the distinctively classical laws. Now, this is exactly the view which comes under fire from the presented uniqueness results. There simply cannot be such a special intuitionistic reading of $\neg$ in addition to a classical reading of it.

To make sense of Kripke's proposal, one has to give up the idea of special readings and concentrate on the first part of the suggestion: "intuitionism is best viewed as a view that takes mathematical sentences to assert the existence of mental constructions". Here is one way to understand this: whenever an intuitionist assertively utters some mathematical sentence S, she wanted to assert that she has performed a mental act which proves S. For example, if an intuitionist assertively utters the sentence "$2+2 = 4$", she wanted to assert that she has performed a mental act which proves "$2+2 = 4$". Such a view is compatible with logical monism: no special intuitionistic meanings of the logical constants have to be invoked. However, I will not consider this radical view according to which intuitionists never say what they want to say.[18]

What are the consequences of the falsity of logical dualism? As far as I can see, the most plausible alternative to logical dualism is inclusive logical monism. Intuitionists and classicists understand the standard logical constants in the

---

[18] In 3.4, I will return to intuitionistic theses about mental constructions and proofs.

same way. They do so although they have conflicting opinions about which of the rules that govern this vocabulary are valid.

The case for adopting inclusive logical monism rather than the exclusive view according to which either intuitionists or classicists do not understand logical vocabulary can be illustrated as follows. Consider the following two alternative possible situations. In the first case, everyone is committed to the rules of the classical calculus $\mathcal{R}^C$, mathematics is only done classically, and all intuitionists have converted to classicism. Otherwise this situation differs as little as possible from the actual situation. In the second case, everyone is only committed to the rules of the intuitionistic calculus $\mathcal{R}^I$, mathematics is only done intuitionistically, and all classicists have converted to intuitionism. Otherwise this situation differs as little as possible from the actual situation.

Assume now that either classicists or intuitionists do not understand logical vocabulary. First, suppose that classicists do not understand logical vocabulary. Then, I take it, they also do not understand logical vocabulary in the first possible situation. Consequently, no native speaker of English in this situation understands words like "not", "or", "and", "if", "some", and "every". Although these words are used as frequently in this situation as they are actually used, no one uses them competently. No one who utters the sentence "I am not hungry." makes an assertion because no meaning is attached to the word "not". Similarly, no one who utters the sentence "Is someone at home?" asks a question because no meaning is attached to the word "some".

Second, suppose that intuitionists do not understand logical vocabulary. Then, I take it, they also do not understand logical vocabulary in the second possible situation. Consequently, no native speaker of English in this situation understands words like "not", "or", "and", "if", "some", and "every". Although these words are as frequently used in this situation as they are actually used, no one uses them competently. No one who utters the sentence "She lives in Canada or in Australia." makes an assertion because no meaning is attached to the word "or". Similarly, no one who utters the sentence "Everyone be quiet!" gives a command because no meaning is attached to the word "every".

As far as I can see, the sketched consequences of the two suppositions are not acceptable. Therefore, the disjunction of the two suppositions, i.e. exclusive logical monism, has to be rejected. Intuitionists and classicists understand the logical constants, and they understand them in the same way despite having conflicting opinions about which of the rules that govern them are correct.[19]

In the next section, I will try to extend the previous arguments to the case of mathematical vocabulary; i.e., I will try to present considerations which

---

[19]   Although it seems clear that inclusive logical monism is intuitively the most plausible alternative to logical dualism, there are theoretical considerations based on certain theses about the relation between use and understanding which seem to speak against this position. I will consider these arguments in 2.4. The considerations against exclusive monism presented there equally apply to the logical and to the mathematical variant of this view. However, since the initial implausibility of exclusive logical monism is so high (and higher than the initial implausibility of exclusive mathematical monism) I take it to be justified to proceed here from the assumption that intuitionists and classicists understand the standard logical constants in the same way.

undermine mathematical dualism. To this end, I will demonstrate certain results regarding mathematical terms which somewhat resemble the logical uniqueness results presented in this section. In 2.4, I will then try to show why the failure of dualism does not necessitate the adoption of exclusive monism but rather leads to inclusive monism.

2.3 Mathematical Vocabulary

Is it plausible to deny that intuitionists and classicists use mathematical vocabulary like "natural number", "real number", "set", and "function" in the same way? Might it be that although they understand in the same way the logical vocabulary they use, this does not apply to the mathematical vocabulary they use? Consider the following two pairs of sentences:

Cl$_2$ Every non-empty set of natural numbers has a least element.

In$_2$ Not every non-empty set of natural numbers has a least element.

Cl$_3$ Not every total function from $\mathbb{R}$ to $\mathbb{R}$ is continuous.

In$_3$ Every total function from $\mathbb{R}$ to $\mathbb{R}$ is continuous.

When a classical mathematician utters the first and third one assertively and an intuitionist utters the second and fourth one assertively, do they contradict each other? Is there something that is affirmed by one and denied by the other? Or should the charitable interpreter conjecture that the two parties do not understand mathematical expressions in the same way, with the consequence that they do not really make incompatible assertions?

I will now consider the prospects of the view that *inclusive mathematical monism*, which states that intuitionists and classicists understand mathematical vocabulary in the same way, is false. There are two variants of this view. According to *exclusive mathematical monism*, either classicists or intuitionists do not understand the mathematical expressions they use. According to *mathematical dualism*, each side has its own understanding of phrases like "natural number", "real number", "set", and "function". It is claimed that one can disambiguate mathematical expressions into those with classical meanings and those with intuitionistic meanings. This dualist view is motivated by its implications for the two practices of classical and intuitionistic mathematics. Consider two of the above sentences after disambiguation:

Cl$_2^*$ Every non-empty$_1$ set$_1$ of natural numbers$_1$ has a least$_1$ element$_1$.

In$_2^*$ Not every non-empty$_2$ set$_2$ of natural numbers$_2$ has a least$_2$ element$_2$.

Mathematical dualists claim that the meanings of the indexed terms can be such that both sentences are true.

In what follows, I will try to defend inclusive mathematical monism. First, I will present considerations which speak against mathematical dualism: there is no room for two radically different ways of understanding mathematical expressions. Second, I will argue that there is no need to adopt exclusive mathematical

monism. One can stick to the initially plausible thesis that classicists and intuitionists understand mathematical expressions in the same way.

2.3.1   Theories About Numbers

According to mathematical dualism, intuitionists and classicists understand
mathematical vocabulary differently: intuitionists attach intuitionistic meanings
to mathematical terms, and classicists attach different classical meanings to
mathematical terms. I will assume that mathematical dualism takes a certain
strong form, according to which it states that the different types of mathematical meanings do not licence the same assertions and argumentative moves.
Forms of mathematical dualism which are compatible with the claim that the
meanings of mathematical terms lead to indistinguishable forms of mathematics
do not play a role in what follows.[20] The basic idea is that mathematical dualism
is motivated by the claim that classical and intuitionistic mathematics are two
independent but equally legitimate enterprises, and that the different meanings
of the employed mathematical terms at least partly explain this state of affairs.
To emphasise this point, I will also speak about *radical mathematical dualism*.
Radical mathematical dualists believe that mathematical terms are ambiguous
in such a way that the legitimacy of some mathematical argumentations depend
on which understanding of these terms are relevant.

By way of motivation, let me present an argument in favour of radical mathematical dualism. To this end, let $\mathcal{L}$ be some mathematical fragment of English
and assume that $\mathcal{L}$ contains only the standard logical constants and a set MT
of mathematical terms, say arithmetical or set-theoretic terms. Now, consider
the following argument:

P$_1$ There is a class c_1 of classical argumentative moves in $\mathcal{L}$ and a *non-
equivalent* class c_2 of intuitionistic argumentative moves in $\mathcal{L}$ (either not
every member of c_1 is derivable from members of c_2 or not every member
of c_2 is derivable from members of c_1), each of which determines meanings
for the terms in MT.

P$_2$ If a class c of argumentative moves in $\mathcal{L}$ determines meanings for the
terms in MT, then the following two conditions are satisfied:

(i) the members of c are valid with respect to the determined meanings;

(ii) the argumentative moves in $\mathcal{L}$ which are valid with respect to the
determined meanings are derivable from the members of c.

C Therefore, not every argumentative move in $\mathcal{L}$ is valid with respect to the
meanings determined by c_1 iff it is valid with respect to the meanings
determined by c_2.

The argument is based on the assumption that mathematical meanings are determined by mathematical practice. This assumption is a variant of the so-called
manifestation thesis, which is a central premise in an argument of Dummett's

[20] I will mention such a modest form of mathematical dualism at the end of this subsection.

that will be discussed in Chapter 4. The first premise is then closely related to the rejection of exclusive mathematical monism: if mathematical meaning is determined by mathematical use and if both intuitionists and classicists employ meaningful mathematical terms, then both intuitionists and classicists use mathematical terms in a way which determines meanings for them. The second premise tries to spell out what it takes for meaning to be determined by use if the use consists of mathematical argumentations. The idea is that meaning-determining uses fix standards of validity: the meaning-determining uses are valid and other uses involving the relevant terms are only valid if their validity follows from the validity of the meaning-determining uses.

Now, as they stand, the premises of this argument are in need of further clarification, but I will not carry out this task here. Rather, I will show directly that radical mathematical dualism, and, therefore, the conclusion of this argument, cannot be correct. I will argue that there is no room for radically different kinds of meanings of mathematical terms. In this subsection, I will begin by considering the case of arithmetical vocabulary. In the following subsections, I will deal with set-theoretic vocabulary (2.3.2) and function-theoretic vocabulary (2.3.3).

Arithmetical dualists claim that intuitionists and classical mathematicians understand arithmetical vocabulary in radically different ways, so that arithmetical argumentations may or may not be correct depending on whether the involved arithmetical terms are to be understood classically or intuitionistically. Consider the following situation. There are two persons who reason in accordance with the rules of the intuitionistic calculus $\mathcal{R}^I$. Furthermore, both persons use arithmetical vocabulary, but they do so in different ways. The first person intends to use this vocabulary to speak about abstract objects whose existence and nature is independent of human activity. As a consequence of that, she accepts all those arithmetical sentences which are logically true according to classical logic, for example, the following one (cp. 3.3):

Goldbach's Disjunction (GD) Every or not every even number larger than 2 is the sum of two primes.

She does not consider this to be a logical truth but to be a sentence which is true partly in virtue of the meanings of the involved arithmetical terms. The other person, by contrast, intends to use arithmetical vocabulary to speak about mental entities created through her mathematical activity and which have their properties in virtue of this activity. As a consequence, she does not accept all arithmetical sentences which are logically true according to classical logic. In particular, she does not accept the sentence GD.[21] Now, might it not be that

---

[21] The idea that it depends on the subject matter of the discourse whether one can rely on classical logic or whether one has to confine oneself to the principles of intuitionistic logic has been put forward by Dummett (1963). It has been taken up, for example, by Rumfitt (2000: 817-20). A somewhat related position is held by Field. Field (2008: 296-8 and 353-7) claims that classical logic can be used in mathematical (set-theoretic) contexts but, in general, he does not believe that sentences of the form "S or not-S" are (logically) true. (One significant difference between Field and the mathematician mentioned above is, of course, that Field does not accept the theory of intuitionistic logic in non-mathematical contexts.)

both persons are correct because they use arithmetical vocabulary with different meanings? Might it not be that there are classical arithmetical meanings
with which one can correctly reason classically and intuitionistic arithmetical
meanings with which one cannot correctly reason classically? It is the purpose
of this subsection to explain why such forms of arithmetical dualism are to be
rejected.

To make the pertinent notions more precise, I will again speak about formal
languages and about calculi for these languages. In the same way as before
these formal languages should be understood as simplified models of fragments
of English, and the rules of the calculi should be understood as corresponding
to ordinary rules applicable to expressions of English.

To begin with, I will consider a formal language $\mathcal{L}^{Ar}$ corresponding to a
simple fragment of English. The language $\mathcal{L}^{Ar}$ has the following symbols:

Ar $\neg, \vee, \wedge, \rightarrow, \forall, \exists, (,), x, y, z, \ldots, a, b, c, \ldots, N, ID, SC, SM, PR.$

Apart from the logical constants, brackets, parameters, and bound variables,
there are five arithmetical predicates: a unary predicate symbol N, two binary
predicate symbols ID and SC, and two ternary predicate symbols SM and PR.
These arithmetical predicates correspond to predicates of English:

- "Nx" $\sim$ "x is a natural number";

- "$ID(x,y)$" $\sim$ "x is the same number as y";

- "$SC(x,y)$" $\sim$ "the number y is a successor of the number x";

- "$SM(x,y,z)$" $\sim$ "the number z is a sum of the numbers x and y";

- "$PR(x,y,z)$" $\sim$ "the number z is a product of the numbers x and y".

It may be noted that the predicate ID is not taken to correspond to the simpler
predicate "is identical to"; this is not essential for the following results but it
simplifies the presentation of the arithmetical rules (see Appx. A.1). Furthermore, it should be noted that $\mathcal{L}^{Ar}$ does not contain function symbols for the
successor function, addition, and multiplication but relation symbols instead.
The main reason for this is that the relevant functions are *partial* functions:
functions which are only applicable to (pairs of) natural numbers. The use of
relation symbols for the graphs of these partial functions is a convenient way
to deal with this complication.[22]

The language $\mathcal{L}^{Ar}$ permits the formation of sentences which are not arithmetical in a narrow sense, e.g. the sentence $\exists a \neg Na$. For what follows, the

---

[22] A second reason is that natural languages like English apparently do not allow for words
corresponding to function symbols. In English, there is no syntactic category of words
which might contain a word w such that "w the number 2" is a phrase corresponding to
"$S2$", in which a function symbol "S" is applied to the singular term "2". In English, one
has to use definite descriptions like "the successor of the number 2" in these contexts.
However, in these descriptions the first three words "the successor of" do not form a
constituent.

sentences of $\mathcal{L}$ that are arithmetical in a narrow sense are of special importance. I will call them *Ar-sentences*. I take an Ar-sentence to be a sentence of
$\mathcal{L}^{Ar}$ containing only quantifiers bounded by the predicate N: every token of a
universal quantifier occurs in a context of the form $\forall a(Na \rightarrow \ldots)$, and every
token of an existential quantifier occurs in a context of the form $\exists a(Na \wedge \ldots)$.
I will write $\mathcal{L}^{Ar}_+$ for the set of Ar-sentences.

Furthermore, I will assume that intuitionists and classicists are committed to
certain rules governing the expressions of $\mathcal{L}^{Ar}$. As regards the logical constants,
I will assume that both accept the rules of the intuitionistic calculus $\mathcal{R}^I$. As
regards the arithmetical constants, I will assume that both accept a standard
collection of rules corresponding to those of *Peano/Heyting Arithmetic* (see
Troelstra & van Dalen 1988: 126). These rules are given in Appendix A.1. I will
refer to the calculus consisting of the logical and the arithmetical rules by $\mathcal{R}^{Ar}$.

There is one peculiarity of $\mathcal{R}^{Ar}$ that I would like to point out here. The
calculus $\mathcal{R}^{Ar}$ does not contain the following rule:

$$\frac{\Gamma \succ \varphi(t) \quad \Delta \succ ID(t, t^*)}{\Gamma, \Delta \succ \varphi(t^*)} \ (ID_E)$$

This is important for two reasons. First, this rule is problematic regardless of
the dispute between classicists and intuitionists.[23] Second, the notion of identity
is a bone of contention between intuitionists and classicists (see Dummett 1977:
15-7). As a consequence of leaving out this rule, the calculus $\mathcal{R}^{Ar}$ contains only
rules which are equally accepted by intuitionists and by classicists.

Let me now introduce radical arithmetical dualism. According to arithmetical dualists, the predicates N, ID, SC, SM, PR are ambiguous. Depending on
whether an intuitionist or a classical mathematician uses them, they have different meanings. Consequently, there are two arithmetical variants of $\mathcal{L}^{Ar}$. On
the one hand, there is a classical arithmetical language ${}^1\mathcal{L}^{Ar}$ containing arithmetical predicates $N_1, ID_1, SC_1, SM_1, PR_1$ with classical meanings. And there is
an intuitionistic arithmetical language ${}^2\mathcal{L}^{Ar}$ containing arithmetical predicates
$N_2, ID_2, SC_2, SM_2, PR_2$ with intuitionistic meanings. Corresponding to these
languages there are then arithmetical calculi ${}^1\mathcal{R}^{Ar}$ and ${}^2\mathcal{R}^{Ar}$ corresponding to
the calculus $\mathcal{R}^{Ar}$, which capture some of the basic classical and intuitionistic
arithmetical commitments. Note that there is an obvious translation function
between the expressions of ${}^1\mathcal{L}^{Ar}$ and ${}^2\mathcal{L}^{Ar}$: to obtain the translation of an expression of ${}^1\mathcal{L}^{Ar}$ one only has to switch the subscripts of the arithmetical predicates.

Now, according to radical arithmetical dualism, the difference in meaning of
the arithmetical terms shows itself in differences in arithmetical commitments.
This can be spelled out differently, but, as an absolutely minimal assumption, I
will take radical arithmetical dualists to be committed to the following thesis:

Arithmetical Dualism (AD) Not for every sentence S_1 in ${}^1\mathcal{L}^{Ar}_+$ with translation S_2 in ${}^2\mathcal{L}^{Ar}_+$: intuitionists and classicists are committed to the biconditional $S_1 \leftrightarrow S_2$.

---

[23]   The following invalid argument seems to instantiate this rule: "Peter believes that Ali
was a boxer. Ali is identical to Clay. Therefore, Peter believes that Clay was a boxer."

If everyone had to accept that every classical arithmetical sentence is materially equivalent to its intuitionistic counterpart, then radical arithmetical dualism would be false. If everyone had to agree that a classical arithmetical sentence is true iff its intuitionistic counterpart is true, then no one could claim that there are two independent arithmetical practices each with its own distinctive results.

However, given fairly uncontentious assumptions about the commitments of intuitionists and classicists, the thesis of which AD is the negation can be proven. Apart from the commitments captured by ${}^1\mathcal{R}^{Ar}$ and ${}^2\mathcal{R}^{Ar}$ one only has to assume that intuitionists and classicists allow for the possibility of making recursive definitions, and this assumption is as unproblematic as the assumption that they accept the rule of induction. Given this assumption, let $\mathcal{L}_*$ be the language containing the symbols of both languages ${}^1\mathcal{L}^{Ar}$ and ${}^2\mathcal{L}^{Ar}$ plus a new binary predicate symbol π. By a recursive definition it can be enforced that π stands for an 'isomorphism' between the objects which satisfy N_1 and the objects which satisfy N_2. Furthermore, this can be captured by a simple set of rules $\mathcal{R}$ for the language $\mathcal{L}_*$ (see Appx. A.1). Now, let $\mathcal{R}_*$ be the union ${}^1\mathcal{R}^{Ar} \cup {}^2\mathcal{R}^{Ar} \cup \mathcal{R}$. Then, the following result can be established:

Theorem $\mathbf{T}^{Ar}$ For every sentence S_1 in ${}^1\mathcal{L}_+^{Ar}$ with translation S_2 in ${}^2\mathcal{L}_+^{Ar}$:
$\qquad S_1 \leftrightarrow S_2$ is $\mathcal{R}_*(\mathcal{L}_*)$-derivable.

I give a proof of T^{Ar} in Appendix A.1. The result T^{Ar} implies the thesis of which AD is the negation; therefore, it shows radical arithmetical dualism to be false.

It is assumed that the relevant arithmetical commitments (the rule of induction and the possibility of definition by recursion) are essentially open-ended (cp. 2.2.4). That is, it is assumed that the arithmetical commitments are not confined to what can be expressed by means of a restricted language. Consider the principle of induction which roughly says that if the smallest number satisfies some open sentence S and if every successor of a number which satisfies S also satisfies S, then every number satisfies S. What is assumed here is that the intuitionistic and the classical commitment to this principle is not restricted to a limited class of open sentences S. Their commitment holds with respect to every possible open sentence S.

It has to be admitted that the result T^{Ar} is weaker than the result T_4^- on the negation operator and the corresponding results on the other logical constants. Roughly put, the result T_4^- can be expressed as follows. There is a set of rules $\mathcal{R}$ which is accepted by intuitionists and classicists to apply to the negation operator and which uniquely characterises a unary sentential operator: with respect to every language $\mathcal{L}$ containing two operators $\neg_1$ and $\neg_1$ satisfying $\mathcal{R}$, the result of applying $\neg_1$ is $\mathcal{R}(\mathcal{L})$-equivalent to the result of applying $\neg_2$. Such a result does not hold in the arithmetical case. In particular, the set of rules $\mathcal{R}_*$ mentioned in T^{Ar} does not uniquely characterise the arithmetical vocabulary. For example, the following is not true: for every language $\mathcal{L}$ containing the symbols of $\mathcal{L}_*$, the result of applying N_1 is $\mathcal{R}_*(\mathcal{L})$-equivalent to the result of applying N_2. If, for example, $\mathcal{L}$ is obtained from $\mathcal{L}_*$ by adding a constant symbol t, then the sentences $N_1 t$ and $N_2 t$ are not $\mathcal{R}_*(\mathcal{L})$-equivalent. (The reason for this is that the arithmetical rules $\mathcal{R}^{Ar}$ do not determine the identity of the natural numbers. For example, the sentence $\forall a(N_1 a \leftrightarrow \neg N_2 a)$ can be consistently added (as a rule without premises) to the specialised calculus $\mathcal{R}_*(\mathcal{L}_*)$. To see this, one may consider classical models of the language $\mathcal{L}_*$ in which the predicates N_1 and N_2 stand for disjoint sets of objects.)

This explains why there is no hope of refuting non-radical forms of arithmetical dualism by means of results like T^{Ar}. Suppose, for example, that the classical predicate N_1 is satisfied by certain abstract entities and that the intuitionistic predicate N_2 is satisfied by certain mental entities. Then, given the assumption that a difference in extension implies a difference in meaning, it follows that the predicates N_1 and N_2 differ in meaning; therefore, it follows that arithmetical dualism is true. Such a position cannot be refuted by relating intuitionistic and classical arithmetical commitments since these only concern Ar-sentences. However, for present purposes it suffices to have refuted radical arithmetical dualism. There cannot be two kinds of arithmetical meanings which differ to such an extent that they give rise to different arithmetical results.

2.3.2 Theories About Sets of Numbers

It might be said that the significance of the result T^{Ar} is limited by the fact that there is no 'mathematical' reason, as opposed to, for example, an ontological reason, for believing that intuitionists and classicists use any of the arithmetical terms N, ID, SC, SM, PR in different ways. In particular, the fundamental argumentative moves involving these arithmetical terms accepted by intuitionists seem to correspond precisely to those accepted by classicists.[24] The fact that they do not accept the same Ar-sentences rather seems to be a consequence of the fact that they do not accept the same fundamental argumentative moves involving logical terms.

This line of thought leads to the consideration of richer mathematical languages containing terms for mathematical objects about whose basic mathematical properties intuitionists and classicists are in greater conflict. I will now consider two such languages. In this subsection, I will consider a language for speaking about sets of numbers. In the next subsection, I will deal with a language for speaking about sequences of numbers. In both cases, intuitionists and classicists are in disagreement about the fundamental mathematical sentences and argumentative moves involving the terms of the respective languages.

To begin with, consider mathematical vocabulary for speaking about sets of natural numbers. Mathematical dualists with respect to such vocabulary claim that intuitionists and classical mathematicians understand this vocabulary in different ways. They assume that such vocabulary can be disambiguated into classical and intuitionistic versions. Furthermore, according to the radical version of this variety of mathematical dualism, the different kinds of meanings of the relevant mathematical terms at least partly explain why intuitionists and classicists employ different argumentative moves involving these terms.

---

[24] One might propose that these fundamental assumptions are captured by the non-logical rules of $\mathcal{R}^{Ar}$ that are given in Appendix A.1.

Consider the following situation. There are two persons who reason in accordance with the rules of the arithmetical calculus $\mathcal{R}^{Ar}$. Furthermore, both use set-theoretic vocabulary, but they do so in different ways. Only the first person accepts the following *Least Number Principle*:

LP Every non-empty set of numbers has a least element.

The second person explicitly rejects this principle but accepts the following *Uniformity Principle* (3.1.1) for every extensional formula φ:

UP If every set of numbers is φ-related to some number, then there is a number to which every set of numbers is φ-related.

Here the Uniformity Principle is explicitly rejected by the first person. Now, might it not be that both persons are correct because they use set-theoretic vocabulary with different meanings? Might it not be that the first person is correct because she attaches classical meanings to set-theoretic vocabulary, while the second person is correct because she attaches intuitionistic meanings to the set-theoretic terms? I will now explain why such forms of mathematical dualism are to be rejected.

To make this precise, I will again use a formal language and an associated calculus. The relevant formal language will be denoted by $\mathcal{L}^{Cl}$. It is an extension of the arithmetical language $\mathcal{L}^{Ar}$ and corresponds to a simple fragment of English. The language $\mathcal{L}^{Cl}$ has the following symbols:

Cl $\neg, \vee, \wedge, \rightarrow, \forall, \exists, (,), x, y, z, \ldots, a, b, c, \ldots, N, ID, SC, SM, PR, C, \in$.

Apart from the symbols of $\mathcal{L}^{Ar}$, it contains two set-theoretic symbols: a unary predicate symbol C and a binary predicate symbol $\in$. These symbols correspond to predicates of English:

- "Cx" $\sim$ "x is a set of numbers";

- "$x \in y$" $\sim$ "the number x is an element of the set of numbers y".

It may be noted that the predicate C only applies to sets of *natural numbers* and that a similar restriction holds for $\in$.

The language $\mathcal{L}^{Cl}$ permits the formation of sentences which do not only concern 'mathematical properties' of natural numbers and sets of them. For example, the truth of the sentence $\exists a(Ca \wedge Na)$, which expresses that some object is both a set of numbers and a number, is irrelevant for the mathematical theory about numbers and their sets. For what follows, it is important to specify a fragment of $\mathcal{L}^{Cl}$ whose sentences only concern the relevant properties of numbers and their sets. An open sentence S of $\mathcal{L}^{Cl}$ which satisfies the following properties will be called an open *Cl-sentence*: (i) every token of a quantifier is bounded by the predicate N or by the predicate C; (ii) for every parameter and variable v (which is not directly adjacent to $\exists$ or $\forall$): either every token of v is in an *arithmetical position* or every token of v is in a *set-theoretic position*.[25] I will write $\mathcal{L}_+^{Cl}$ for the set of Cl-sentences.

Furthermore, I assume that intuitionists and classicists are committed to certain rules governing the expressions of $\mathcal{L}^{Cl}$. I will assume that both accept the arithmetical rules of $\mathcal{R}^{Ar}$ and a comprehension principle for sets of numbers:

Comprehension $\exists a(Ca \wedge \forall b(b \in a \leftrightarrow (Nb \wedge \varphi(b))))$.

This principle says that there is a set of numbers which contains some number b iff b satisfies a condition given by some open sentence. (In fact, for the following result it suffices to assume that both accept a *predicativist* version of this principle. In this weaker principle, only those open sentences φ that solely contain 'arithmetical quantifiers' are allowed, i.e. quantifiers that are bounded by the predicate N.) The resulting calculus will be referred to as $\mathcal{R}^{Cl}$ (see Appx. A.2).

Now, the pertinent form of mathematical dualism can be expressed as follows. The predicates C, $\in$ are ambiguous and can be replaced by different versions C_1, $\in_1$ and C_2, $\in_2$ with classical and intuitionistic meanings. This yields a classical language ${}^1\mathcal{L}^{Cl}$ with a classical calculus ${}^1\mathcal{R}^{Cl}$ and an intuitionistic language ${}^2\mathcal{L}^{Cl}$ with an intuitionistic calculus ${}^2\mathcal{R}^{Cl}$. As in the arithmetical case, there is an obvious translation function between ${}^1\mathcal{L}^{Cl}$ and ${}^2\mathcal{L}^{Cl}$. Again, I will assume that set-theoretic dualists of the relevant form are minimally committed to the following thesis:

Set-Theoretic Dualism (SD) Not for every sentence S_1 in ${}^1\mathcal{L}_+^{Cl}$ with translation S_2 in ${}^2\mathcal{L}_+^{Cl}$: intuitionists and classicists are committed to the biconditional $S_1 \leftrightarrow S_2$.

The difference in meaning between C_1, $\in_1$ and C_2, $\in_2$ has to show up in different kinds of commitments expressed by Cl-sentences.

However, SD is false. To see this, let $\mathcal{L}_*$ be the language containing the symbols of both languages ${}^1\mathcal{L}^{Cl}$ and ${}^2\mathcal{L}^{Cl}$, and let $\mathcal{R}_*$ be the union of the two calculi ${}^1\mathcal{R}^{Cl}$ and ${}^2\mathcal{R}^{Cl}$. Then, the following result can be established:

Theorem $\mathbf{T}^{Cl}$ For every sentence S_1 in ${}^1\mathcal{L}_+^{Cl}$ with translation S_2 in ${}^2\mathcal{L}_+^{Cl}$:
$S_1 \leftrightarrow S_2$ is $\mathcal{R}_*(\mathcal{L}_*)$-derivable.

The proof of this result is given in Appendix A.2.

This result implies the thesis of which SD is the negation; therefore, it shows the relevant form of set-theoretic dualism to be false. As in the case of logical and arithmetical vocabulary, it is here assumed that the relevant set-theoretic commitments are essentially open-ended: classicists and intuitionists are committed to believing in the existence of sets containing those numbers

---

[25] The notions of an arithmetical position and of a set-theoretic position are to be understood in the intuitive way. In particular, in the following (open) sentences, the parameter x is in an arithmetical position:

$$Nx, ID(x,y), ID(y,x), SC(x,y), SC(y,x), SM(x,y,z), SM(y,x,z), SM(y,z,x),$$

$$PR(x,y,z), PR(y,x,z), PR(y,z,x), x \in y.$$

In the sentences Cx and $y \in x$, the parameter x is in a set-theoretic position.

that satisfy some open sentence no matter which vocabulary is used in this sentence. Furthermore, it has to be admitted that the result T^{Cl} is weaker than the result $T_4^{\neg}$ about the negation operator. The rules of $\mathcal{R}^{Cl}$ for C and $\in$ do not uniquely characterise these predicates. If, for example, $\mathcal{L}$ is obtained from $\mathcal{L}_*$ by adding a constant symbol t, then the sentences C_1t and C_2t are not $\mathcal{R}_*(\mathcal{L})$-equivalent.[26]

However, for present purposes it suffices to have refuted the relevant radical form of set-theoretic dualism. There cannot be two kinds of meanings for C and $\in$ which differ to such an extent that they give rise to different set-theoretic results.

2.3.3 Theories About Sequences of Numbers

In this subsection, I will transfer the result about sets of numbers from 2.3.2 to sequences of numbers.[27] Since the involved notions, claims, and argumentations are very similar in these cases, I will be rather brief.

There are two motivations for considering theories about sequences of numbers in addition to theories about sets of numbers. First, a basic intuitionistic assumption about sequences of natural numbers will figure prominently in an anti-classical argument discussed in the following chapter. Second, it is not *obvious* that theories of sequences of numbers are similar to theories of sets of numbers as regards the question of mathematical dualism. It might be that the intuitionistic conception of a sequence differs more radically from its classical counterpart than the intuitionistic conception of a set differs from its classical counterpart.

Classical mathematicians assertively utter the following *Sequential Least Number Principle*:

LP* Every sequence of numbers has a least item.

Intuitionists assertively utter the negation of LP*. In addition, they accept the following *Continuity Principle* (3.1.2) for every extensional formula φ:

CN If every sequence of numbers is φ-related to some number, then there is an initial segment of every sequence and a number such that every sequence extending this initial segment is φ-related to that number.

This sentence is explicitly rejected by classical mathematicians. Might it be that both are correct because they attach different senses to the function-theoretic

---

[26] The set-theoretic rules $\mathcal{R}^{Cl}$ do not determine the identity of the sets of numbers or of the relation of being an element. For example, the sentence $\forall a \forall b (a \in_1 b \leftrightarrow \neg a \in_2 b)$ can be consistently added (as a rule without premises) to the specialised calculus $\mathcal{R}_*(\mathcal{L}_*)$. (To see this, consider a classical model of the language $\mathcal{L}_*$ in which the arithmetical predicates and the predicates C_1, C_2 and $\in_1$ are interpreted in the standard way, while $\in_2$ is interpreted to be satisfied by those pairs (x, y), for which x is a number, y is a set of numbers, and x is not an element in y.)

[27] Throughout this book "sequence" should always be understood as referring to infinite sequences (of length ω); i.e., they can be identified with functions whose domain is the set of natural numbers.

vocabulary? I will now explain why such forms of mathematical dualism are to be rejected.

Corresponding to a simple fragment of English, consider the formal language $\mathcal{L}^{Seq}$ containing the following symbols:

Seq $\neg, \vee, \wedge, \rightarrow, \forall, \exists, (,), x, y, z, \ldots, a, b, c, \ldots, N, ID, SC, SM, PR, SQ, AP.$

Apart from the symbols of $\mathcal{L}^{Ar}$, it contains two function-theoretic symbols: a unary predicate symbol SQ and a ternary predicate symbol AP. These symbols correspond to predicates of English:

- "$SQ\,x$" $\sim$ "x is a sequence of natural numbers";

- "$AP(x, y, z)$" $\sim$ "the sequence x maps the number y to the number z".

Similar to the previous cases, it may be noted that SQ is only satisfied by sequences of *natural numbers* and that a similar restriction applies to AP.

An open sentence S of $\mathcal{L}^{Seq}$ which satisfies the following properties will be called an open *Seq-sentence*: (i) every token of a quantifier is bounded by the predicate N or by the predicate SQ; (ii) for every parameter and variable v (which is not directly adjacent to a token of $\exists$ or $\forall$): either every token of v is in an arithmetical position or every token of v is in a function-theoretic position. I will write $\mathcal{L}^{Seq}_+$ for the set of Seq-sentences.

Furthermore, I assume that intuitionists and classicists are committed to certain rules governing the expressions of $\mathcal{L}^{Seq}$. I will assume that both accept the arithmetical rules of $\mathcal{R}^{Ar}$ and a comprehension principle for sequences of numbers. If $\varphi(x, y)$ is arithmetical and functional in x and y,[28] then the following is assumed to hold:

Comprehension* $\exists a(SQ\,a \wedge \forall b \forall c(AP(a, b, c) \leftrightarrow \varphi(b, c)))$.

This principle says that there is a sequence of numbers which maps a number b to a number c iff b and c are φ-related. (Again, it suffices to assume that both accept a *predicativist* version of this principle, in which only those open sentences φ that solely contain arithmetical quantifiers are allowed.) The resulting calculus will be referred to as $\mathcal{R}^{Seq}$ (see Appx. A.3).

The pertinent form of mathematical dualism leads again to two languages $^1\mathcal{L}^{Seq}$ and $^2\mathcal{L}^{Seq}$ with corresponding calculi $^1\mathcal{R}^{Seq}$ and $^2\mathcal{R}^{Seq}$, and there is again an obvious translation function between $^1\mathcal{L}^{Seq}$ and $^2\mathcal{L}^{Seq}$. I will assume that function-theoretic dualists of the relevant form are minimally committed to the following thesis:

---

[28] This means that the following conditions are satisfied:

(a) $\forall a \forall b(\varphi(a, b) \rightarrow (Na \wedge Nb))$;

(b) $\forall a \exists b \varphi(a, b)$;

(c) $\forall a \forall b \forall b^*(\varphi(a, b) \wedge \varphi(a, b^*) \rightarrow ID(b, b^*))$.

The first one states that only numbers are φ-related. The second one states that every number is φ-related to at least one number. And the third one states that every number is φ-related to at most one number.

Function-Theoretic Dualism (FD) Not for every sentence S_1 in $^1\mathcal{L}_+^{Seq}$ with
translation S_2 in $^2\mathcal{L}_+^{Seq}$: intuitionists and classicists are committed to the
biconditional $S_1 \leftrightarrow S_2$.

The difference in meaning between SQ_1, AP_1 and SQ_2, AP_2 has to show up in
different kinds of commitments expressed by Seq-sentences.

To see that FD is false, let $\mathcal{L}_*$ be the language containing the symbols of
both languages $^1\mathcal{L}^{Seq}$ and $^2\mathcal{L}^{Seq}$, and let $\mathcal{R}_*$ be the union of the two calculi $^1\mathcal{R}^{Seq}$
and $^2\mathcal{R}^{Seq}$. Then, the following result can be established:

Theorem T^{Seq} For every sentence S_1 in $^1\mathcal{L}_+^{Seq}$ with translation S_2 in $^2\mathcal{L}_+^{Seq}$:
$S_1 \leftrightarrow S_2$ is $\mathcal{R}_*(\mathcal{L}_*)$-derivable.

The proof of this result is given in Appendix A.3. This result implies the thesis
of which FD is the negation; therefore, it shows the relevant form of function-
theoretic dualism to be false.

2.3.4 Second-Order Theories

Before I turn to the discussion of exclusive mathematical monism, I would like
to answer a certain question suggested by the previous results. As regards the
standard logical constants, classicists accept a proper superset of the set of rules
accepted by intuitionists: the intuitionistic calculus $\mathcal{R}^I$ is properly contained in
the classical calculus $\mathcal{R}^C$. However, it was possible to show that the rules of the
intuitionistic calculus $\mathcal{R}^I$ are uniquely characterising. For example, the rules
A, E, raa$_\neg$, and efq$_\neg$ uniquely characterise $\neg$. As regards set-theoretic and
function-theoretic vocabulary, classicists and intuitionists accept incomparable
sets of rules: some rules are accepted by intuitionists and rejected by classicists
and other rules are accepted by classicists and rejected by intuitionists. Now,
in these cases it was not possible to show that rules accepted by both parties
are uniquely characterising. All that could be shown was that in a combined
calculus it can be proven that a classical sentence is materially equivalent to its
intuitionistic counterpart.

One might now ask whether this is an instance of a general connection
between shared commitments and unique characterisability, or whether there
is also vocabulary uniquely characterised by a set of rules $\mathcal{R}$ which is accepted
by intuitionists and classicists such that classicists accept a superset of rules
$\mathcal{R}_1 \supset \mathcal{R}$ rejected by intuitionists and such that intuitionists accept a superset
of rules $\mathcal{R}_2 \supset \mathcal{R}$ rejected by classicists. The aim of this subsection is to draw
attention to the fact that there is indeed such vocabulary, namely so-called
higher-order logical vocabulary. For simplicity, I will restrict attention to monadic second-order logic here. (The following discussion can be easily extended
to other types of higher-order logical resources.)

First, I will introduce the formal resources and the simple result of unique
characterisability. Then, I will quickly discuss the question of how monadic
second-order resources can be found in natural languages like English.

Suppose that $\mathcal{L}$ is some standard formal language of first-order logic, say
the arithmetical language $\mathcal{L}^{Ar}$. The corresponding language of monadic second-
order logic $\mathcal{L}_{II}$ is obtained by adding two new quantifier symbols: $\forall^{II}$ and $\exists^{II}$,

a countably infinite set of bound variables $(A, B, C, \dots)$, and a countably infinite set of parameters $(X, Y, Z, \dots)$. The notion of an open sentence of $\mathcal{L}_{\mathrm{II}}$ is defined in such a way that the only new atomic open sentences are of the form Xt, in which a second-order parameter X is applied to a singular term t.[29]

The new quantifiers are governed by rules corresponding to those for the first-order quantifiers:[30]

$$\frac{\Gamma \succ \varphi}{\Gamma \succ \forall^{\mathrm{II}} A \; \varphi_X^A} \; (\forall_I^{\mathrm{II}})$$

$$\frac{\Gamma \succ \forall^{\mathrm{II}} A \; \varphi}{\Gamma \succ \varphi_A^{\lambda a \psi(a)}} \; (\forall_E^{\mathrm{II}})$$

$$\frac{\Gamma \succ \varphi_A^{\lambda a \psi(a)}}{\Gamma \succ \exists^{\mathrm{II}} A \; \varphi} \; (\exists_I^{\mathrm{II}})$$

$$\frac{\Gamma \succ \exists^{\mathrm{II}} A \; \varphi \quad \Delta, \varphi_A^X \succ \psi}{\Gamma, \Delta \succ \psi} \; (\exists_E^{\mathrm{II}})$$

There are restrictions for the rules $(\forall_I^{\mathrm{II}})$ and $(\exists_E^{\mathrm{II}})$ corresponding to those for the first-order quantifiers (see 1.2.3): in $(\forall_I^{\mathrm{II}})$, the parameter X must not occur in Γ, and in $(\exists_E^{\mathrm{II}})$, the parameter X must not occur in $\Gamma, \Delta, \exists^{\mathrm{II}} A \; \varphi, \psi$.

At the end of 2.2.3, it has been remarked that not only $\neg$ but also the other logical constants $\lor, \land, \to, \forall, \exists$ are uniquely characterised by their rules from the intuitionistic calculus $\mathcal{R}^I$. Now, in a perfectly analogous manner, it can be shown that the second-order quantifiers $\forall^{\mathrm{II}}$ and $\exists^{\mathrm{II}}$ are uniquely characterised by the rules (A), $(\forall_I^{\mathrm{II}})$, $(\forall_E^{\mathrm{II}})$ and (A), $(\exists_I^{\mathrm{II}})$, $(\exists_E^{\mathrm{II}})$.

To formulate this result, consider the following notions corresponding to the notions for unary sentential operators presented in 2.2.3:

$\mathcal{R}(\mathcal{L})$-**Equivalent Quantifiers** Two monadic second-order quantifiers Q_1 and Q_2 are called $\mathcal{R}(\mathcal{L})$-equivalent iff $Q_1 A \; S \succ Q_2 A \; S$ and $Q_2 A \; S \succ Q_1 A \; S$ are $\mathcal{R}(\mathcal{L})$-derivable for every S such that $Q_1 A \; S$ is a sentence of $\mathcal{L}$.

$\mathcal{R}$-**Equivalent Quantifiers** Two monadic second-order quantifiers Q_1 and Q_2 are called $\mathcal{R}$-equivalent iff they are $\mathcal{R}(\mathcal{L})$-equivalent for every language $\mathcal{L}$ with monadic second-order parameters and bound variables which contains Q_1 and Q_2.

If two $\mathcal{R}$-equivalent operators are applied to the same input of $\mathcal{L}$, then the resulting sentences are "$\mathcal{R}(\mathcal{L})$-equivalent". Then, there are the following results:

Theorem T$^{\forall^{\mathrm{II}}}$ For every collection of rules $\mathcal{R}$ containing (A), $(^1\forall_I^{\mathrm{II}})$, $(^1\forall_E^{\mathrm{II}})$, $(^2\forall_I^{\mathrm{II}})$, $(^2\forall_E^{\mathrm{II}})$: $^1\forall^{\mathrm{II}}$ and $^2\forall^{\mathrm{II}}$ are $\mathcal{R}$-equivalent.

Theorem T$^{\exists^{\mathrm{II}}}$ For every collection of rules $\mathcal{R}$ containing (A), $(^1\exists_I^{\mathrm{II}})$, $(^1\exists_E^{\mathrm{II}})$, $(^2\exists_I^{\mathrm{II}})$, $(^2\exists_E^{\mathrm{II}})$: $^1\exists^{\mathrm{II}}$ and $^2\exists^{\mathrm{II}}$ are $\mathcal{R}$-equivalent.

---

[29] Recall from 1.2.1 that it is assumed that every quantifier binds some variables and that every variable is bound by some quantifier.

[30] In $(\forall_I^{\mathrm{II}})$, φ_X^A results from φ by replacing every token of X by a token of A, where it is required that every such new token of A is bound by the newly introduced initial quantifier. In $(\forall_E^{\mathrm{II}})$ and $(\exists_I^{\mathrm{II}})$, $\varphi_A^{\lambda a \psi(a)}$ results from φ by replacing every token of the form At in which A bound by the initial quantifier by a token of ψ_a^t; cp. Troelstra & Schwichtenberg 2000: 345-6. In $(\exists_E^{\mathrm{II}})$, φ_A^X results from φ by replacing every token of A bound by the initial quantifier by some token of X.

Proof. For each of the two assertions I will present one of the two required derivations. The other can be obtained by switching the indices:

$$
\cfrac{\cfrac{\cfrac{}{{}^2\forall^{II}A\ \varphi \succ {}^2\forall^{II}A\ \varphi}\ (\mathrm{A})}{{}^2\forall^{II}A\ \varphi \succ \varphi_A^X}\ ({}^2\forall_E^{II})}{{}^2\forall^{II}A\ \varphi \succ {}^1\forall^{II}A\ \varphi}\ ({}^1\forall_I^{II})
$$

$$
\cfrac{\cfrac{}{{}^2\exists^{II}A\ \varphi \succ {}^2\exists^{II}A\ \varphi}\ (\mathrm{A}) \qquad \cfrac{\cfrac{}{\varphi_A^X \succ \varphi_A^X}\ (\mathrm{A})}{\varphi_A^X \succ {}^1\exists^{II}A\ \varphi}\ ({}^1\exists_I^{II})}{{}^2\exists^{II}A\ \varphi \succ {}^1\exists^{II}A\ \varphi}\ ({}^2\exists_E^{II})
$$

In the two derivations, the second-order parameter X should not be contained in ${}^1\forall^{II}A\ \varphi$ and ${}^1\exists^{II}A\ \varphi$ respectively. $\qquad\square$

Thus, the monadic second-order quantifiers $\forall^{II}$ and $\exists^{II}$ are uniquely characterised by rules accepted by classicists and intuitionists alike.

As in the case of set-theoretic and function-theoretic languages, a second-order language permits the formation of sentences about which classicists and intuitionists have contradictory opinions. For example, the sentence

(5) $\forall^{II}A\ \forall a\ (Aa \vee \neg Aa)$

is accepted by classicists and rejected by intuitionists. Given that there are such sentences in set-theoretic languages, it is not very surprising that there are such sentences in monadic second-order languages since these two types of languages are rather similar. But then it might be asked how it is possible that second-order resources and set-theoretic resources behave differently as regards unique characterisability.

The set-theoretic sentences of $\mathcal{L}^{Cl}$ which are mathematically most important, namely the Cl-sentences, behave like sentences of a two-sorted language in which the different sorts of terms (those for numbers and those for sets) *cannot* be applied to each other. The connection between the two sorts of terms is effected by a bridge predicate: $\in$. By contrast, the sentences of a second-order language like $\mathcal{L}_{II}^{Ar}$ behave like sentences of a two-sorted language in which the different sorts of terms (the second-order terms and the first-order terms) *can* be applied to each other. And the fact that the gulf between the two sorts of terms in $\mathcal{L}_+^{Cl}$ has to be bridged by an additional predicate is responsible for the failure of unique characterisability because the needed bridge predicate is not uniquely characterised by a set of rules accepted by intuitionists and classicists. Since there is no bridge predicate in the case of monadic second-order logic, there is no such obstacle to unique characterisability.

Up to this point, I have only presented a mathematical result about a certain formal language and a certain formal calculus. However, as I stressed in 2.1.2, such a result can have significance for questions about logical validity only if the formal language corresponds to a fragment of a natural language like English.

I would like to close with some remarks about monadic second-order resources in English. According to an influential suggestion from Boolos (1984, 1985), monadic second-order quantifiers correspond to plural quantifiers of English. I will give a very brief introduction of Boolos's idea.

English not only contains expressions which can be used as singular terms ("Peter", "I", "this book"); it also contains expressions which can be used as plural terms ("Peter and Paul", "we", "these books"). Similarly, English not only contains expressions which can be used as singular predicates ("sleeps", "is wise", "is a philosopher"); it also contains expressions which can be used as plural predicates ("sleep", "are wise", "are philosophers"). Furthermore, English also contains *plural quantifier phrases* and expressions which behave like *plural bound variables* in addition to singular quantifier phrases and expressions which behave like singular bound variables:

(6) *Some critics* admire only one another.

(7) If there are *some things* such that every set is one of them, then there is no set which contains all of them.[31]

Boolos had the idea that sentences of a formal language with monadic second-order quantifiers can be taken to correspond to sentences of natural languages with plural quantifiers and expressions which function like plural bound variables; e.g:

(8) $\exists^{II}A \ \forall a \ (Na \rightarrow Aa) \approx$ There are some things such that every natural number is one of them.

(9) $\forall^{II}A \ \forall a \ (Aa \lor \neg Aa) \approx$ If there are some things then every object is such that it is or is not one of them.

It is possible to present recursive clauses by means of which sentences of monadic second-order languages can be seen to correspond to sentences of English, but it is not necessary to present them here.

It has to be acknowledged that there are two problems with Boolos's proposal. The first one is of minor importance. Obviously, if there are some things, then there is at least one object which is one of them:[32]

(10) $\forall^{II}A \ \exists a \ Aa$

However, the negation of this sentence is logically true according to the theories of intuitionistic and classical second-order logic. In fact, the sentence $\exists^{II}A \ \forall a \ \neg Aa$ is a theorem of these theories.[33] It is not difficult, however, to adjust the correspondence between sentences of a monadic second-order language and sentences of English to account for this anomaly.

---

[31] Note that in (7) *universal* plural quantification is expressed by means of the plural quantifier phrase "some things". One may also note that (6) is "a sentence whose quantificational structure cannot be captured by means of first-order logic" (Boolos 1984: 432).

[32] Arguably, it even holds that if there are some things, then there are at least two (distinct) objects x and y such that x is one of them and y is one of them (cp. Boolos 1984: 443).

[33] It can be obtained by means of the rule $(\exists_I^{II})$ from the logical truth $\forall a \neg (Xa \land \neg Xa)$.

The second problem with Boolos's proposal is more far-reaching (cp. Rayo & Yablo 2001: 78-9). In monadic second-order languages, second-order variables are directly applied to singular terms; for example, if the language contains the singular term 0, then $\exists^{II} A\ A0$ is one of its sentences. However, in the corresponding fragment of English, plural 'variables' are not directly applied to singular terms; for example, in the English sentence "there are some things such that the number 0 is one of them", the expression "is one of" is used to make the plural pronoun "them" applicable to the singular term "the number 0". This might invite the objection that Boolos's proposal fails for general syntactic reasons.

I will not try to solve this problem here.[34] I only wanted to illustrate that there is no principled obstacle to the existence of vocabulary uniquely characterised by a set of rules $\mathcal{R}$ accepted by intuitionists and classicists such that classicists accept a superset of rules $\mathcal{R}_1 \supset \mathcal{R}$ rejected by intuitionists and such that intuitionists accept a superset of rules $\mathcal{R}_2 \supset \mathcal{R}$ rejected by classicists.

2.4 The Simplicity of Understanding

I will now discuss the consequences of the failure of logical and mathematical dualism. The question is whether one should embrace *inclusive* or *exclusive* monism. Do intuitionists and classicists understand logical and mathematical vocabulary in the same way? Or do the adherents of at least one side fail to understand the logical and mathematical expressions they use?

I have already stressed that the exclusive view according to which a large number of native speakers of English do not understand words like "not", "two", "number", or "set" seems very implausible. What motivates someone to adopt such a view? It might be suggested that we know about similar cases from non-logical and non-mathematical contexts. In particular, it has been claimed that we know about similar cases from physics:

> Before special relativity, it was surely intrinsic to the meaning of "before" that the question of whether one event occurred before another had an answer which it made no sense to qualify as from a particular point of view (frame of reference); once we have accepted the theory we cannot attach that meaning to the word or any other word (Dummett 1991: 11).

Dummett rejects the claim that native speakers of English from the 19th century understood the word "before" in the same way as those of their descendants who accept Einstein's theory of special relativity. And he holds an analogous thesis regarding logical and mathematical vocabulary. According to Dummett, it took 20th century physics to endow words like "before" with coherent meanings, and it took 20th century intuitionistic mathematics to endow words like "not" and "number" with coherent meanings.

It should be acknowledged, I think, that the claim that a large number of native speakers of English do not understand words like "not" or "number" is

---

[34] See Rayo & Yablo 2001 for an alternative suggestion for a correspondence of monadic second-order languages and fragments of English.

no more implausible than the claim that a large number of native speakers of English do not understand words like "before". Unlike Dummett, however, I do not find it clear that persons who did not accept the theory of special relativity, perhaps because they lived in the 19th century, did not attach the same senses to words like "before" as, for example, modern physicists do. Dummett believes that there is a close connection between the acceptance of physical, logical, and mathematical theories on the one hand and the understanding of physical, logical, and mathematical vocabulary on the other hand. I take this assumption to be the basis of the doubts about inclusive logical and mathematical monism. Consequently, this assumption will be discussed here further.

2.4.1 An Argument Against Shared Understanding

To begin with, I will try to spell out what I take to be Dummett's idea. I will do so with respect to a Cl-sentence about which intuitionists and classicists have contradictory opinions:

TND^{Cl} $\forall a(Ca \rightarrow \forall b(Nb \rightarrow (b \in a \vee \neg b \in a)))$.

This sentence, a set-theoretic version of the principle *tertium non datur*, says that for every set of numbers a and for every number b, the number b is or is not an element of a. This sentence is logically true according to the theory of classical logic, and its negation is true according to the intuitionistic theory of sets of natural numbers. I will use $\neg$TND^{Cl} to refer to the negation of TND^{Cl}. In what follows, I will use "set-theoretic expression" and similar phrases for all expressions of the set-theoretic language $\mathcal{L}^{Cl}$, including the standard logical constants and the arithmetical predicates of $\mathcal{L}^{Ar}$. In the same way, I will use "set-theoretic practice" and similar phrases for all mathematical activity in which sentences of $\mathcal{L}^{Cl}$ are used, including purely arithmetical argumentations. Now, consider the following *Dummettian argument*:

P$_1$ Either partaking in classical set-theoretic practice or partaking in intuitionistic set-theoretic practice can ground an understanding of set-theoretic vocabulary.

P$_2$ It is not the case that partaking in classical set-theoretic practice and partaking in intuitionistic set-theoretic practice can ground an understanding of set-theoretic vocabulary.

P$_3$ If partaking in classical but not in intuitionistic set-theoretic practice can ground an understanding of set-theoretic vocabulary, then a person who understands set-theoretic vocabulary accepts TND^{Cl}.

P$_4$ If partaking in intuitionistic but not in classical set-theoretic practice can ground an understanding of set-theoretic vocabulary, then a person who understands set-theoretic vocabulary accepts $\neg$TND^{Cl}.

P$_5$ Intuitionists do not accept TND^{Cl}, and classicists do not accept $\neg$TND^{Cl}.

C It is not the case that intuitionists and classicists understand set-theoretic vocabulary.

The first premise states that someone's understanding of set-theoretic vocabulary can be grounded in her set-theoretic activity, leaving it open whether this activity has to follow classical or intuitionistic lines. The second premise excludes the possibility that both kinds of activity, classical and intuitionistic set-theoretic practice, can account for someone's set-theoretic understanding. It follows that either classical but not intuitionistic practice, or intuitionistic but not classical practice can ground set-theoretic understanding. In the former case, the third premise implies that set-theoretic understanding requires acceptance of TND^{Cl}; under the assumption that set-theoretic understanding can only be grounded in *classical* set-theoretic activity, not also in intuitionistic set-theoretic activity, such understanding requires the acceptance of fundamental assumptions of *classical* set theory. In the latter case, the fourth premise implies that set-theoretic understanding requires acceptance of $\neg TND^{Cl}$; under the assumption that set-theoretic understanding can only be grounded in *intuitionistic* set-theoretic activity, not also in classical set-theoretic activity, such understanding requires the acceptance of fundamental assumptions of *intuitionistic* set theory. But then the fifth premise directly implies the falsity of inclusive monism: it is not the case that intuitionists *and* classicists understand set-theoretic vocabulary.

Now, the first four premises of this argument are surely in need of clarification. What are the requirements for participating in a certain mathematical practice? And what does it take for such activity to ground someone's understanding of the involved expressions? However, I will not discuss these questions here. (They will be a central topic of Chapter 4.) A preliminary understanding has to suffice for the following discussion.

The first premise of the Dummettian argument is clearly not incontestable. Why should it be possible that someone's set-theoretic understanding can be grounded in her linguistic behaviour. And even if it can be grounded in all her linguistic behaviour: why should it be possible that someone's understanding of set-theoretic expressions can be grounded in her set-theoretic activity alone, given that these expressions also figure prominently in other kinds of linguistic behaviour (e.g. in non-mathematical contexts). Despite its being not indubitable, however, I will not question the premise P_1 here. It has some initial plausibility, and I think it would be remarkable if inclusive monism could only be defended by denying that any kind of set-theoretic activity can ground an understanding of the involved expressions. (The premise P_1 is closely related to the central premise of Dummett's *manifestation argument*; see the thesis MT in 4.1.1.)

Assuming that the first premise correctly states that either classical or intuitionist set-theoretic practice can account for set-theoretic understanding, and given that the fifth premise is uncontroversial (it only reports set-theoretic assertions of classicists and intuitionists), there only remain three further premises. The second premise excludes that both set-theoretic activities can account for someone's set-theoretic understanding. The third and fourth premise develop the supposition that only one of the two activities can account for someone's set-theoretic understanding. These remaining three premises will be examined in what follows. Given that the conclusion of the Dummettian argument, i.e. exclusive monism, is very counter-intuitive, the pertinent question will be whether there are plausible accounts of set-theoretic understanding and its relation to set-theoretic practice which allow for a rejection of at least one of these premises.

2.4.2 A Causal Theory of Understanding

The premises P_3 and P_4 of the Dummettian argument express Dummett's view on the close relation between the acceptance of sentences and rules on the one hand and the understanding of the terms that are contained in these sentences and rules on the other hand. This view has been criticised by Williamson:

> What binds together uses of a word by different agents or at different times into a common practice of using that word with a given meaning? This is an instance of a more general type of question: what binds together different events into the history of a single complex object [...]? In brief, what makes a unity out of diversity? [...] [I]t is the complex interrelations of the constituents, above all, their causal interrelations. [...] The claim that a shared understanding of a word requires a shared stock of platitudes depends on the assumption that uses of a word by different agents or at different times can be bound together into a common practice of using that word with a given meaning only by an invariant core of beliefs. But that assumption amounts to one of the crudest and least plausible answers to the question of what makes a unity out of diversity. In effect, it assumes that what animates a word is a soul of doctrine (Williamson 2007a: 123).

According to Williamson, the complex of (causal) interrelations of certain uses of a word ensures that the same meaning is attached to it in these uses. Applied to the present case, this means that intuitionists and classicists attach the same meanings to words like "not" and "number", not because they accept similar sentences and rules which contain these words, but because their uses of these words stand in suitable causal relations.

Williamson's considerations can directly be used to argue against exclusive logical and mathematical monism by way of the following *causal argument*:

P_1^W That classicists partake in classical set-theoretic practice can ground their understanding of set-theoretic vocabulary.

P_2^W Intuitionists make uses of set-theoretic vocabulary which are suitably related to uses of this vocabulary by classicists.

P_3^W If intuitionists make uses of set-theoretic vocabulary which are suitably related to uses of this vocabulary by persons who understand it, then they too understand it.

C^W Intuitionists and classicists understand set-theoretic vocabulary.

The underlying idea would be that, considered in isolation, understanding of set-theoretic vocabulary can only be accounted for by partaking in *classical* set-theoretic practice. But this partly negative view is then combined with the claim that intuitionists nevertheless understand set-theoretic expressions in virtue of the fact that their uses of them are causally linked to 'classical' uses of them.[35] The premise P_3 is thus rejected. Although partaking in classical but not in intuitionistic set-theoretic practice can ground an understanding of set-theoretic vocabulary, such an understanding does not require acceptance of fundamental assumptions of classical set theory. Someone's understanding of set-theoretic vocabulary can obtain in virtue of suitable causal relations between her own uses of it and the uses of others, where these other uses might belong to a set-theoretic practice based on assumptions which differ considerably from her own set-theoretic commitments. (Corresponding to this 'classical' version of the argument, there is, of course, also an 'intuitionistic' version of it in which the roles of intuitionists and classicists are reversed.)

Now, while I endorse the claim that a shared understanding of a word may be based on causal relations of its uses rather than on shared beliefs expressed by sentences containing it, I doubt that the causal argument is the most convincing critique of Dummett's view. As I said, the underlying idea of the causal argument coheres with the second premise of the Dummettian argument: considered in isolation, it is not the case that partaking in the classical set-theoretic practice *and* partaking in the intuitionistic set-theoretic practice can account for an understanding of set-theoretic vocabulary. I take this to be a highly questionable assumption. My thesis is that even if considered in isolation, the classical practice can ground an understanding of set-theoretic vocabulary iff the intuitionistic practice can do so. And since I (provisionally) accept the claim that at least one of the two practices can ground an understanding of set-theoretic vocabulary, this amounts to the claim that both practices can do so. A proponent of the Dummettian argument takes mathematical understanding to be grounded in mathematical practice, while she takes intuitionistic and classical mathematics to be dissimilar to such an extent that a shared understanding is precluded. A proponent of the causal argument takes mathematical understanding to be grounded either in mathematical practice or in causal relations to persons who already have mathematical understanding (with the consequence that a shared understanding is possible because suitable causal relations can compensate for the differences between intuitionistic and classical practice). A proponent of the position that is defended here, by contrast, allows for a shared mathematical understanding to be grounded in partly incompatible mathematical practices.

---

[35] Since I do not know whether Williamson accepts the thesis P_1^W, I do not attribute this argument to him. In a way, the underlying idea of this argument is similar to Putnam's (1975: 145-6) idea about what he calls the "division of linguistic labour". Classical mathematicians would function as the *experts* whose understanding of logical and mathematical vocabulary is reflected in their mathematical uses of it, while intuitionists owe their understanding to causal relations between their uses of logical and mathematical vocabulary and the uses of such vocabulary by the experts. (A difference to the case described by Putnam would be, of course, that intuitionists do not accept the authority of the classical 'experts'.)

In the following subsection, I will spell out this third position in more detail. My aim is to show that one can resist the counter-intuitive conclusion of the Dummettian argument without invoking causal relations to compensate for someone's non-standard mathematical views. To this end, I will assume that partaking in classical set-theoretic practice, considered in isolation, can ground understanding of set-theoretic vocabulary. I will then argue that partaking in intuitionistic set-theoretic practice, considered in isolation, can also ground understanding of set-theoretic vocabulary. Thus, in the discussion to follow, the symmetric treatment of intuitionism and classicism is temporarily given up. I would like to emphasise that this has been an arbitrary choice and that everything that follows could be modified in such a way that intuitionism and classicism switch their roles.

2.4.3 Low Requirements for Understanding

Consider the following possible future events. In the next years, the groups of classical mathematicians and intuitionists gradually stop conversing with members of the other group. At some point classicists only speak with other classicists and intuitionists only speak with other intuitionists. In the course of time, the separation between the two groups becomes more severe. They no longer read texts written by members of the other group, they only listen to audio recordings of speech from members of their own group, and they only watch films produced by members of their own group. Eventually, everyone is only confronted with linguistic products that have been produced by members of his own group. Then, due to some terrible events, only the members of the intuitionistic group survive. Everyone living then argues in accordance with intuitionistic logical and mathematical rules; no one feels any commitment to distinctively classical rules. As time passes, the intuitionists forget that there ever had been a group of persons who reasoned differently from the way they do. Everyone accepts the intuitionistic rules, and there is no one who even considers the possibility that this was or could be different.

Now, as far as I can see, the claim that no member of the intuitionistic group in the final state understands the words "not" or "number" is as implausible as the claim that intuitionists in the present state or in any of the intermediate states do not understand these words. This suggests that the plausibility of the claim that intuitionists and classicists all understand the logical and mathematical expressions they use is not based on the assumption that the uses of these expressions are causally related.[36]

It might be replied that the described form of causal separation is still not strong enough. After all, the future intuitionistic uses of the words "not" and "number" are descendants of uses which came from a time in which classical

---

[36]   Keep in mind that dualism is out of contention. It is only investigated whether it is possible to avoid the counter-intuitive position of exclusive monism, and, if this is possible, whether a shared understanding of logical and mathematical expressions by classicists and intuitionists has to be based on causal relations between their uses of these expressions, or whether a shared understanding of logical and mathematical vocabulary can also be grounded in partly incompatible mathematical practices.

uses of them also existed. Might it not be that long ago in the past classicists endowed logical and mathematical vocabulary with classical meanings and that even those intuitionists living in a distant future in which all signs of former classicism have vanished would still attach classical meanings to logical and mathematical vocabulary because their uses originate from past classical uses?[37]

Imagine an alternative situation which is physically indistinguishable from the final state of the future development sketched above, but where there never were any classical mathematicians or other people who argued in accordance with distinctively classical rules of inference. In this case there never was any classical practice which could have endowed logical and mathematical vocabulary with classical meanings. Either those uses which are neutral together with those that are distinctively intuitionistic can account for an understanding of these terms in this situation, or no uses of them can do so. Now, in my view, even in this situation it is more plausible to adopt the former view. If it is assumed that set-theoretic terms now have such meanings that classical set-theoretic argumentations are correct, and if it is also assumed that someone can have an understanding of these terms because she produces such classical set-theoretic argumentations, then it should also be claimed that someone can have an understanding of these terms because she produces intuitionistic set-theoretic argumentations. Classical set-theoretic practice can ground an understanding of set-theoretic terms iff intuitionistic set-theoretic practice can do so.

Apart from being intuitively plausible, this view also has important theoretical merits. The basic idea is that an understanding of set-theoretic vocabulary is not very difficult and that classical and intuitionistic set-theoretic practice are rather similar. First, both classical and intuitionistic mathematics are strong enough to fix unique meanings for the set-theoretic terms. And second, they agree on the vast majority of basic principles and rules which govern these terms.

As to the first point, we saw that classical as well as intuitionistic set-theoretic practice uniquely determines the logical and mathematical behaviour of the involved logical and mathematical expressions. As was shown in 2.2, there are sets of rules for the standard logical operators which are accepted by intuitionists and classicists alike and which uniquely characterise these operators. Furthermore, as was shown in 2.3, there are sets of rules for the pertinent arithmetical, set-theoretic, and function-theoretic vocabulary which are accepted by intuitionists and classicists alike and which suffice to preclude conflicting results: if there were two versions of such vocabulary, possibly associated with additional rules, it can be shown that a mathematical theorem formulated by means of either version always corresponds to a theorem formulated by means of the other version. This suggests that to argue in accordance with these 'neutral' logical and mathematical rules can already account for an understanding of the involved logical and mathematical expressions.

Now, it has to be acknowledged that it is not necessarily true that someone who argues in accordance with uniquely characterising rules understands the involved terms. To take the most extreme case, suppose that some crazy logician

---

[37]   Cp. Kripke's (1972) causal theory about the meaning and denotation of proper names.

is committed to *every* rule of inference. Then, she is *a fortiori* committed to sets of uniquely characterising rules. This does not show, of course, that she understands the vocabulary she uses.[38]

Here, however, the second point comes into play: classical and intuitionistic set-theoretic practice are not very different. To begin with, consider classical and intuitionistic collections of rules for the arithmetical language $\mathcal{L}^{Ar}$. It can be assumed that every basic rule which governs only expressions of $\mathcal{L}^{Ar}$ and is accepted by intuitionists is also accepted by classicists, and that every basic rule which governs only expressions of $\mathcal{L}^{Ar}$ and is accepted by classicists is derivable from rules accepted by intuitionists together with the rule dne$_\neg$. From an intuitionistic perspective, classicists employ one unwarranted basic rule for an expression of $\mathcal{L}^{Ar}$, namely $\neg$, while, from a classical perspective, intuitionists fail to employ one warranted basic rule for this expression. It seems to be an overreaction if one takes this to indicate that either intuitionists or classicists do not understand logical or arithmetical vocabulary (lest they stand in a suitable causal relation). Less dramatically, it might be taken to indicate that intuitionists and classicists have conflicting logical and arithmetical beliefs, and, in addition, that semantically interested intuitionists and classicists may have conflicting beliefs about the meaning of "not": semantically interested classicists normally believe while semantically interested intuitionists normally disbelieve that a commitment to the informal rule corresponding to dne$_\neg$ is justified in virtue of the meaning of "not".[39] But the claim that one of them has a false belief about the meaning of "not" seems to be far more plausible than the claim that either classicists or intuitionists do not understand "not", or that one of the two sides can only understand "not" by standing in suitable causal relations to the other side.

Furthermore, the situation does not change dramatically when a richer mathematical language, such as the set-theoretic language $\mathcal{L}^{Cl}$, is taken into account. Even in such a language the vast majority of fundamental principles and rules for its expressions are accepted by classicists and intuitionists alike. One or two sentences which one side treats as expressing fundamental truths while the other side rejects them as false do not justify the claim that either side does not understand the expressions it uses without the help of the other side. The initial plausibility of the view that ordinary native speakers of English, no matter whether they are intuitionists or classicists, and no matter whether they are surrounded by intuitionists or classicists, understand words like "not", "number", and "set" rates so high that only a very strong argument could undermine this view. As far as I can see, the differences in the intuitionistic and the classical set-theoretic commitments are too marginal to form the basis of such an argument. Instead of saying that either intuitionists or classicists only

---

[38] I take it to be most plausible to assume that such a person does not understand the vocabulary she uses. If she does, this can only be explained by recourse to considerations about appropriate (causal) relations of her uses to the uses of others.

[39] I have inserted the qualification "normally" to stress that nothing in their respective mathematical practices forces classicists and intuitionists to have such beliefs. For example, a classical mathematician might believe that there are no meanings and, therefore, disbelieve that a certain commitment is justified in virtue of such an entity.

produce meaningless sounds when they engage in their preferred mathematical practices, it seems far more plausible to say that they make meaningful but contradictory assertions.

(Similarly, I would defend the view that native speakers of English from the 19th century understood words like "before". This view is also adopted, for example, by Field. After he raises the question of whether Einstein has changed the meaning of "simultaneous", Field makes the following remark:

> If we do say there has been a change of meaning, say in "simulta-neous", what are we to say that the term prior to Einstein "meant"? If it "meant" a relation obeying all the properties that Newton assumed, then since there is no such relation, virtually everything speakers said using the word was false. If it "meant" some two-place relation that is relativistically kosher, then some very surprising claims come out true. [...] Given this, it may be best to interpret Newton as having used the word "simultaneous" in a way that actually was frame-relative, though he didn't know it (Field 2009b: 255).

This suggests that Field believes that native speakers of English who lived before the 20th century attached the same meaning to words like "before" and "simultaneous" as physically informed people do now.)

To sum up, I hope to have made plausible the following thesis: intuitionists and classicists understand logical and mathematical vocabulary, and they understand this vocabulary in the same way. The standard logical constants are governed by uniquely characterising rules accepted by intuitionists and classicists alike. If there are versions of the logical constants which behave classically, then there are no versions of them which behave intuitionistically. Similarly, standard arithmetical, set-theoretic, and function-theoretic vocabulary is governed by rules accepted by intuitionists and classicists alike and which, in a sense, determine the mathematical truths which can be expressed by means of this vocabulary. If there were two versions of such vocabulary, then intuitionists and classicists would have to concede that a sentence formulated by means of vocabulary of either version is true iff the corresponding sentence formulated by means of vocabulary of the other version is true. Finally, I have argued that there is no need to embrace the counter-intuitive claim that either intuitionists or classicists do not understand the logical vocabulary they use. Their respective mathematical practices are similar enough to stick to the claim that classicists utter meaningful mathematical sentences iff intuitionists do so as well. Together with the presupposed view that either intuitionists or classicists understand the logical and mathematical expressions they use, it follows that all of them do.

In the following chapter, I will turn to the discussion of arguments against the theory of classical logic which are inspired by or related to mathematical intuitionism. From now on, I will assume that intuitionists and classicists understand the logical and mathematical terms which figure in these arguments in the same way.

Chapter 3

Counterexamples to Classical Logic

How is intuitionistic criticism of the theory of classical logic to be understood? On the one hand, intuitionists accept mathematical sentences which directly conflict with classical logic:

In_1 Not every set of numbers does or does not contain at least one element.

The sentence In_1 is the negation of a sentence which is logically true according to the theory of classical logic. On the other hand, they accept epistemic sentences which conflict with classical logic only indirectly:

In_2 The following is not known: every or not every even number larger than 2 is the sum of two primes.

The sentence In_2 states something expressed by a simple arithmetical sentence which is logically true according to the theory of classical logic to be unknown. On what basis do intuitionists accept sentences like In_1 and In_2? And how are these two anti-classical attitudes related: is one of them more fundamental than the other?

In 1.4 and 1.5, it was seen that relevantists and freedom fighters question the theory of classical logic on the basis of fairly accessible considerations: relevantists appeal to widely shared intuitions about the unacceptability of certain conditionals and implications, while freedom fighters draw attention to the phenomenon of empty terms which is usually neglected by classicists. The reasons for intuitionists to abandon the theory of classical logic are less easily understood. They do not claim to account for commonsensical intuitions which conflict with principles of classical logic, and they do not appeal to a familiar phenomenon ignored by many classical logicians. In the following three chapters, the peculiarities of the intuitionistic attack on classical logic will be explored. Here, the intuitionistic views of mathematics and of the relation between mathematics and logic turn out to be most crucial.

By and large, the more complex a mathematical domain is, the stronger are the differences between the corresponding intuitionistic and classical theories. In the first two sections of this chapter, I will discuss arguments against classical logic based on principles about moderately complex mathematical domains: the domain of *sets of natural numbers* and the domain of *sequences of natural numbers*. In the remaining two sections, I will deal with intuitionistic arguments concerning the simplest infinite mathematical domain: the domain of *natural numbers*.

3.1   A Mathematical Conflict

In the study of mathematical domains such as the domain of sets of numbers or the domain of sequences of numbers, intuitionists put forward claims which contradict basic principles of classical logic. Sentences which express such claims are usually called *strong counterexamples* (to classical logic). That is, a true negation "not-S" is a strong counterexample iff S is logically true according to classical logic. To put forward an argument whose conclusion is a strong counterexample, if it is true, is to argue against a fundamental consequence of a canonical logicality theorem of classical logic; it is an instance of *basic criticism* of classical logic in the sense of 1.3.2.

In this and the next section, I will discuss two intuitionistic mathematical arguments whose conclusions are strong counterexamples to classical logic, if they are true. The first argument is based on the *Uniformity Principle* about relations between sets of numbers and numbers. The second argument is based on the *Continuity Principle* about relations between sequences of numbers and numbers.

3.1.1   The Uniformity Principle (UP)

The Uniformity Principle (UP) deals with relations between sets of natural numbers and natural numbers. It was introduced by Troelstra (1973), and, unlike the Continuity Principle presented in 3.1.2, it did not play an important role in the development of intuitionistic mathematics. Because of its conceptual simplicity, however, it is especially suitable for philosophical discussion.

I will present UP in a definitional extension of the formal set-theoretic language $\mathcal{L}^{Cl}$ from 2.3.2. Recall that $\mathcal{L}^{Cl}$ contains the standard logical constants, a standard collection of arithmetical predicate symbols, and two set-theoretic predicate symbols, C and $\in$, which correspond to the predicates "is a set of numbers" and "is an element of".

In what follows, I will use n and m as variables for natural numbers and X and Y as variables for sets of numbers. I will use $n = m$ and $X = Y$ to express identity between numbers and sets.[1] Furthermore, if φ is a two-place open sentence, then I will use $\mathrm{Ext}(\varphi)$ to express that φ is governed by the

---

[1]   Formally, I take the predicate $=$ to be defined as follows here:

$$\forall a \forall b \left(a = b \leftrightarrow \left((Na \wedge Nb \wedge ID(a,b)) \vee (Ca \wedge Cb \wedge \forall z\,(z \in a \leftrightarrow z \in b)) \right) \right).$$

That is, the predicate $=$ applies to a pair of objects a and b iff a and b are identical numbers or a and b are sets of numbers with the same elements.

following rule (in whose statement X, Y, n, m are used as schematic terms for (sets of) natural numbers):

$$\frac{X = Y \quad n = m \quad \varphi(X, n)}{\varphi(Y, m)}$$

It expresses that φ is *extensional*: if φ applies to a set X and a number n and if Y and m are identical to X and n, then φ also applies to Y and m. The Uniformity Principle is the following schema for extensional open sentences φ:

UP $\forall X \exists n \varphi(X, n) \rightarrow \exists n \forall X \varphi(X, n).$

This principle states for every extensional φ that if every set of numbers is φ-related to some number, then there is a number to which every set of numbers is φ-related.

To grasp the significance of UP, one should take a look at principles with the same structure. Consider the following schema in which the variable a ranges over some domain D_1 and the variable b ranges over some domain D_2:

$\forall^{D_1}\exists^{D_2}$ $\forall a \exists b \varphi(a, b) \rightarrow \exists b \forall a \varphi(a, b).^2$

If one writes $\mathbb{N}$ for the domain of natural numbers and $\wp\mathbb{N}$ for the domain of sets of natural numbers, then UP amounts to $\forall^{\wp\mathbb{N}}\exists^{\mathbb{N}}$. Many principles of the form $\forall^{D_1}\exists^{D_2}$ are uncontroversially false. The following are two examples:

$\forall^{\mathbb{N}}\exists^{\mathbb{N}}$ $\forall m \exists n \varphi(m, n) \rightarrow \exists n \forall m \varphi(m, n);$

$\forall^{^{\mathbb{N}}\mathbb{N}}\exists^{\mathbb{N}}$ $\forall \alpha \exists n \varphi(\alpha, n) \rightarrow \exists n \forall \alpha \varphi(\alpha, n).$

Here $^{\mathbb{N}}\mathbb{N}$ stands for the domain of sequences of numbers, and α is used as a variable for sequences of numbers.

An instance of $\forall^{\mathbb{N}}\exists^{\mathbb{N}}$ states that if every number is φ-related to some number, then there is a number to which every number is φ-related. Clearly, the principle $\forall^{\mathbb{N}}\exists^{\mathbb{N}}$ is false: although every number is identical to some number, there is no number which is identical to every number; i.e., the principle is false with respect to $\varphi(m, n) := m = n$. More generally, the principle $\forall^{D_1}\exists^{D_2}$ is uncontroversially false if D_1 is a subset of D_2 and if there is a reflexive relation that is not universal on D_1 (e.g. the identity relation of D_1).

An instance of $\forall^{^{\mathbb{N}}\mathbb{N}}\exists^{\mathbb{N}}$ states that if every sequence of numbers is φ-related to some number, then there is a number to which every sequence of numbers is φ-related. This principle is also false: although every sequence has some first member, there is no number which is the first member of every sequence; i.e., the principle is false with respect to a two-place open sentence φ that applies to a sequence α and a number n iff α maps 0 to n.

This shows that the intuitionistic acceptance of the Uniformity Principle is based on specific features of the pair of domains $(\wp\mathbb{N}, \mathbb{N})$. As will be further

---

2   More precisely, the principle should be stated with predicates F for D_1 and G for D_2:

$$\forall a(Fa \rightarrow \exists b(Gb \wedge \varphi(a, b))) \rightarrow \exists b(Gb \wedge \forall a(Fa \rightarrow \varphi(a, b))).$$

However, I will use the simpler expression here.

explored in 3.2.1, it is essential that the first domain exhibits a certain 'fuzzi-ness', while the second domain comprises 'simple objects' in a 'discrete way'. (As McCarty (2005: 361-2) notes, the relevant feature of the first domain seems to be shared by other non-trivial power sets. That is, if $\wp M$ is the power set of some inhabited set M, then the principle $\forall^{\wp M}\exists^{\mathbb{N}}$ seems to be as plausible as UP.)

It is not difficult to see that UP directly implies the negation of a classical logical truth (cp. McCarty 2005: 359):

SC$_1$ $\neg\forall X(\exists n\, n \in X \ \lor \ \neg\exists n\, n \in X)$.

This potential strong counterexample states that not every set of numbers is or is not inhabited.

Proof. Suppose that every set of numbers is or is not inhabited:[3]

$$\forall X(\exists n\, n \in X \ \lor \ \neg\exists n\, n \in X).$$

Then, for every set of numbers X there is a number n such that the following holds: either n is an element of X, or nothing is:

$$\forall X\exists n(n \in X \ \lor \ \neg\exists n\, n \in X).$$

Relying on the Uniformity Principle, one can then infer that there is some number n such that for every set of numbers X, n is an element of X, or X is not inhabited:

$$\exists n\forall X(n \in X \ \lor \ \neg\exists n\, n \in X).$$

But this is false. For any given number n, let X be the singleton of some different number. Then, n is not contained in X but X is inhabited:

$$\forall n\exists X(\neg n \in X \ \land \ \exists n\, n \in X).$$

This contradicts the previous sentence.   □

Note that apart from its reliance on UP this argument is acceptable by clas-sical as well as by intuitionistic standards. It is therefore obvious how classical mathematicians will react to it. While intuitionists claim that not every set is or is not inhabited because UP is true, classicists assert that UP is false because every set is or is not inhabited.

3.1.2   The Continuity Principle (CP)

Continuity principles are the oldest intuitionistic principles used to demonstrate strong counterexamples to classical logic. The Continuity Principle (CP) deals with relations between sequences of natural numbers and natural numbers. It has been used in Brouwer 1918, and it has been explicitly formulated in Heyting

---

3   I will present an informal proof. It is easy to transform it into a derivation of the calculus obtained from $\mathcal{R}^{Cl}$ (see Appx. A.2) by adding UP as a rule without premises.

1930a. There are various stronger continuity principles, but for present purposes
these do not have to be considered (see Troelstra & van Dalen 1988: 206-17).

I will present CP in a definitional extension of the formal language $\mathcal{L}^{Seq}$ from
2.3.3. Recall that $\mathcal{L}^{Seq}$ contains the standard logical constants, a standard collection of arithmetical predicate symbols, and two function-theoretic symbols:
$SQ(x)$ corresponds to "x is a sequence of numbers", and $AP(x, y, z)$ corresponds
to "x maps y to z". I will use α and β as variables for sequences, 0 and 1 for
the first two natural numbers, and $<$ for the natural ordering of the natural
numbers. Furthermore, I will use $\alpha = \beta$ to express identity between sequences
of numbers,[4] and I will use a binary function symbol "$(\cdot)$" for application:
$\alpha(n) = m$ corresponds to $AP(\alpha, n, m)$. Finally, I will use $\bar{\alpha}m$ to denote the
initial segment of α that consists of the first m steps; I will use this expression
only in the context $\bar{\alpha}m = \bar{\beta}m$, in which it can be taken to be an abbreviation of
$\forall n\, (n < m \rightarrow \alpha(n) = \beta(n))$. As in the case of the Uniformity Principle, if φ is a
two-place open sentence, then I will use $\mathrm{Ext}(\varphi)$ to express that φ is governed by
the following rule (in whose statement α, β, n, m are used as schematic terms
for (sequences of) natural numbers):

$$\frac{\alpha = \beta \quad n = m \quad \varphi(\alpha, n)}{\varphi(\beta, m)}$$

It expresses that φ is *extensional*: if φ applies to a sequence α and a number n
and if β and m are identical to α and n, then φ also applies to β and m. The
Continuity Principle is the following schema for extensional open sentences φ:

CP $\forall \alpha \exists n \varphi(\alpha, n) \rightarrow \forall \alpha \exists m \exists n \forall \beta (\bar{\alpha}m = \bar{\beta}m \rightarrow \varphi(\beta, n))$.

This principle states for every extensional φ that if every sequence is φ-related to
some number, then for every sequence there is an initial segment of it such that
for some number every sequence that extends this initial segment is φ-related
to that number.

It is easy to see that CP implies the negation of a classical logical truth (cp.
Troelstra & van Dalen 1988: 209):

SC$_2$ $\neg \forall \alpha (\exists n\, \alpha(n) = 0 \ \vee \ \neg \exists n\, \alpha(n) = 0)$.

This potential strong counterexample states that not every sequence of numbers
is or is not at some point identical to 0.

Proof. Assume that every sequence is or is not at some point identical to 0:

$$\forall \alpha (\exists n\, \alpha(n) = 0 \ \vee \ \neg \exists n\, \alpha(n) = 0).$$

---

[4]   Thus, in this subsection, I take the predicate $=$ to be defined as follows:

$$\forall a \forall b \left(a = b \leftrightarrow ((Na \wedge Nb \wedge ID(a, b)) \vee (SQa \wedge SQb \wedge \forall y \forall z\, (AP(a, y, z) \leftrightarrow AP(b, y, z)))) \right).$$

That is, the predicate $=$ applies to a pair of objects a and b iff a and b are identical
numbers or a and b are sequences of numbers which are identical in every coordinate.

Then, for every sequence there is a number such that the following holds: either
the sequence maps the number to 0, or the sequence is not at some point
identical to 0:

$$\forall\alpha\exists n(\alpha(n) = 0 \ \vee \ \neg\exists n\,\alpha(n) = 0).$$

Relying on the Continuity Principle, one can then infer that there is some initial
segment of α and some number n such that every sequence extending the initial
segment maps n to 0 or is not at some point identical to 0:

$$\forall\alpha\exists m\exists n\forall\beta(\overline{\alpha}m = \overline{\beta}m \ \rightarrow \ (\beta(n) = 0 \ \vee \ \neg\exists n\,\beta(n) = 0)).$$

But this is false. Let α be the sequence which is constant 1, let m and n be any
numbers, and let β be the sequence which is 1 for the first $\max(m, n) + 1$ steps
and is then constant 0. Then, β agrees with α for the first m steps, β maps n
not to 0, and β is at some point identical to 0:

$$\exists\alpha\forall m\forall n\exists\beta(\overline{\alpha}m = \overline{\beta}m \ \wedge \ \beta(n) \neq 0 \ \wedge \ \exists n\,\beta(n) = 0).$$

This contradicts the previous sentence.   □

Note that apart from its reliance on CP this argument is acceptable by classical
as well as by intuitionistic standards. It is therefore obvious how classical mathematicians will react to it. While intuitionists claim that not every sequence is
or is not at some point identical to 0 because CP is true, a classicist will say
that CP is false because every sequence is or is not at some point identical to
0.

3.2   Justifications of UP and CP

In the previous section, it was seen that classicists and intuitionists have contradictory mathematical beliefs. There are mathematical sentences S such that
classicists accept S while intuitionists accept not-S. Is there any possibility of
resolving this mathematical conflict? Here is an example of a pessimistic viewpoint:

> I would not deign to argue that an intuitionistic approach to math-
> ematics is the *only* correct one. For starters, it is difficult to see how
> such an argument could proceed. The standard of correctness for the
> steps of the argument would have to be determined at the outset.
> But, to do that, we would already have to decide whether intuition-
> istic or classical standards are the more appropriate (McCarty 1987:
> 537).

I think that the underlying thought can be spelled out as follows. Mathematical
knowledge is either fundamental or derived from fundamental mathematical
knowledge in accordance with mathematical methods of proof. Suppose now
that two persons accept contradictory mathematical claims on the basis of their
basic mathematical commitments. Then, neither of them can correctly argue

in favour of her view on the basis of non-mathematical premises since mathematical knowledge is not based on non-mathematical knowledge. Furthermore, neither of them can argue in favour of her view on the basis of mathematical assumptions and in accordance with argumentative rules which the other one accepts. Each of them can only produce 'proofs' of mathematical sentences that she accepts and 'refutations' of mathematical sentences that she rejects which involve mathematical assumptions or argumentative moves which the other one does not accept. Consequently, the conflict between intuitionists and classicists is unresolvable.

In what follows, I will consider the prospects of responses to such pessimistic considerations. In particular, I will discuss arguments of intuitionists who claim to be able to refute principles of classical logic (which are part of classical mathematics). In this section, I will deal with intuitionistic justifications of the anti-classical principles UP and CP. As will be seen, they lie somewhat at the border between mathematical and non-mathematical argumentations. Afterwards, I will turn to considerations of a more philosophical nature.

3.2.1 A Justification of UP

I will now discuss a justification of the Uniformity Principle. Recall that this principle states for every extensional φ that if every set of numbers is φ-related to some number, then there is a number to which every set is φ-related. The most explicit justification of UP that I know of stems from McCarty. (See Troelstra & van Dalen 1988: 234-5 for a very condensed justification.) I will present it here in full, inserting bracketed letters for future reference:[5]

> To see how such a principle might be made plausible, [a] let R be any extensional, binary relation between sets of natural numbers X and natural numbers n for which $R(X, n)$. For this liaison between sets and numbers to subsist, there should be a discernible association in virtue of which R links sets to numbers. That association is expressible as a list of instructions ρ for determining, from sets X suitable n for which $R(X, n)$. [b] In contrast to the natural numbers and integers, the collection of all sets of natural numbers is not the trace of some recursive generation process. The relation R and the sets themselves are all extensional. Hence, the action of ρ should not depend upon the fine points of a set's possible specification in language. [c] Further, since ρ is a rule with which one can *act* on *all* sets X of numbers, the action of ρ should not depend upon the membership conditions for any particular X. Those conditions might well be so complicated as to elude capture in anything one would rightly call a "rule." [d] The application of ρ to sets should therefore be uniform: what ρ does to one set, it does to all. [e] The identity badge

---

[5] To avoid confusion I have replaced the symbol α with the symbol ρ. (The symbol α will only be used as a variable for sequences of natural numbers in this chapter.) Note also that McCarty speaks about a relation: R, while I speak about an open sentence: φ. Nothing of what follows depends on this choice.

of intuitionism as a branch of constructive mathematics is the in-
sistence that every rule underwriting an existential statement about
numbers $\exists n P(n)$ must provide, if implicitly, an appropriate numerical term t and knowledge that $P(t)$ holds. [f] Therefore, since ρ is
constructive and labels each set X uniformly with *some* number, ρ
must yield a designation for some particular number m uniformly
in terms of the Xs. Obviously, for the association to be uniform,
m must be the same for every set of numbers X. Hence, there is a
number related by R to every set, and UP is seen to hold (McCarty
2005: 358).

Since McCarty does not comment on his argumentation, my discussion will have
a partly speculative character.

The first part of the argumentation, [a], consists of three claims:

(1) For every set X, there is a number n such that (X, n) satisfies φ.

(2) There is a discernible association A in virtue of which for every set X,
there is a number n such that (X, n) satisfies φ.

(3) There is a rule ρ which can be used, by applying it to a set X, to determine
a number n such that (X, n) satisfies φ.

Note that a rule is here a list of instructions which can be applied to sets of
numbers, and whose application to such a set can be used to determine at least
one number with a certain property. Regard the following example of a rule
with two instructions which is applicable to sets of numbers:

I_1^1 Form the union of x and $\{0\}$: $x_1 := x \cup \{0\}$!

I_2^1 Take the least member of x_1: $x_2 := \min(x_1)$!

The first instruction takes one from a set X to the possibly different set $X \cup \{0\}$,
and the second instruction takes one from this set to the number 0. The instructions I_1^1 and I_2^1 constitute a rule which can be applied to sets of numbers and
which can be used to determine a number, 0 to wit, if they are applied to any
such set.

Two points about [a] are worth mentioning. First, it should be noted that
McCarty uses *epistemic* vocabulary in (2): there is something *discernible* in
virtue of which sets are φ-related to numbers. His justification of UP thus
involves epistemological considerations. Second, the passage [a] contains what
I take to be the most important step of McCarty's argumentation. In (3), it
is claimed that there is a *finite rule* which accounts for the fact that every set
is φ-related to some number. This rule can be applied to sets of numbers and
can be used, by applying it to a set X, to determine at least one number to
which X is φ-related. This assumption is then exploited in the remaining steps
of the argumentation, in which McCarty especially deals with two questions:
(i) What is a rule that captures an extensional relation and can be applied to
arbitrary sets of numbers? (ii) What is a rule which can be used to determine
numbers with certain (relational) properties? By combining his answers to these

questions, he reaches the conclusion that there has to be a number to which every set is φ-related.

The second part of the argumentation, [b], has again three items:

(4) The collection of all sets of natural numbers is not recursively generated.

(5) Sets with the same members are identical, and φ is extensional.

(6) For every set X, how ρ acts on X is independent of how exactly X can be specified in a language.

I take (4) to be uncontroversial: while the natural numbers can be obtained by repeated applications of the successor operation to the initial number 0, the sets of natural numbers (like the sequences of natural numbers) cannot be obtained by a recursive procedure. Similarly, (5) is unproblematic (cp. 3.1.1). The import of (6), however, is less obvious.

What would be a rule whose action on an input depends on how exactly the input can be specified in a language? Suppose that $\mathcal{L}$ is an arithmetical language which contains the numerals $0, S0, SS0, \ldots$ for the natural numbers, and consider the following rule that is applicable to natural numbers:

$\mathbf{I}^2$ Take the number of tokens of symbols in the numeral of x in $\mathcal{L}$!

This rule takes one from a number to its successor. Its universal applicability depends on the existence of a numeral in $\mathcal{L}$ for every natural number. It seems plausible to assume that (6) is somehow based on the fact that there is no corresponding *set-theoretic* language, which contains a canonical term for every set of natural numbers.[6] But this observation cannot capture the whole point of (6), which is presented as a consequence of (5).

To understand the significance of (6), one should contemplate the question of how to conceive of an act of applying a rule to a mathematical object. After all, mathematical objects are unlike ordinary concrete objects on which one can act physically. It is intuitively clear how to apply the following instruction to a knife and an apple:

$\mathbf{I}^3$ Take the knife x and cut the apple y into small pieces!

It is less clear in which sense certain instructions can be applied to numbers, sets, or sequences.

I think that the idea has to be that one acts on a mathematical object by acting on a representation of this object. Examine the instruction $\mathbf{I}_1^1$ to take the union of the input set and the singleton of 0. If a set X of numbers is represented by a term, then one can use $\mathbf{I}_1^1$ to act on X by acting on this term. Here is an illustration of such an application:

$$\emptyset \quad \rightsquigarrow \quad \emptyset \cup \{0\}.$$

[6]   If one acknowledges only languages whose terms can be enumerated, then Cantor's Theorem (which is accepted by intuitionists) shows that there cannot be a language whose terms denote every set of numbers (since it shows that the sets of numbers cannot be enumerated).

The instruction I_1^1 is applied to the empty set given by the constant symbol $\emptyset$, and it yields the singleton of 0 by producing the complex term $\emptyset \cup \{0\}$. It might be objected that terms are also abstract objects with which one cannot causally interact. But since they (can) have tokens with which one can interact, they are less difficult to deal with than mathematical objects like numbers, sets, or sequences. In particular, one might say that an application of I_1^1 is effected by copying the token which represents the input set and then appending a token of the expression $\cup \{0\}$.

To make this precise, one would have to say more about representations of mathematical objects. However, I will not pursue this task here. For what follows, I will make the simplifying assumption that these representations are linguistic expressions and not, for example, mathematical diagrams. The present proposal then amounts to the claim that rules act on sets of numbers by acting on linguistic expressions representing these sets. Note that this proposal does not violate the above mentioned observation that there is no language which contains a term for every set of numbers. For example, the presentation of the instruction I_1^1 does not require a specification of some language which contains terms for all sets. Rather, in any *application* of this instruction, a set of numbers has to be given by some representation. For this to be feasible one does not have to fix in advance a system of representations for all sets of numbers.

With this understanding, it is now possible to explain the significance of (6). My hypothesis is that McCarty draws attention to the fact that the action of a rule on a set given by some term is not allowed to depend on features of the term which are not shared by all possible co-referring terms. Scrutinise the following rule:

I^4 Take the number of tokens of symbols of the term by which x is given!

This rule yields, if it is applied to a set given by a term t, the number of tokens of symbols of t. For example, a set given by the term $\emptyset$, leads to 1, a set given by the term $\{0\}$ leads to 3, and a set given by the term $\{1,3\}$ leads to 5. (Note that this rule yields a natural number for every set of numbers but that is does not yield the same number for all such sets. It can therefore be used to construct a potential counterexample to UP.) However, apparently, this cannot be a rule of the relevant type because it violates the extensionality condition that is stated in (5). For example, if it is applied to a set given by the term $\{0,0\}$, the instruction I^4 yields the number 5, although it maps the same set to the number 3 if this set is given by the term $\{0\}$. Since a rule of the relevant type expresses an *extensional* relation, it cannot act differently on two co-referring representations.

With (6), interpreted in the suggested way, McCarty has reached his first more specific conclusion about the nature of the rule ρ: although the action of ρ on a set X relates to a representation of X, it is not allowed to depend on features of this representation which are not shared by other representations of X. Understood in this way, (6) seems plausible (given the assumption that ρ is a rule which accounts for an extensional relation between sets and numbers and which acts on sets by acting on representations of them).

The third part of the argumentation, [c], contains two statements:

(7) Some sets have membership conditions which cannot be captured by ρ.

(8) For every set X, how ρ acts on X is independent of the membership conditions of X.

What does is mean to say that a rule captures the membership conditions of some set, and why should one assume that ρ acts on a set independently of its membership conditions? Regard the rule with the following two instructions:

$\mathbf{I}_1^5$ Form the union of x and $\{1\}$: $x_1 := x \cup \{1\}$!

$\mathbf{I}_2^5$ Take the least member of x_1: $n := \min(x_1)$!

One might think that this rule yields a natural number for every set of numbers, but that it does not yield the same number for all such sets (again, leading to a potential counterexample to UP). However, for example, if one tries to apply these instructions to a set which contains 0 iff Goldbach's Conjecture is true, then it turns out that this is presently impossible: as long as one does not know whether Goldbach's Conjecture is true, one does not know whether one has to take 0 or 1 if one wants to follow the second instruction.

The problem might be described as follows. Rules have to be *applicable*, while there are sets for which one cannot apply a rule if that requires knowledge as to whether a certain number satisfies its membership conditions. Correspondingly, I assume that (7) amounts to the claim that there are sets with membership conditions $\mathbf{m}$ and numbers n such that ρ is only applicable if it is not required that it is known whether n satisfies $\mathbf{m}$.

However, why does the existential claim (7) justify the universal claim (8)? The reason has to be that ρ is applicable to *every* set of numbers. Suppose that the application of ρ were to require knowledge of whether some given number n satisfies the membership conditions of the input set. Then, ρ could not be applied to those sets for which it is presently unknown whether n satisfies their membership conditions. And it is not difficult to see that for every number n there is some set for which it is presently not known whether n satisfies its membership conditions.

With (8), McCarty has reached his second more specific conclusion about the nature of ρ: ρ acts on an input independently of the input's membership conditions. Now, I have to admit that I do not find this claim completely clear, but the previous considerations seem to indicate that it is not altogether unreasonable to suppose that a precise version of this claim expresses a plausible desideratum for rules of the kind that McCarty is interested in.

The fourth part of the argumentation, [d], consists of the following claim:

(9) For all sets X and Y, ρ acts on X is the same way as it acts on Y.

McCarty draws this temporary conclusion from (6) and (8), according to which the action of ρ is independent of the exact linguistic specifications and membership conditions of its inputs.

The sentence corresponding to (9) is introduced with the claim that the application of ρ is *uniform*. How exactly is uniformity to be understood? There are two possible and somewhat opposite misunderstandings of this notion. First, uniformity does *not* mean that the rule produces the same result no matter to which set it is applied. Consider the following rule:

I[6] Take the complement of x in $\mathbb{N}$: $x_1 := \mathbb{N} - x$!

This rule yields distinct results for any pair of distinct sets of numbers. Thus, if (9) is not absurd, the rule **I**[6] has to be a rule which acts on all sets in the same way. Second, uniformity is not a trivial property already secured by the fact that the same list of instructions is used for every input. Consider the following rule, which is applicable to sequences of numbers:

I[7] Take the first item of x!

In one sense, one might say that this rule acts on all sequences of numbers in the same way because it expresses a single and simple operation on the set of all sequences of numbers. However, this cannot be the pertinent sense of uniformity. If the application of this rule were uniform, then McCarty would be committed to the plainly false claim that every sequence of numbers has the same first item.[7] Consequently, if (9) is not absurd, the rule **I**[7] has to be a rule which does *not* act on all sequences of numbers in the same way.

The notion of uniformity expressed by (9) has to be understood in the light of (6) and (8). The application of ρ to some input set X is, according to (6), independent of the exact linguistic representation of X and, according to (8), does not require knowledge of whether some given number satisfies the membership conditions of X. Probably, this shows that ρ acts uniformly on all inputs in the sense of not requiring knowledge about anything which is specific to their identities (since all that *is* specific to their identities is captured by their linguistic representations and membership conditions). Again, it has to be acknowledged, I think, that the relevant notion of uniformity and, therefore, the significance of (9) have not been made completely clear. However, it does not seem to be entirely unreasonable to think that there is a coherent conception of rules for which it is plausible to assume that if they capture some *extensional* relation, and if they are applicable to *all sets of numbers*, then they have to act on their inputs uniformly in the sense that McCarty has given an indication of here.

The fifth part of the argumentation, [e], consists of two claims:

(10) For every set X, the rule ρ underwrites the existential open sentence $\exists n \varphi(x, n)$ with respect to X.

(11) For every set X, the rule ρ must provide an appropriate numerical term t and knowledge that $\varphi(x, t)$ is true of X.

---

[7]   To see this, suppose that McCarty were to say that **I**[7] labels each sequence α *uniformly* with some number. Corresponding to what he says in part [f] of his justification of UP, he would then have to say that **I**[7] "must yield a designation for some particular number m uniformly in terms of the $[\alpha]$s". And then he would also have to acknowledge that "for the association to be uniform, m must be the same for every [sequence] of numbers $[\alpha]$".

The relation between (3) and (10) is not entirely clear. What is the connection between the claim that ρ, if it is applied to a set X, allows one to determine a number n such that (X, n) satisfies φ, and the claim that ρ underwrites the existential open sentence $\exists n \varphi(x, n)$ with respect to every set X? Since McCarty treats (10) as a trivial observation at this stage of his argumentation, I assume that he takes it to be just a reformulation of (3): underwriting $\exists n \varphi(x, n)$ with respect to X amounts to the same thing as allowing someone to determine a number n such that (X, n) satisfies φ.[8]

In (11), McCarty gives information about how a number is supposed to be determined by the application of ρ: the application must provide an appropriate term that denotes this number. But which numerical terms are appropriate? Contemplate the following instruction:

$\mathbf{I}^8$ Take the least element m of x such that (x, m) satisfies φ: $n := \mu_m(\varphi(x, m))$!

This instruction is not applicable in general, because its application to a set X requires knowledge of whether an arbitrary number m is such that (X, m) satisfies φ (cp. the instructions I_1^5 and I_2^5 from above). Consequently, μ-terms cannot in general be considered appropriate here. I will not try to figure out precisely what counts as appropriate. Roughly speaking, I conjecture that a numerical term t is appropriate iff an algorithm is known that could in principle be used to calculate the value of t. If such a conjecture was true, then problematic rules such as I^8 would be excluded.

In addition to providing an appropriate numerical term t, the application of ρ is also meant to provide knowledge that the open sentence $\varphi(x, t)$ is true of the input set. I take this to mean that if someone follows the instructions of ρ with respect to a set X, then these acts put her in a position to know that the relevant open sentence is true of X. (Note that another *epistemic* term thus enters the justification of UP.) However, as far as I can see, this additional assumption does not play an essential role here. For what follows, it seems to suffice that ρ (uniformly) provides an appropriate numerical term t such that $\varphi(x, t)$ is true of the input set; it does not have to be taken into account that this can be known by someone who applies ρ to this input set.

With (11), McCarty has reached his third more specific conclusion about the nature of ρ: for every input X, ρ must provide an appropriate numerical term t such that $\varphi(x, t)$ is true of X. While the first two specific conclusions about the nature of ρ were inferred from the assumption that ρ can be applied to *sets of numbers*, the third conclusion is derived from the supposition that ρ can be used to determine *natural numbers* which satisfy some (relational) condition.

The final part of the argumentation, [f], consists of two claims:

(12) For every set X, the rule ρ must uniformly provide an appropriate numerical term t and knowledge that $\varphi(x, t)$ is true of X.

---

[8] Alternatively, one could propose that (10) amounts to the claim that ρ can be used to derive $\exists n \varphi(x, n)$ from certain other open sentences (in combination with certain additional rules). But such a claim could not simply be treated as a trivial observation here.

(13) There is a number n such that for every set X, (X, n) satisfies φ.

Now, (12) follows from (11) and the uniformity assumption that is captured by (9). Thus, the final question is why (12) implies (13).

McCarty takes it to be obvious that if the appropriate numerical terms are obtained uniformly, they all have to be coreferential. Given the suggested understandings of uniformity and appropriateness, the inference from (12) to (13) indeed seems to have some plausibility. Recall that the action of ρ on a set X was assumed to be mediated by a corresponding action on a representation of X, where (6) guarantees that this underlying action does not depend on details of the representation which are not shared by other representations of X. Now, if this underlying action is also not allowed to depend on facts about which numbers satisfy the membership conditions of X, i.e., if no application of ρ depends on knowledge of whether some number n satisfies the membership conditions of X, it is difficult to see how different input sets might lead to appropriate numerical terms that are not coreferential.

By way of summary, I think that McCarty's justification of UP can be structured as follows.

MC$_1$ For every set X, there is a number n such that (X, n) satisfies φ.

MC$_2$ Therefore, there is a rule ρ which can be used to determine, for every set X represented by a term t_X, a number n represented by an appropriate numerical term $t_{\mathbb{N}}^X$, such that $\varphi(t_X, t_{\mathbb{N}}^X)$ is true.

MC$_3$ The action of ρ is independent of the exact choice of t_X and independent of which numbers are members of X.

MC$_4$ Consequently, for all sets X and Y represented by t_X and t_Y, the produced numerical terms $t_{\mathbb{N}}^X$ and $t_{\mathbb{N}}^Y$ are coreferential.

MC$_5$ Thus, there is a number n such that for every set X, (X, n) satisfies φ.

In short, if every set is φ-related to some number, then there is a certain finite object, a rule, which grounds this fact. But then the finiteness of this object, in combination with the fuzziness of the inputs, the discreteness of the outputs, and the extensionality of φ, implies that there is some number to which every set is φ-related.

I would like to make four remarks about McCarty's justification of UP. First, in my view, the crucial premise of the argument is the assumption that if every set of numbers is φ-related to some number, then there is a rule which can be used to determine, for every set of numbers, a number that is φ-related to the set. The remaining steps of the argumentation are surely in need of further clarification and defence, but, as I have tried to show, it is not out of the question that they might be made convincing. As a consequence to this, the correctness of classical logic hinges on the non-existence of certain finite 'quasi-mathematical' objects: rules which account for relations between sets of numbers and numbers.

Second, if my reconstruction of McCarty's justification is on the right lines, then its crucial existence assumption presupposes certain views about *representations* of mathematical objects. According to this interpretation, it is essential for McCarty's justification that rules act on mathematical objects by acting on representations of mathematical objects. This suggests that considerations of a broadly *semantic* nature might be relevant for an evaluation of UP and, consequently, for a choice between intuitionistic and classical mathematics. In Chapters 4 and 5 such semantic considerations will be the central topic.

Third, if one concentrates on the representations which I hypothesised to underlie the action of ρ and, therefore, the truth of the sentence "every set of numbers is φ-related to some number", then it stands out that they are *terms* for various mathematical objects. It will be seen that other justifications of intuitionistic mathematics are based on the claim that complexes of representations of another kind underlie the truth of mathematical sentences, namely *proofs* of these sentences. In McCarty's justification of UP, however, underlying proofs of the involved sentences do not seem to play any role.

Fourth, McCarty's justification seems to lie somewhat at the border between (a sketch of) a mathematical proof and (a sketch of) a non-mathematical argumentation. On the one hand, one might propose that the pertinent rules are mathematical objects and that the justification might be developed into a mathematical proof founded on fundamental mathematical principles about these rules. (Compare McCarty's (2005: 376-8) remarks about the relation between his justification of UP, computability theory, and Kleene's (1945) so-called *realisability interpretation*, and consider also the following statement:

> [U]ltimately, the intuitionist erects a mathematical edifice, and thereby rejects conventional mathematics, on mathematical foundations and on mathematical foundations alone (McCarty 2005: 356-7).

This suggests that McCarty conceives of an adequate justification of UP as having a mathematical character.) If this mathematical project were to be completed, then one would obtain an intuitionistic 'refutation' of classical logic which takes its starting point not in a mathematical principle about relations between sets of numbers and numbers but in mathematical principles about mathematical rules.

On the other hand, one might consider the possibility that McCarty's justification of UP must be understood as having a non-mathematical nature; in particular, one might think that it is bound to involve certain epistemic and semantic theses. If this proposal was correct and one was to complete his argumentation, then one would obtain an intuitionistic 'refutation' of classical logic which is ultimately based on non-mathematical principles.[9] The question of whether the dispute between intuitionists and classicists is ultimately to be understood as a purely mathematical dispute or whether it is based on conflicting non-mathematical principles will continue to play an important role in what follows.

---

[9]   Cp. Sundholm 1983: 159-60 about the difference between mathematical projects such as those of Kleene (1945), Gödel (1958), and Kreisel (1962) and semantic projects such as those of Heyting (1930b, 1931, 1934) and Kolmogorov (1932).

3.2.2   A Justification of CP

Recall that CP states for every extensional φ that if every sequence of numbers is φ-related to some number, then, for some initial segment $\overline{\alpha}m$ of any given sequence α, there is a number to which every sequence which extends $\overline{\alpha}m$ is φ-related.

I will concentrate on a justification given by de Swart (1992). Before I turn to his reasoning, however, I would like to mention a justification of CP given by McCarty, which is very similar to his justification of UP.[10] Suppose that φ is extensional and that every sequence of numbers is φ-related to some number. Then, McCarty claims that there has to be a rule which accounts for this fact:[11]

> As above, there should be a rule ρ attaching numbers to sequences
> in accord with R (McCarty 2005: 365).

Now, however, the rule ρ does not have to act uniformly. It can be sensitive to initial segments of its inputs:[12]

> [T]he members s of Seq can be represented in ρ only through the numerical values of the terms that comprise their extensions [...]. Because rules are finite and not infinite bearers of information, ρ can contain and operate on only the information presented by some finite number of those terms, perhaps the finite initial segment of s ending with s_m [...]. It should be on the sole basis of such an initial segment that ρ attaches an n to s so that $R(s, n)$. Moreover, since ρ can take only finite initial segments of input sequences into consideration, if t is another sequence sharing with s exactly this initial segment [...], ρ should attach to t the very same n that it attaches to s (McCarty 2005: 365-6).

McCarty's justification of CP is very similar to his justification of UP. As before, he assumes that there is a certain finite object, a rule, which grounds the assumption that every sequence is φ-related to some number. But since sequences of numbers are less fuzzy than sets of numbers, one can only reach a weaker conclusion. The finiteness of the underlying rule, in combination with the limited representability of sequences, the discreteness of numbers, and the extensionality of φ, implies that for some initial segment $\overline{\alpha}m$ of any given sequence α there is a number to which every sequence extending $\overline{\alpha}m$ is φ-related.

I will now turn to de Swart's justification of CP. His reasoning has to be understood against the background of a certain conception of the set $^{\mathbb{N}}\mathbb{N}$ of sequences of natural numbers. According to de Swart (1992: 203-4), the set $^{\mathbb{N}}\mathbb{N}$ is a certain *construction-project*, where the notion of a construction-project is taken to be primitive (p. 203). Construction-projects are mental projects for

---

[10]   See van Atten & van Dalen 2002 for yet another justification of CP.
[11]   In this and the following quotation I have replaced the symbol α with the symbol ρ.
[12]   McCarty uses "Seq" as a term for the set of all sequences of natural numbers, and he uses the symbol s as a variable for sequences of numbers.

"generating the elements of a set in the course of time" (ibid.). The specific construction-project $^N\mathbb{N}$ is explained as follows:[13]

> By choosing an element from $[\mathbb{N}]$ at successive moments or stages, potentially infinite sequences of [natural numbers] come into being. These sequences are generated in the course of time by a simple precept, called the choice-law: at each state choose [a natural number]. We identify the (intuitionistic) set $[^N\mathbb{N}]$ with this precept; and we call the potentially infinite sequences of [natural numbers] the elements of this set, since they are generated in accordance with the corresponding choice-law (de Swart 1992: 205).

The elements of $^N\mathbb{N}$ are called *choice-sequences*. De Swart then presents the following justification of the Continuity Principle:

> Suppose $\forall\alpha\exists n\varphi(\alpha,n)$,[14] i.e., I can carry out for each α a finite construction n and a finite construction of the truth of $\varphi(\alpha,n)$. A construction of α can be thought of as given step by step. While such an α is produced I can make the finite construction n and the finite construction of the truth of $\varphi(\alpha,n)$. As these are finite constructions, I already make them after finitely many steps $\alpha(0),\ldots,\alpha(m-1)$ of the construction of α. Hence $\exists m\exists n\forall\beta\big(\overline{\beta}m = \overline{\alpha}m \rightarrow \varphi(\beta,n)\big)$ (de Swart 1992: 207).

In the first step of the argumentation, de Swart moves from the claim that every sequence of numbers is φ-related to some number to a statement about his abilities:

(1) For every sequence α, there is a number n such that (α,n) satisfies φ.

(2) For every sequence α, I can carry out a finite construction n and a finite proof that (α,n) satisfies φ.

I have replaced the notion of a *construction of the truth* with the notion of a *proof*, which seems to yield a more accessible formulation of what de Swart wants to convey here.

Notice that de Swart uses epistemic vocabulary. Unlike McCarty, who refers to a rule which he takes to underlie the fact that every sequence of numbers is φ-related to some number, de Swart refers the abilities of an agent. From the assumption that α is φ-related to n, he moves to the claim that it can be proven that α is φ-related to n. One might contemplate that de Swart accepts a corresponding general principle about proofs of mathematical sentences:

---

[13] In the following quote, I have replaced the references to the two-element set $\{0,1\}$ with references to the set of natural numbers $\mathbb{N}$ since I am here only interested in the latter set.

[14] In the original text one finds the sentence $\forall\alpha \in \sigma\ \exists k \in N\ A(\alpha,k)$, in which N stands for the set of natural numbers and σ stands for a so-called "spread", where the set of sequences of numbers is one example of a spread. Restricting attention to a specific spread, I am, therefore, only concerned with an instance of the principle considered by de Swart. (Furthermore, as before, I leave the restrictions on the variables implicit, and I use n instead of k and φ instead of A.)

Provability Principle If ψ, then it can be proven that ψ.

The significance of considerations about proofs for the conflict between classicists and intuitionists will be discussed in 3.3 and 3.4.

The second part of the argumentation consists of three claims:

(3) Every sequence is a construction which proceeds by successive choices of its members.

(4) For every sequence α, while α is constructed I can carry out a finite construction n and a finite proof that (α, n) satisfies φ.

(5) For every sequence α, while some finite number m of steps in the construction of α have been performed, I can carry out a finite construction n and a finite proof that (α, n) satisfies φ.

In this second part, de Swart brings constructions of choice-sequences into play, and he relates them to his abilities which he had introduced in (2).

Apart from the transition from (1) to (2), I take the assumption stated in (3) to be the most crucial step in de Swart's justification. For the sake of the argument, I will assume that every mathematical object is a mental construction: either a finite construction or an infinite construction-project. On the face of it, however, it is not clear that the construction of an infinite sequence of natural numbers cannot be finite. Think about sequences of numbers which can be defined by (finite) expressions, e.g. constant sequences. What more does it take to mentally construct such a sequence than to grasp its definition? And since grasping a (finite) definition is surely a finite mental process, does this not show that at least some sequences can be constructed in finite mental processes? However, as has been quoted above, according to de Swart, sequences of natural numbers are, *by definition*, the elements of a construction-project in which they are generated by successively choosing natural numbers. It seems that according to de Swart, (3) expresses a *conceptual truth* about sequences of numbers.

The transitions from (2) and (3) to (4) and from (3) and (4) to (5) seem to be less problematic. First, it is claimed that the construction of the relevant number and the proof (of which it is already known that they can be carried out) can be carried out while the sequence α is constructed. If one accepts the assumption that mathematical objects are given by mental constructions, then this seems to be a relatively unproblematic assumption. Second, (5) can plausibly be taken to be a consequence of (3) and (4). If a finite construction, such as a number or a proof, is carried out, then it is completed at some point in time. Therefore, since α is constructed by infinitely many steps, where it may plausibly be assumed that there is some lower bound for the amount of time needed for one step, the two finite processes are completed at some time when only a finite number of the steps in the construction of α have been made.

The final part of the argumentation consists of two claims:

(6) For all sequences α and β and for all numbers m and n: if I can carry out a finite proof that (α, n) satisfies φ while m steps in the construction of α have been performed, and if β coincides with α on the first m steps, then I can carry out a finite proof that (β, n) satisfies φ.

(7) For every sequence α there are numbers m and n such that for every β which coincides with α on the first m steps, (β, n) satisfies φ.

The transition from (6) to (7) is clearly unproblematic. Thus, it remains to be seen whether (6) is plausible. The idea is the following: suppose that you witness the successive choices of natural numbers and that you know that this process will continue on forever, thereby producing the infinite sequence α. Furthermore, suppose that you are able to come up with a number n and with a proof that α is φ-related to n during the first m steps of the construction of α. Then, neither the construction of n nor the proof that α is φ-related to n can depend on choices that succeed the first m choices in the construction of α. Whatever features of α allow you to carry out the two finite processes, these features have to 'supervene' on the initial segment you are confronted with while carrying out the two processes; i.e., every sequence β that coincides with α on this initial segment also has to have those features that allow you to come up with n and a proof that β is φ-related to n.

By way of summary, I think that de Swart's justification of CP can be structured as follows.

DS$_1$ For every sequence α, there is a number n such that (α, n) satisfies φ.

DS$_2$ Consequently, a (finite) agent Ag can, for every α, produce a number n and a proof that (α, n) satisfies φ.

DS$_3$ Therefore, Ag can, drawing only on an initial segment $\overline{\alpha}m$ of α, produce a number n and a proof that (α, n) satisfies φ.

DS$_4$ It follows that Ag can produce a number n and a proof that (β, n) satisfies φ, for every sequence β which extends $\overline{\alpha}m$.

DS$_5$ Thus, for every sequence α, there are numbers m and n such that for every sequence β that extends $\overline{\alpha}m$, (β, n) satisfies φ.

In short, if every sequence is φ-related to some number, then for every sequence α one can find a number n and a proof π that α is φ-related to n. But since α is always in the process of construction, there has to be a finite part $\overline{\alpha}m$ which suffices for one's activity. Therefore, for every sequence β which extends $\overline{\alpha}m$, the proof π also shows that β is φ-related to n.

I would like to make two remarks about de Swart's justification of CP. In my view, there are two crucial premises: (a) the assumption that if α is φ-related to n, then it can be proven that α is φ-related to n, and (b) the assumption that sequences of natural numbers are to be thought of as forever unfinished constructions of a certain construction-project. I will comment on these two assumptions in turn.

By making the first assumption, de Swart makes it more explicit than McCarty that he takes CP to rest on *epistemic* considerations. An evaluation of his argumentation crucially depends on theses about mathematical proofs. This relates to another difference to McCarty's justification: while McCarty's reasoning is based on the idea that operations on *terms* account for operations on mathematical objects which, in turn, account for the truth of CP, de Swart's reasoning is based on the idea that *proof processes*, i.e. mental processes which involve *propositional* mental acts, can capture the relation between sequences and numbers and which, ultimately, account for the truth of CP. Epistemic theses and theses of a broadly semantic nature are, therefore, important ingredients of de Swart's justification. I will treat mathematical proofs in 3.4 and investigate semantic 'foundations' of intuitionistic arguments against classical logic in Chapters 4 and 5.

The second assumption is of a more 'metaphysical' or 'conceptual' nature. It is an assumption about what certain mathematical objects are or how they must be thought of. The argumentations of McCarty were seen to implicitly rely on theses about *representations* of mathematical objects, but the nature of these representations was not spelled out. By contrast, de Swart is more explicit about how one must conceive of sequences of numbers. They are forever unfinished mental 'products' of a mathematical construction-project. They come into existence over an infinitely long period of time by successive choices of natural numbers. Furthermore, this claim about what sequences are or how they must be thought of was seen to be of crucial significance for de Swart's argumentation.

Despite the differences between McCarty's and de Swart's justifications of CP, one can also see various similarities. In particular, it is assumed in both argumentations that there is a certain finite object, a rule or an agent, which accounts for the fact that a given sequence of numbers is φ-related to some number. This finite entity can only use a finite amount of information about the given sequence, and both agree that this information can only relate to an initial segment of it. In what follows, I will discuss the crucial notion of de Swart's version of this assumption: the notion of a mathematical proof. However, before I come to that, I will take a look at another way to argue against classical logic, which will also turn out to be based on considerations about mathematical proofs.

3.3   Weak Counterexamples

In 3.1.1 and 3.1.2, I introduced two potential strong counterexamples to classical logic: SC_1 and SC_2. Both of these sentences are negations of classical logical truths which contain only logical and mathematical vocabulary, and they are direct consequences of two basic intuitionistic principles about sets and sequences of numbers: the Uniformity Principle and the Continuity Principle. As I have pointed out, the justifications of UP and CP can either be interpreted as sketches of mathematical proofs based on mathematical principles about a certain class of finite mathematical objects, or they can be interpreted as sketches of argumentations based on principles of a more philosophical nature. In the rest of this chapter, I will discuss some such philosophical principles which have been taken to ground intuitionistic mathematics. For simplicity, I will consider these principles in relation to a mathematical domain that is simpler than domains of sets or sequences, namely the domain of natural numbers.

3.3.1 What are Weak Counterexamples?

Intuitionists do not claim that there are sentences of a first-order arithmetical language which are strong counterexample to classical logic. Nevertheless, they diagnose the classical treatment of some such sentences as a symptom of the inadequacy of classical logic. Instead of saying that some classical logical truths of a first-order arithmetical language are not true, they only say that some of them lack a certain *epistemic* feature, namely the property of being known to be true.

The sentences whose epistemic status is under discussion are the sentences of the first-order arithmetical language $\mathcal{L}^{Ar}$ (2.3.1), which correspond to simple arithmetical sentences of English. The epistemic claims put forward by intuitionists are expressed in a slight extension of this language. First, there are resources for speaking about the expressions of $\mathcal{L}^{Ar}$ (see 1.2.1). Second, there is a unary predicate symbol T corresponding to the predicate "is a true sentence of $\mathcal{L}^{Ar}$". Third, there is a unary sentential operator K for our present knowledge; i.e., "KS" may be read as "it is (presently) known that S".[15] Then, a true sentence of the form $\neg KT(t)$ is called a *(first-order arithmetical) weak counterexample* iff t stands for a sentence of $\mathcal{L}^{Ar}$ that is logically true according to classical logic. Intuitionists put forward arguments whose conclusions are weak counterexamples, if they are true.

It is important to note that weak counterexamples are sentences which express *universal* claims about human knowledge. (I will not consider the possibility of non-human mathematical knowledge here.) The sentences that are mentioned in weak counterexamples are such that *no one* presently knows them to be true. In addition, it should be emphasised that weak counterexamples express claims about the *present* time. Intuitionists do not present classical logical truths of $\mathcal{L}^{Ar}$ about which they say that no one *ever* knows them to be true; presently unsolved problems may be solved tomorrow.

It might be objected that weak counterexamples are not really incompatible with classical logic. After all, we are not logically omniscient; we do not know every logical truth to be true. Even a classical logician must accept that not every classical logical truth is known to be true. But although this is correct, it is irrelevant because in the pertinent cases it is shared knowledge that the mentioned sentences are classical logical truths; therefore, classicists are committed to their truth.

I will assume that mathematical knowledge is always based on mathematical proof: a mathematical sentence is known to be true iff it has been proven. Strictly speaking, this does not seem to be correct. On the face of it, there are *basic* mathematical truths of which we can have immediate knowledge, i.e. truths about which we can have knowledge that is not based on further knowledge. An example might be the proposition that 0 is a natural number.

---

[15]   I will make the idealised assumption that mathematical knowledge is never lost: if a mathematical truth is known at a time t, then it is also known at any time later than t. If someone is not willing to make this assumption, then she can replace the notion of present knowledge with the notion of present or past knowledge throughout the following discussion.

Strictly speaking, it seems that such basic mathematical truths can be known to be true without being proven. (Here it is important to keep in mind that *informal* mathematical proofs are at issue. In the usual theories of formal proofs, every axiom can be used in a 'proof' of itself.) For simplicity, however, I will make the assumption that if a basic mathematical truth is recognised as true, then this act of recognition counts as a proof of it. This assumption will allow for simpler formulations in the following, but nothing substantial depends on it.

In what follows, I will restrict attention to simple arithmetical instances of the principle *tertium non datur*

$$\varphi \vee \neg\varphi$$

as targets of potential weak counterexamples. If the corresponding rule $\overline{\vdash \varphi \vee \neg\varphi}$ is added to the intuitionistic calculus $\mathcal{R}^I$, then one obtains a calculus which produces the same arguments as the classical calculus $\mathcal{R}^C$. This rule is therefore on a par with the rule of *double negation elimination*.

Recall that Goldbach's Conjecture is the claim that every even number larger than 2 is the sum of two primes. I will write G for the corresponding sentence of $\mathcal{L}^{Ar}$, $\neg G$ for its negation, and $G \vee \neg G$ for the disjunction of G and $\neg G$. The sentence $\neg G$ will be called *Goldbach's Negation*, and the sentence $G \vee \neg G$ will be called *Goldbach's Disjunction*. I will focus on Goldbach's Disjunction as a classical logical truth which intuitionists refer to in a possible weak counterexample:

$$\neg KT(G \vee \neg G).$$

Being a universal sentence whose instances are decidable by a simple algorithm, Goldbach's Conjecture gives rise to a potential weak counterexample that is especially simple. (One may notice that Goldbach's Disjunction cannot be a strong counterexample: every instance of the schema $\neg\neg(\varphi \vee \neg\varphi)$ is logically true according to intuitionistic logic.)

Brouwer claims that classical logic is generally not to be trusted. A short paper published in 1908 can be seen as the first intuitionistic attack on classical logic.[16] Brouwer (1908: 110) presents sentences which concern the decimal expansion of π as providing weak counterexamples. I speak about Goldbach's Disjunction rather than about these examples only because it is a simple first-order arithmetical sentence.

Before he presents the sentences about π as giving rise to weak counterexamples, Brouwer criticises a more abstract 'classical' principle:

> [The principium *tertii exclusi*] claims that every supposition is either
> true or false; in mathematics this means that for every supposed em-
> bedding of a system into another, satisfying certain given conditions,
> we can either accomplish such an embedding by a construction, or

---

[16]   The theses of this paper must be understood against the background of Brouwer's dis-
       sertation (1907). However, in his dissertation he had not yet drawn the conclusion that
       the principles of classical logic are not generally to be relied on (see Brouwer 1907: 74 &
       88, and cp. Franchella 1995: 307-9).

we can arrive by a construction at the arrestment of the process which would lead to the embedding. It follows that the question of the validity of the principium tertii exclusi is equivalent to the question whether unsolvable mathematical problems can exist. There is not a shred of a proof for the conviction which has sometimes been put forward that there exist no unsolvable mathematical problems (Brouwer 1908: 109).

What Brouwer here calls the *principium tertii exclusi*, I will call the *Propositional Principle of Bivalence*:

PB$_{\mathrm{Prop}}$ Every proposition is either true or false.

I assume that Brouwer's use of "supposition" corresponds to my use of "proposition".

It may be noted that Brouwer asserts that the universal claim that every mathematical supposition is either true or false is equivalent to the negative existential claim that there are no unsolvable mathematical problems. However, this seems to be questionable from an intuitionistic viewpoint. I assume that Brouwer infers this equivalence from the assumption that there is a correspondence f between mathematical suppositions and mathematical problems such that a supposition a is true or false iff the problem $f(a)$ is solvable. It then follows that the claim that every mathematical supposition is either true or false is equivalent to the claim that every mathematical problem is solvable. But although it might be plausible by intuitionistic standards to question the claim that every mathematical problem is solvable, it seems less clear that it is plausible for intuitionists to question the claim that there are no unsolvable mathematical problems. Making use of the above assumption, this latter claim is equivalent to the claim that there is no mathematical supposition which is neither true nor false, a claim that intuitionists typically do not deny.

I will return to a sentential version of the Principle of Bivalence (see 4.1.2), but here I will concentrate on the question of how intuitionists try to establish that a sentence like Goldbach's Disjunction is presently not known to be true.

3.3.2 Establishing Weak Counterexamples

On what basis might an intuitionist claim that Goldbach's Disjunction $(G \vee \neg G)$ is presently not known to be true? Consider the following proposal:

> [W]e have at present experienced neither its truth nor its falsity, so intuitionistically speaking, it is at present neither true nor false, and hence we cannot at present assert "Goldbach's conjecture is true, or it is false" (van Atten 2011: supp. "weakcounterex.html").

Reformulated by means of the notion of present knowledge, van Atten's idea is that Goldbach's Disjunction is not known to be true, because its two disjuncts are not known to be true:

P_1^* $\neg(KT(G) \vee KT(\neg G))$;

P_2^* $\neg(KT(G) \vee KT(\neg G)) \rightarrow \neg KT(G \vee \neg G)$.

The premise P_1^* states the uncontroversial fact that we neither know Goldbach's Conjecture nor Goldbach's Negation to be true, and the premise P_2^* states that if neither Goldbach's Conjecture nor Goldbach's Negation is known to be true, then Goldbach's Disjunction is also not known to be true. The combination of P_1^* and P_2^* then trivially implies that Goldbach's Disjunction is not known to be true.

Unfortunately, it is quite unclear why anyone who does not already accept the consequent of P_2^*, should accept P_2^* itself. Consider the generalisation of this conditional:[17]

$\mathrm{P}^?$ $\forall S\, (\neg(KT(S) \vee KT(\neg S)) \rightarrow \neg KT(S \vee \neg S))$.

According to $\mathrm{P}^?$, whenever we fail to know the arithmetical sentences S and $\neg S$ to be true, then we also fail to know their disjunction to be true. Now, it seems that claim expressed by P_2^* could only be the basis of someone's knowledge that Goldbach's Disjunction is not known to be true if its generalisation $\mathrm{P}^?$ was acceptable. However, $\mathrm{P}^?$ is not acceptable. To see this, let H be a sentence of $\mathcal{L}^{Ar}$ which expresses that some large number, say $1000^{1000} + 1$, is prime, such that neither it not its negation is presently known to be true, and consider the following instance of $\mathrm{P}^?$:

$$\neg(KT(H) \vee KT(\neg H)) \rightarrow \neg KT(H \vee \neg H)$$

This conditional is uncontroversially false: it is common ground between intuitionists and classicists that we know that every number is or is not prime. There are simple algorithms for finding out whether an arbitrary natural number is or is not prime. However, assuming that we have not applied such a primality algorithm to the number $1000^{1000} + 1$, we neither know that it is prime nor that it is not prime. Consequently, $\mathrm{P}^?$ is false. But then it is unclear why someone should accept P_2^* if she has not already accepted its consequent.

Intuitionists have to proceed from a stronger assumption about Goldbach's Conjecture than the assumption that we neither know it nor its negation to be true. A stronger assumption suggested by the primality example concerns our knowledge of decision methods. We know that we have a decision method for the primality sentence H, but we do not know that we have a decision method for Goldbach's Conjecture.[18] Instead of directly proceeding from the assumption that no decision method for Goldbach's Conjecture is known, however, it turns out to be more fruitful to proceed directly from a claim that arguably follows from this assumption: the claim that it is not known that either Goldbach's Conjecture or Goldbach's Negation is provable. I will use the unary predicate symbol P for provability: "Pt" corresponds to "t is a provable sentence of $\mathcal{L}^{Ar}$". Then, the pertinent claim can be stated as follows:

---

[17]   In this and the following principles the quantifiers range over the sentences of $\mathcal{L}^{Ar}$.

[18]   Note that it is not claimed that there is no decision method for G. For all we know, there might be some even number n which is not the sum of two primes, and then an algorithm for testing whether n is the sum of two primes would yield a decision method for G.

P_1 $\neg K(P(G) \vee P(\neg G))$.

According to P_1, it is not known that Goldbach's Conjecture is either provable or refutable. (A sentence is referred to as refutable iff its negation is provable.) The premise P_1 is accepted by many philosophers independently of whether they have classical or intuitionistic inclinations, and it will not be questioned here.[19]

In what follows, I will assume that a mathematical sentence is provable iff there is a proof of it, that there is a proof of it iff it can be proven, and that it can be proven iff it is possible that someone sometimes proves it. Furthermore, given the assumption that a mathematical sentence is known to be true iff it has been proven, it follows that provability coincides with knowability for mathematical sentences: a mathematical sentence is knowable iff it is provable.

The next premise in the argument to the conclusion that Goldbach's Disjunction is not known to be true is the following conditional:

P_2 $\neg K(P(G) \vee P(\neg G)) \rightarrow \neg KP(G \vee \neg G)$.

According to P_2, if it is not known that Goldbach's Conjecture or Goldbach's Negation is provable, then it is also not known that their Disjunction is provable. The combination of P_1 and P_2 directly entails that Goldbach's Disjunction is presently not known to be provable: $\neg KP(G \vee \neg G)$.

Now, unlike P_2^*, the conditional P_2 is an instance of a general principle which does not have instances that are uncontroversially false: the principle that it is not known that a disjunction is provable if it is not known of at least one disjunct that it is provable. This general principle is based on the assumption that if there is a mathematical proof of some disjunction, then there also has to be a mathematical proof of one of the disjuncts:

$P_\vee$ $\forall S_1 \forall S_2 \left(P(S_1 \vee S_2) \rightarrow (P(S_1) \vee P(S_2)) \right)$.[20]

According to $P_\vee$, a disjunction of $\mathcal{L}^{Ar}$ is provable only if at least one of its disjuncts is provable. (Note that $P_\vee$ follows from the thesis that an arithmetical sentence of $\mathcal{L}^{Ar}$ is true iff it is provable, in combination with the assumption that an arithmetical disjunction is true only if at least one of its disjuncts is true.)

The final premise of the argument is the following conditional:

P_3 $\neg KP(G \vee \neg G) \rightarrow \neg KT(G \vee \neg G)$.

According to P_3, if Goldbach's Disjunction is not known to be provable, then it is not known to be true. I will not question this assumption here. The combination of P_1, P_2, and P_3 directly entails that Goldbach's Disjunction is presently not known to be true: $\neg KT(G \vee \neg G)$.

---

[19] It is worth noting that even a stronger thesis than P_1 might be compatible with classical arithmetic: $\neg(P(G) \vee P(\neg G))$. This thesis states that Goldbach Conjecture is neither provable nor refutable. Leitgeb (2009: 292) calls the question of whether there are mathematical sentences that are neither provable nor refutable the Holy Grail of the philosophy of mathematics. Most intuitionists disbelieve that there is a Holy Grail.

[20] The inference from $P(G \vee \neg G) \rightarrow (P(G) \vee P(\neg G))$ to $\neg K(P(G) \vee P(\neg G)) \rightarrow \neg KP(G \vee \neg G)$ is perhaps not indubitable, but I will not question it here.

In sum, there is the following short argument that establishes a weak counter-example to classical logic if its premises are true:

P$_1$ $\neg K(P(G) \vee P(\neg G))$

P$_2$ $\neg K(P(G) \vee P(\neg G)) \rightarrow \neg KP(G \vee \neg G)$

P$_3$ $\neg KP(G \vee \neg G) \rightarrow \neg KT(G \vee \neg G)$

C $\neg KT(G \vee \neg G)$

In this argument, the premises P$_1$ and P$_3$ are rather unproblematic. The critical premise is P$_2$. As indicated above, intuitionists usually not only accept P$_2$ but also its generalisation and the underlying assumption P$_\vee$ according to which an arithmetical disjunction is provable only if one of the disjuncts is provable. On what basis do intuitionists accept P$_\vee$? This question will be pursued in the following section.

3.4   A Logical Basis of Mathematics?

The findings up to this point suggest that considerations about (mathematical) proofs are important for an evaluation of the intuitionistic criticism of classical logic and classical mathematics. First, de Swart's justification of the Continuity Principle (3.2.2) is based on the thesis that if every sequence of numbers is φ-related to some number, then for every sequence α there is a proof that establishes that α is φ-related to some number. Considerations about proofs might therefore be used to show that there are *strong* counterexamples to classical logic. Second, a justification of the claim that Goldbach's Disjunction is presently not known to be true (3.3.2) crucially depends on the thesis that an arithmetical disjunction is provable only if one of the disjuncts is provable. Considerations about proofs thus also seem to be important for the question of whether there are *weak* counterexamples to classical logic.

How is the mathematical status of potential proofs related to the logical status of mathematical arguments? Do proofs owe their correctness, at least in part, to the logical validity of associated linguistic expressions? Or does a plausible conception of logical validity have to build on more fundamental principles about the nature of proofs? Several intuitionists think that opposing views about the relation between acts of constructing/following proofs on the one hand and logically valid mathematical arguments on the other hand lie at the heart of the conflict between intuitionists and classicists. Some of them claim that classical logic and classical mathematics are based on an erroneous view about the nature of mathematical activity and its relation to logic. Others think that classicists are committed to an untenable reduction of the correctness of logical and mathematical inferences to properties of associated linguistic expressions.

In this section, it will be explored to what extent classical logic and classical mathematics depend on a questionable conception of mathematical activity and its relation to the logical status of mathematical arguments. To begin with

(3.4.1), I will introduce Brouwer's view that classical mathematics involves a misconception of the relation between mathematical and logical activity. Then (3.4.2), I will introduce the view, prominently held by Sundholm, that classicism incorporates an erroneous reduction of the correctness of logical and mathematical inferences to properties of associated linguistic and propositional entities. Finally (3.4.3), I will discuss Prawitz' account of correct inferences and consider the thesis that intuitionism is superior to classicism because only the former allows for such an account. It will be seen that considerations of a broadly semantic nature underlie the most plausible variants of the intuitionistic criticism mentioned in this section. These considerations will be dealt with in the remaining Chapters 4 and 5.

3.4.1   Brouwerian Mathematics

In Brouwer's view, classical mathematics is, to some extent, a *purely linguistic* activity, in which logical principles without any mathematical justification are used to derive mathematical sentences (cp. Brouwer 1908: 109). But in fact, according to Brouwer, legitimate mathematical activity is an essentially non-linguistic mental activity, and he thinks that an adequate understanding of the nature of this activity leads to a rejection of classical logic and to the adoption of intuitionistic logic instead. In this subsection, I will give a brief introduction to Brouwer's views about mathematics and logic. This will allow for a clearer understanding of two possible construals of intuitionistic criticism that is based on considerations about mathematical activity and logical validity. The more plausible (and less radical) construal of this kind of criticism will then be explored in 3.4.2 and 3.4.3.

Brouwer believes that mathematical activity is based on temporal awareness, and he thinks that mathematical objects are generated in mental abstraction processes which are grounded in such temporal experiences. He takes intuitionistic mathematics to be based on two acts:

> [T]he first act of intuitionism separates mathematics from mathematical language [...] and recognizes that intuitionist mathematics is an essentially languageless activity of the mind having its origin in the perception of a move of time [...] (Brouwer 1952: 509-10).

> [T]he second act of intuitionism [...] recognizes the possibility of generating new mathematical entities: firstly in the form of infinitely proceeding sequences [...]; secondly in the form of mathematical species [...] (Brouwer 1952: 511).

(Here, species may be thought of as sets.) Brouwer's basic picture is that mathematics is a purely mental activity, that the use of mathematical language is a secondary phenomenon whose only functions are to assist one's own memory and to influence other mathematicians, and that logic is a tertiary activity, namely the study of certain regularities in the use of the mathematical language (see Brouwer 1918: 174 and 1947: 477; cp. Troelstra & van Dalen 1988: 23).

Now, this picture might indicate one possible sense of *logical activity*, relative to which it has to be posterior to certain mathematical activity: the sense in which it is a study of linguistic by-products of a (temporarily) prior mathematical activity. If one considers mathematical proofs, however, one may well think that there is also a different sense of *logical activity*, relative to which mathematical activity is not independent of it, simply because mathematical proofs involve inferences some of which seem to be correct solely in virtue of their structure and the involved logical notions. Does Brouwer's conception of mathematics allow for such inferential acts, so that, in some sense, mathematics is partly a logical activity itself? To become clearer about this question, it is helpful to take a closer look at the nature of mathematical proofs.[21]

According to Gödel (1953: 341 n. 20), in its original meaning the term "proof" signifies "a sequence of thoughts convincing a sound mind". I will basically stick to this meaning. *Proofs* are certain types of mental processes, and by following a proof a 'sound mind' can obtain mathematical knowledge. (To *follow* a proof is to carry out a mental process that instantiates the proof.) A proof consists of a sequence of *proof steps*, and it ends with a mathematical judgement.

On the face of it, there are different kinds of proof steps that involve propositional mental acts, i.e. proof steps that are justified in different kinds of ways.[22] In particular, there seem to be what I will call *trivial steps, (quasi-)perceptual steps*, and *inferential steps*. Trivial steps are judgements that are not based on previous judgements or assumptions of the proof and also not based on (quasi-)perceptual awareness of some mathematical structure. A relatively uncontroversial example might be someone's judgement that 0 is a natural number.[23] (Quasi-)Perceptual steps are non-trivial judgements that are not based on previous judgements or assumptions of the proof. One may think of judgements based on one's perception of mathematical diagrams, or, alternatively, of judgements based on one's introspective awareness of mental counterparts of such diagrams.[24] Inferential steps are based on previous judgements or assumptions of the proof. The simplest inferential steps are *inferences*, which can be conceived of as sequences of judgements. (More complex inferential steps involve assumptions and conditional judgements; see Ch. 5.) If someone *presents*

---

[21]   Compare Detlefsen 1990: 514-25 for a characterisation of Brouwer's conception of mathematical activity, and see Franchella 1995 for Brouwer's view on the relation between mathematics and logic.

[22]   There are also non-propositional steps in proofs (cp. Leitgeb 2009: 266), e.g. the construction of mathematical objects. But such steps do not have to be considered here.

[23]   One suggestion would be that trivial judgements are grounded in a grasp of the involved notions, e.g. that someone's trivial judgement that 0 is a natural number is based on her grasp of the notion of 0 and the notion of a natural number. (Note that this suggestion does not imply that anyone who grasps these notions can recognise that 0 is a natural number, a kind of claim that has been criticised at length by Williamson (2006 and 2007a: 73-133). The idea is only that it is *possible* to recognise that 0 is a natural number on the basis of one's grasp of the involved notions, and that in a trivial judgement this possibility is realised.)

[24]   For the importance of mathematical diagrams, from a non-intuitionistic perspective, see Leitgeb 2009: 273-82 and Feferman 2012.

a simple inference, then she performs a sequence of assertions which produces a token of a complex linguistic expression, an *argument* to wit.

Now, as was said above, the existence of certain kinds of inferential steps seems to imply that mathematical activity is not wholly independent of logic, because some such steps seem to be correct solely in virtue of their structure and the involved logical notions. Two different intuitionistic reactions to this alleged observation have been suggested. Remarkably, there are some traces in Brouwer's writings suggesting the radical view that mathematical proofs have to be non-inferential. According to Brouwer (1948: 488), mathematical theorems are established "exclusively by means of introspective construction". Mathematical knowledge is claimed to be based on introspection and not on (logical) inference:

> [T]here are no non-experienced truths and [...] logic is not an absolutely reliable instrument to discover truths [...] (ibid.).

Brouwer rejects

> the possibility of extending one's knowledge of truth by the mental process of thinking, in particular thinking accompanied by linguistic operations independent of experience called 'logical reasoning' [...] (Brouwer 1955: 551).

If the mental processes of thinking comprise what I have called "inferential steps", then Brouwer here rejects the possibility of mathematical proofs which involve inferential steps. This suggests that Brouwer conceives of a mathematical proof process as a combination of non-propositional acts, in which mathematical objects are created and operated on, and acts of judgements that are based on introspective awareness of these non-propositional acts and their (mental) products.

I am not sure that Brouwer really believes that inferential acts are not part of mathematical activity. This radical view seems to be strikingly implausible, and I will not consider it here. Perhaps the more charitable interpretation is that Brouwer only objects to a certain conception of inferential acts, a conception according to which they owe their correctness to the (logical) status of associated linguistic expressions. I take this latter view to be far more plausible, and most intuitionists explicitly allow for inferential acts in mathematical proof processes. Restricting attention to such 'moderate' intuitionistic views, I will now turn to the relation between (mental) mathematical inferences and corresponding (linguistic) arguments and pursue the question of whether an intuitionistic argument against classical logic can proceed from plausible theses about this relation.

3.4.2   The Normativity of Logic and Mathematics

Many intuitionists think that classicism differs from intuitionism in involving a misconception of the relation between mathematical activity and the status of certain linguistic expressions. Brouwer's student Arend Heyting, for example, follows his teacher in this regard:

> While you think in terms of axioms and deductions, we think in
> terms of evidence; that makes all the difference (Heyting 1956: 13).

(Heyting presents a dialogue between persons with different conceptions of
mathematics, and the quoted words are said by INT, an intuitionist, to CLASS,
a classical mathematician.) Similarly, McCarty claims that intuitionism stands
out in its conception of mathematics as an independent mental activity that is
prior to logic:

> [In intuitionism there] is no set of normatively binding principles
> for right thinking that (1) hold sway over all scientific domains,
> including mathematical ones, and (2) can be set up once and for all,
> before we start to pursue mathematics. As for (1), the doctrines of
> intuitionistic logic are one and all statements of mathematics, largely
> of an unexciting sort. Ultimately, they owe their correctness to vivid
> intuition rather than to pale generality. As we have seen, the claim
> that TND [*tertium non datur*] fails of validity is, as with all other
> logical results, a theorem of mathematics admitting of mathematical
> demonstration. As for point (2), one learns of logical principles only
> after the fact, from truths and proofs in mathematics. Intuitionists
> intuit, construct, and reason first, and legislate for these activities
> later (McCarty 2005: 369-70).

Part of what intuitionists such as Brouwer, Heyting, and McCarty want to
deny is that the correctness of a mathematical inferential act is based on the
logical status of a corresponding argument. (An argument whose n-th sentence
expresses the content of the n-th item of an inference is said to *correspond* to
the inference.) They reject the claim that some correct mathematical inferences
are correct *because* corresponding arguments are logically valid, and they accuse
classical mathematicians of being committed to this claim.

How plausible is the thesis that an act of mathematical inference is correct
if a corresponding argument is logically valid, and that such an act is then
correct *in virtue of* corresponding to a logically valid argument? Similarly, how
plausible is the claim that an act of mathematical judgement is correct if a
corresponding sentence is logically true, and that such an act is then correct
because it corresponds to a logically true sentence? To begin with, consider
positive answers to the first half of each question:

Logical Validity Implies Correctness (LVIC) An act of mathematical inference is correct if some corresponding argument is logically valid.

Logical Truth Implies Correctness (LTIC) An act of mathematical judgement is correct if some corresponding sentence is logically true.

Are at least these conditionals true? And how do they relate to the conflict
between intuitionists and classicists?

I will introduce here Göran Sundholm's criticism of 'classical' conceptions
of inference which incorporate these principles. But first I have to mention one
ingredient of his reasoning, which I will set aside. When Sundholm discusses

linguistic or mental acts, he often speaks about what he calls the *products* of these acts. For example, according to Sundholm, an act of assertion has an *assertion* as its product, and an act of judgement has a *judgement* as its product. These products are not to be identified with the acts themselves, and they are neither types of such acts nor linguistic expressions nor propositional contents. Like Künne (2008: 390), I am skeptical about these hypothetical entities, and following him (2003: 250), I use terms like "assertion" or "judgement" to stand for linguistic or mental acts, for types of such acts, or for the propositional contents of such acts. Now, as far as I can see, nothing essential is lost by framing the discussion in terms of acts, types of acts, and their linguistic and propositional counterparts. In particular, I will speak about types of mental and linguistic acts (*judgements, inferences, proofs, assertions*) instead of speaking about products of such acts. However, it should be noted that Sundholm (1983: 164-7, 1994: 120-1, 1997: 204-5, 1998, 2004: 439-41 & 453-5) heavily stresses the importance of products of mental acts and speech acts.

To be precise, Sundholm levels his criticism against propositional variants of the theses LVIC and LTIC:

Logical Validity^p Implies Correctness (LV^{p}IC) An act of mathematical inference is correct if the corresponding propositional argument is logically valid.

Logical Truth^p Implies Correctness (LT^{p}IC) An act of mathematical judgement is correct if its (propositional) content is logically true.

Here a propositional argument corresponds to an inference iff its n-th member is the content of the n-th judgement of the inference. The theses LV^{p}IC and LT^{p}IC presuppose a notion of logical validity for propositional arguments and a notion of logical truth for propositions. Such notions have been developed, for example, by Bolzano, who is one of the targets of Sundholm's criticism:

> [U]nder the classical Bolzano reduction of validity, we could find ourselves in the position that we knew the premises of an inference, and, furthermore, that, unknowingly to us, logical consequence does obtain between the relevant propositional contents of premises and conclusion. In such a position one would be allowed to carry out the inference - because under the Bolzano reduction the inference *is* valid - but still we would not know the conclusion (Sundholm 2004: 454).

According to Sundholm's Bolzano, one is allowed to carry out an inference if the corresponding propositional argument is logically valid; it may then happen that someone has knowledge with the content of the initial judgements of an inference and is allowed to carry out the inference although her act does not put her in a position to have knowledge with the content of the final judgement of the inference.[25] Note that Sundholm speaks about *valid* inferences, while I speak

---

[25]   According to Sundholm (1997: 210), the premises and the conclusion of an inference are *judgements*, and he uses formulations like "know a judgement" in this context.

about *correct* inferences; nothing of what follows depends on this terminological choice.

Sundholm thinks that a thesis like LV^{p}IC is clearly unacceptable, and he (2004: 454) claims that "Bolzano was followed by virtually the entire modern tradition in classical logic" in accepting it. Similarly, Sundholm (2004: 441-4) rejects LT^{p}IC and accuses the vast majority of classicists to being committed to this principle. In fact, he (2004: 453) even accuses them of being committed to the stronger thesis that "judgements which are [...] unwarranted are still held to be knowledge, simply in virtue of having a (classically) true proposition as content".

Now, Sundholm's considerations invite two questions. First, are LV^{p}IC and LT^{p}IC really thus implausible? Second, are classicists in some way committed to such theses? I am inclined to regard the first question as a question which 'only' concerns the proper uses of the phrases "correct act of judgement" and "correct act of inference". There one can distinguish between *descriptivist* and *normativist* approaches. According to descriptivist approaches, correct acts of inferences and judgements do not have to be legitimate acts. Following Bolzano and Husserl, Künne adopts a descriptivist approach:

> [W]hether an act of judgement is correct only depends on whether the content of the act is true [...] (Künne 2008: 390).

Unlike Sundholm, who presupposes that one is always allowed to make correct judgements and to carry out correct inferences, Künne denies in his reply to Sundholm (2008) that the pertinent notion of correctness carries such a normative burden:

> In the Bolzano-Husserl sense, correctness is compatible with lack of epistemic merit, for an act of judgement, or a belief, can be correct (=: have a true content) even if it is capricious and ill founded. Presumably, making such a judgement, or having such a belief, is not judging or believe, as one *ought* to judge or believe (Künne 2008: 392).

Notice that descriptivists are not committed to the claim that one is allowed to make any judgement whose content is (logically) true, or that one may carry out every inference whose corresponding propositional argument is (logically) valid. Sundholm, on the other hand, adopts a normativist approach: he presupposes that one is allowed to make correct judgements and to carry out correct inferences. (It should be emphasised that according to the normativist approach, theses like LV^{p}IC and LT^{p}IC are false independently of whether logical validity and logical truth behave classically or intuitionistically. There are, for example, also complex intuitionistic logical truths whose corresponding judgements cannot be legitimately made by everyone at every time.)

For the sake of the argument, I will follow the normativist approach of Sundholm: making a correct judgment is judging as one ought to judge and carrying out a correct inference is inferring as one ought to infer. Consequently, I assume that there are incorrect acts of judgements that correspond to logical truths and incorrect acts of inferences that correspond to logical validities. Given this assumption, there only remains the second of the above mentioned questions: are classicists in some sense committed to a (false) descriptivist account of correct acts of inferences and judgements?

Now, on the face of it, it is difficult to see why classicists should be committed to such a view. Why should the question about the proper use of the phrases "correct act of judgement" and "correct act of inference" be relevant for the question of which arguments are logically valid? Furthermore, given that some classicists explicitly adopt a normativist approach (cp. Williamson 2000: ch. 11 esp. p. 241), it follows that the connection between classical logic and a descriptivist approach would have to be a rather oblique one.

According to what I take to be the most promising strategy for arguing that classicism (obliquely) depends on a false descriptivist account of correct acts of inferences, intuitionists try to establish two things: (a) that specific claims about logical validity (claims to the effect that canonically specified arguments are logically valid) have to be grounded in principles about correct acts of inferences,[26] and (b) that such principles could only ground the specific claims about logical validity of classical logic if they reduced the correctness of an act of inference to properties of a corresponding (linguistic or propositional) argument. In fact, intuitionists can work out a strategy of this kind without claiming that classicists are strictly committed to theses like LV^pIC and LT^pIC. It suffices if they can show that the specific claims about logical validity of a sound logical theory have to be grounded in principles about correct inferential acts and that such principles do not ground the relevant claims of classical logic.

If an intuitionist wants to work out this strategy, then she can proceed in either of two ways. If she follows the *direct approach*, then she puts forward and defends principles about correct inferential acts which have immediate anti-classical consequences, e.g. weak or even strong counterexamples to classical logic. If she follows the *indirect approach*, then she (i) explicates and substantiates the claim that the specific claims about logical validity of a sound logical theory have to be grounded in principles about correct inferential acts, (ii) develops and defends desiderata for collections of such principles, and (iii) shows that no collection of principles which satisfies the desiderata yields classical logic. Both approaches would imply that classical logic is not sound.

In the remaining subsection, I will present an account of correct acts of inferences that, if it is sound, undermines classical logic and, arguably, grounds intuitionistic logic. This will illustrate a possibility to work out the mentioned direct approach. In addition, it will also provide a better understanding of the indirect approach, because it is displays what kind of inferential principles intuitionists suggest to underlie a logical theory. A closer inspection of the presented account of correct acts of inferences leads to what I take to be the most powerful intuitionistic challenges to classical logic.

---

[26] Recall the importance of such *canonical logicality theorems* (1.3) for an evaluation of a logical theory.

3.4.3   Prawitz' Account of Correct Inferences

In this subsection, I will introduce Prawitz's account of correct acts of inferences and display its potential to yield arguments against classical logic.

Prawitz (2009: 183, 2012a: 890) identifies knowledge with possession of (conclusive) ground: someone knows a sentence S to be true iff she has a conclusive ground for accepting S.[27] He (2009: 194) mentions two classes of examples: verifying observations are grounds for accepting observation sentences, and verifying computations are grounds for accepting arithmetical identity-sentences. (The idea must be that one comes in possession of such grounds by making them.) Now, if mathematical knowledge consists in the possession of grounds, and if every ground can become someone's possession (cp. Prawitz 2012a: 893), then a mathematical sentence can be known to be true iff there is a ground for accepting it. From the assumption that mathematical knowability coincides with mathematical provability (see 3.3.1), it then follows that provability consists in the existence of a ground: a mathematical sentence is provable iff there is a ground for accepting it.

Prawitz (2009: 184, 2012a: 890) then poses the question of how one can obtain grounds for accepting sentences by means of acts of inferences. His basic idea is that if one performs a correct inference, then one applies an operation to the grounds for accepting certain sentences, thereby obtaining a ground for accepting another sentence. I will spell this out in more detail for conjunctions and disjunctions.

Suppose that you have a ground g_1 for accepting the sentence S_1 and a ground g_2 for accepting the sentence S_2. You can then make a correct inference with the result that you have a ground for accepting the conjunction $S_1 \wedge S_2$. Prawitz's idea is that there is a binary operation $\wedge G$ and that in making this inference you apply $\wedge G$ to the pair (g_1, g_2) thereby obtaining a ground $\wedge G(g_1, g_2)$ for accepting the conjunction $S_1 \wedge S_2$. More strongly, Prawitz (2009: 192) claims that a ground for accepting a conjunction is always composed of grounds for accepting the conjuncts. That is, if the predicate symbol $\vDash$ corresponds to the binary predicate "is a (conclusive) ground for accepting", then Prawitz puts forward the following claim:

Grounds for Conjunctions (GfC)   For all sentences S_1 and S_2 and for every ground g, $g \vDash S_1 \wedge S_2$ iff for some grounds g_1 and g_2: $g = \wedge G(g_1, g_2)$, and $g_1 \vDash S_1$, and $g_2 \vDash S_2$.

According to this principle, something is a ground for accepting a conjunction iff it is the result of applying $\wedge G$ to grounds for accepting the conjuncts.

---

[27]   Like Prawitz (2009: 184), I will often drop the qualification "conclusive" in what follows: grounds are always conclusive grounds. Prawitz (2012a: 890) uses various formulations, e.g. "grounds for a judgement", "grounds for an assertion", "grounds for holding a proposition to be true". But since the fundamental theses of his account (see below) describe potential grounds in accordance with the syntactic structure of associated sentences, I find it best to adopt a formulation like "grounds for accepting a sentence" that explicitly brings sentences into play. Correspondingly, I will here assume that acts of inferences are made up from acts of accepting sentences.

This principle suggests (2009: 197) that there are two unary operations on grounds, $\wedge R_1$ and $\wedge R_2$, which can be applied to grounds for accepting a conjunction and which then yield grounds for accepting its conjuncts; that is, $\wedge R_1(\wedge G(g_1, g_2)) = g_1$ and $\wedge R_2(\wedge G(g_1, g_2)) = g_2$. Suppose that you have a ground g for accepting a conjunction $S_1 \wedge S_2$. Then, you can obtain a ground for accepting S_1 by applying the operation $\wedge R_1$ to g, and, similarly, you can obtain a ground for accepting S_2 by applying the operation $\wedge R_2$ to g.

If I understand Prawitz correctly, then he would reject the claim that whenever you have a ground for accepting a conjunction, then you have also grounds for accepting its conjuncts. For example, it might be that you have obtained a ground for accepting a conjunction from a ground for accepting a universal sentence of which the conjunction is an instance. In such a case, you have to do something, namely applying the operations $\wedge R_1$ and $\wedge R_2$, to obtain grounds for accepting the conjuncts.

I will now consider grounds for accepting disjunctions. Suppose that you have a ground g_1 for accepting S_1. Then, you can obtain a ground for accepting the disjunction $S_1 \vee S_2$. According to Prawitz's account, there is a unary operation $\vee_1 G$ on grounds which you can apply to the ground g_1 to obtain a ground $\vee_1 G(g_1)$ for accepting the disjunction $S_1 \vee S_2$. Similarly, if you have a ground g_2 for accepting S_2, then you can apply another operation $\vee_2 G$ to obtain a ground $\vee_2 G(g_2)$ for accepting $S_1 \vee S_2$. More strongly, Prawitz claims that a ground for accepting a disjunction is always thus composed out of a ground for accepting one of the disjuncts:

Grounds for Disjunctions (GfD) For all sentences S_1 and S_2 and for every ground g, $g \vDash S_1 \vee S_2$ iff for some ground g_1: $g = \vee_1 G(g_1)$ and $g_1 \vDash S_1$, or for some ground g_2: $g = \vee_2 G(g_2)$ and $g_2 \vDash S_2$.

That is, something is a ground for accepting a disjunction iff it is either the result of applying $\vee_1 G$ to a ground for accepting the first disjunct or the result of applying $\vee_2 G$ to a ground for accepting the second disjunct.

Similar to the case of conjunctions, I think that Prawitz would reject the claim that whenever you have a ground for accepting a disjunction, you also have a ground for accepting one of the disjuncts. Moreover, and in contrast to the case of conjunctions, it is not even clear that whenever you have a ground for accepting a disjunction, you are in a position to obtain a ground for accepting a disjunct. In the conjunctive case, you could always apply the two operations $\wedge R_1$ or $\wedge R_2$ to obtain grounds for accepting the conjuncts from a ground for accepting a conjunction. In the disjunctive case, however, it is not clear whether you can apply an operation to a ground for accepting a disjunction, thereby obtaining a ground for accepting a disjunct. The problem is not that there is no such operation. We may assume that if there is a ground for accepting a disjunction, then there is or is not a ground for accepting the first disjunct. Then, given some relatively weak existential assumptions about operations on grounds, there is an operation bound to yield a ground for accepting a disjunct when applied to a ground for accepting a disjunction. There is, for example, the operation whose application to a ground for accepting a disjunction yields a ground for accepting the first disjunct if there is such a ground and which otherwise yields a ground for accepting the second disjunct. The problem is that it is not clear whether a person who possesses a ground for accepting a disjunction can *apply* such an operation. For all that has been said, there might be a person who possesses a ground for accepting a disjunction although she is in no position to find out which of the two disjuncts is true. Such a person could not apply the operation to her ground for accepting the disjunction because this would require that she then knew for which disjunct she possesses a ground.

The theory also pertains to grounds for accepting conditionals, negations, and quantified sentences. However, the treatment of such sentences is more complex and does not have to be pursued in this chapter. (I will return to a variant of such a treatment in Chapter 5.) Prawitz proposes to identify an inference, what he (2012a: 895-6) calls a "(generic) inference (act)", with four kinds of elements: (i) acts of accepting finitely many sentences $S_1, \ldots, S_n$, (ii) grounds $g_1, \ldots, g_n$ for accepting $S_1, \ldots, S_n$, (iii) an operation Φ that is applicable to $g_1, \ldots, g_n$, and (iv) an act of accepting another sentence S. He then calls such an inference *valid* iff $\Phi(g_1, \ldots, g_n)$ is a ground for accepting S. To make an act of inference that instantiates such an inference is to apply the operation Φ to the given grounds. This then yields an account of correct acts of inferences: an act of inference is correct iff it instantiates an inference that is valid in the specified sense.

Prawitz conceives of his theory of grounds for accepting sentences as a semantic theory (see below), and as such it is inspired by the so-called BHK-Interpretation of intuitionistic logic.[28] In contrast to Prawitz's theory in which the recursive clauses are biconditionals about *grounds* for accepting sentences, the recursive clauses of the BHK-Interpretation are biconditionals about *proofs* of sentences. (Notice that grounds are to be distinguished from proofs: proofs are carried out to get in possession of grounds). The earliest versions of the BHK-Interpretation were given by Heyting (1930b, 1931, 1934) and Kolmogorov (1932). Since then, it has been much discussed and refined; in particular, it is now usually presented as a simultaneous specification of *canonical* and *non-canonical* proofs of sentences (see e.g. Prawitz 1973 and Dummett 1973b).[29] In Prawitz's theory of grounds one does not differentiate between canonical and non-canonical grounds, but it can be useful to differentiate between canonical and non-canonical *terms* for grounds (cp. Prawitz 2009: 195). Since Prawitz's theory of grounds is simpler than the (refined) BHK-Interpretation and, as far as I can see, not less plausible, I will concentrate on it here.

I will now relate Prawitz's theory to the previous discussion of weak counterexamples to classic logic (3.3.2). Recall that the predicate symbol P corresponds to the unary predicate "is a provable sentence of $\mathcal{L}^{Ar}$", and that the claim that Goldbach's Disjunction is presently not known to be true could be established by invoking a thesis about the provability of arithmetical disjunctions:

---

[28]   See Troelstra & van Dalen 1988: 9. In this acronym, "B" stands for "Brouwer", "H" stands for "Heyting", and "K" stands for "Kolmogorov". (The term "BHK-Interpretation" was introduced by Troelstra (1977: 977), in which "K" stood for "Kreisel".)

[29]   For two recent overviews, see van Atten 2014 and Prawitz unpublished.

$$\mathrm{P}_{\vee}\ \forall S_1 \forall S_2\ \big(P(S_1 \vee S_2) \to (P(S_1) \vee P(S_2))\big).$$

According to $\mathrm{P}_{\vee}$, an arithmetical disjunction is only provable if one of its disjuncts is provable.

Now, the combination of the assumption that a mathematical sentence is provable iff there is a ground for accepting it and the principle GfD directly imply $\mathrm{P}_{\vee}$. To see this, suppose that some arithmetical disjunction is provable: $P(S_1 \vee S_2)$. Then, there is a ground for accepting it: $\exists g\, g \vDash S_1 \vee S_2$. Consequently, the principle GfD implies that there is either a ground for accepting the first disjunct or a ground for accepting the second disjunct: $(\exists g\, g \vDash S_1) \vee (\exists g\, g \vDash S_2)$. But then it follows that at least one disjunct is provable: $P(S_1) \vee P(S_2)$. This shows that Prawitz's theses about grounds for accepting mathematical sentences can be used to substantiate the claim that there are weak counterexamples to classical logic.

It requires more effort to extend such a theory of grounds so that it can be used to substantiate the claim that there are *strong* counterexamples to classical logic. McCarty (2008b: 48) has made a suggestion for deriving the Uniformity Principle (3.1.1) from such an extension.[30] I will not discuss McCarty's short sketch here but only remark that it not only depends on the assumption that for every true mathematical sentence there is a ground for accepting it, but also on a strong assumption about the nature of grounds for accepting universal quantifications over sets of numbers.

Prawitz's theory of grounds clearly has anti-classical potential. How can classicists react to the claim that there is a plausible theory of correct mathematical inferences and proofs which leads to weak (and possibly even to strong) counterexamples to classical logic?

Part of a classical response, of course, has to consist in questioning the plausibility of Prawitz's account, since it supports weak (and possibly strong) counterexamples to classical logic. A general worry could be that the proposed notion of ground somehow plays a questionable combination of a subjective epistemic role and an objective metaphysical role. On the one hand, operations on grounds and possession of grounds are meant to account for correct inferences and inferential knowledge. On the other hand, grounds are taken to be governed by structural principles which lead to grounds of arbitrary complexity and comprise principles such as GfD that might seem most plausible for a metaphysical notion of ground (cp. Fine 2012). However, intuitionists might respond by pointing out that it is not to be expected that fundamental principles about correct acts of inferences cohere with the epistemological and metaphysical views of classicists. Given the assumption that a sound logical theory has to be grounded in fundamental principles about correct acts of inferences (see 3.4.2), it seems to speak in favour of intuitionism if only intuitionists can come up with plausible candidates for such principles. Consequently, it is desirable

---

[30]   More precisely, McCarty does not speak about grounds for accepting sentences but about constructions which prove sentences, that is, he presents his proposal in terms of a simple version of the BHK-Interpretation. But his reasoning can be directly transferred to the framework proposed by Prawitz.

for classicists to embed their specific critique of Prawitz' theory of grounds in a more general response.

Apart from criticising specific intuitionistic theories about correct acts of inferences, classicists can follow two routes. First, they can question the claim that a sound logical theory has to be based on a more fundamental account of correct acts of inferences. Why does a true claim to the effect that some argument is logically valid has to be grounded in principles about correct inferential acts?

The most common intuitionistic justification for this grounding claim invokes considerations about *meaning* and *semantic theories*. Similarly to other intuitionists' understanding of the BHK-Interpretation, Prawitz takes his theory of grounds to be a semantic theory:

> [T]he meaning of a sentence is determined by what counts as a ground for the judgement expressed by the sentence (Prawitz 2009: 191-2).[31]

Using the present terminology, this amounts to the claim that the meaning of a sentence is determined by what counts as a ground for accepting it. A common intuitionistic idea now seems to be that a sound logical theory has to be grounded in fundamental principles about correct inferential acts because it has to be grounded in semantic principles, which are directly correlated to principles about correct inferential acts.[32]

Many philosophers think that a sound logical theory has to be validated by a sound semantic theory. In addition, intuitionists think that there is no sound semantic theory that validates classical logic. One of their strategies, a purely negative one, is to show that classical logic is only validated by an implausible version of truth conditional semantics (see Ch. 4). A different and (partly) positive strategy is to show directly that a sound semantics validates intuitionistic but not classical logic. Here intuitionists try to show that a sound semantics has to incorporate a treatment of the logical constants which is linked up with desiderata on a *coherent inferential practice*. Furthermore, they try to establish that a semantic theory such as Prawitz's theory of grounds or the BHK-Interpretation incorporates such a treatment, while no 'classical' semantic theory does (cp. Ch. 5).

A classicist who follows the mentioned first route has to take these semantic considerations into account. If she accepts that a sound logical theory has to be semantically validated, then she has to combine her denial of the claim that

---

[31]   Compare Prawitz 2012a: 893 and 2012b: 12-3. Correspondingly, Prawitz (2012a: 893) characterises a conjunction $S_1 \wedge S_2$ as a sentence whose meaning is such that a ground for accepting it is formed by bringing together a ground for accepting S_1 and a ground for accepting S_2.

[32]   One version of this idea would be that principles about correct inferential acts *are* semantic principles. A second version of this idea would be that they are *grounded in* semantic principles; thus, semantics would provide a common foundation to claims about logical validity and claims about correct inferential activity. A third version of this idea would be that principles about correct inferential acts directly ground semantic principles which then ground claims about logical validity; thus, semantics would provide a link between principles about correct inferential acts and claims about logical validity.

a sound logical theory has to be based on principles about correct inferential acts with a denial of the claim that the fundamental semantic principles are (grounded in) principles about correct inferential acts. On the other hand, if she accepts a close connection between semantic principles and principles about correct inferential acts, then she has to combine her denial of the claim that a sound logical theory has to be based on principles about correct inferential acts with a denial of the claim that a sound logical theory has to be based on semantic principles (cp. the end of 4.5.2).

However, there is also a second possible route that classicists can take if they want to answer the charge that only intuitionists can come up with plausible candidates for fundamental (semantic) principles about correct inferential acts. This route consists in presenting principles about correct acts of inferences which cohere with classical logic while being relevantly similar to the principles that intuitionists have proposed.

Now, *technically* it is not difficult to develop a classical theory of correct inferential acts which resembles, for example, Prawitz's theory of grounds. Here is an indication of an example. Recall that mathematical knowledge is not only obtained by means of inferential acts. There are also trivial steps in mathematical proofs, in which a mathematician might judge, for example, that 0 is a natural number (see 3.4.2). Transferred to the setting suggested by Prawitz, one might say that someone can come to possess a ground for accepting the sentence "0 is a natural number" just by combining her grasp of the meanings of the term "0" and the predicate "is a natural number" in accordance with the significance of the syntactic operation of predication; her knowledge is grounded in an activity that consists in assembling concepts rather than in an inferential activity. A classicist might now claim that someone may come to possess a ground for accepting sentences of the form "S or not-S" by combining her grasp of the meanings of the displayed logical constants in accordance with the significance of the involved syntactic operations. If it is possible to have a ground for accepting the sentence "0 is a natural number" that is not obtained from further grounds for accepting other sentences, why might it not also be possible to have a ground for accepting, say, Goldbach's Disjunction that is not obtained from further grounds for accepting other sentences? To make such a proposal consistent with the remainder of Prawitz's theory, one would have to alter the biconditionals about grounds for accepting the sentences of the different forms. The simplest idea might be to distinguish between two kinds of grounds: *inferential grounds* and, say, *trivial grounds*. One would then reinterpret Prawitz's biconditionals as specifying inferential grounds for accepting complex sentences in terms of grounds for accepting simpler sentences, and one would add principles about trivial grounds for accepting sentences. For example, one might say that every *inferential* ground for accepting a disjunction is obtained from a ground for accepting a disjunct, but one would also allow for additional *trivial grounds* for accepting disjunctions of the form "S or not-S".

Intuitionists will respond by claiming that the proposed classical variant of Prawitz's theory is not an adequate semantics. But the existence of such examples indicates that subtle questions about adequate semantic theories have to be answered if one wants to evaluate the claim that there are principles about correct acts of inferences which cohere with classical logic while being relevantly similar to the principles proposed by intuitionists (see Ch. 5). Consequently, both classical routes for answering the objection (that classical logic fails because it is not founded on a suitable account of correct inferential acts) lead to subtle questions about meaning and semantic theories.

To sum up, the discussed justifications of principles which imply potential strong or weak counterexamples to classical logic are not by themselves compelling. As regards McCarty's justifications of UP (3.2.1) and CP (3.2.2), a classicist will simply reject the assumption about the existence of the pertinent rule, and as regards de Swart's justification of CP (3.2.2), a classicist will simply reject the assumption about the existence of the relevant mathematical proofs. If the justifications of McCarty and de Swart are not supplemented with justifications of these crucial existential assumptions, then they do not seem to present something that a classicist would have to be afraid of. Similarly, in response to the presented justification of the claim that Goldbach's Disjunction is a weak counterexample (3.3.2), a classicist will simply deny that an arithmetical disjunction is provable only if at least one of its disjuncts is provable. Thus, an intuitionist has to present a justification of this provability claim to complete her argument against classical logic.

As was spelled out in this section, intuitionists try to provide completions of their arguments by defending certain claims about mathematical activity and logical validity. The most plausible such claims were seen to come down to the thesis that a sound logical theory has to be grounded in fundamental principles about correct inferential acts, while such principles do not yield classical logic. But again, such a thesis is clearly non-trivial and thus in need of further defence, and the usual strategy for defending it proceeds by relating it to considerations about *meaning* and *semantic theories*. Now, I do not claim that semantic considerations constitute the only possibility of justifying McCarty's or de Swart's crucial existential assumptions, that they constitute the only possibility of justifying the thesis that a provable disjunction necessitates a provable disjunct, or that they constitute the only possibility of justifying the claim that a sound logical theory has to be grounded in fundamental principles about correct inferential acts, while such principles do not yield classical logic. But if the presented arguments are meant to be real obstacles to classical logic and classical mathematics, then they are in need of some supplementation. And a semantic supplementation is not only an intuitively plausible candidate, it is also the candidate which has received the most philosophical attention.

In the following two chapters, I will therefore discuss intuitionistic arguments against classical logic which are based on considerations about meaning and semantic theories. In Chapter 4, I will discuss a famous argument of Dummett's which is meant to undermine classical logic by undermining classical truth conditional semantics. In Chapter 5, I will return to semantic theories of the kind that Prawitz favours. It will then be discussed whether the suggested intuitionistic semantic theories are plausible and, on the assumption that they are, whether they have plausible counterparts that cohere with classical logic.

Chapter 4

The Manifestation Argument

In this chapter, I will discuss an argument of Dummett's which is meant to undermine classical logic by undermining classical truth conditional semantics, i.e. truth conditional semantics based on an epistemically unconstrained notion of truth.

Dummett's strategy for attacking classical logic by attacking classical truth conditional semantics proceeds from the assumption that a sound logical theory is *semantically validated*: every argument that is logically valid according to the logical theory is semantically guaranteed to be truth-preserving.[1] His attack on classical logic can then be reconstructed by the following argument:

Dummett's Semantic Argument (DSA)

P_1 A logical theory $\mathcal{T}$ is sound only if every argument which is logically valid according to $\mathcal{T}$ is semantically guaranteed to be truth-preserving.

P_2 If every argument which is logically valid according to classical logic is semantically guaranteed to be truth-preserving, then sentential meanings are truth conditions, and truth is epistemically unconstrained.

P_3 If sentential meanings are truth conditions, then truth is epistemically constrained.

C The theory of classical logic is not sound.

Suppose that classical logic is sound, i.e., suppose that it contains only true axioms. The first premise of DSA then implies that every argument which is logically valid according to classical logic is semantically guaranteed to be truth-preserving. According to P_2, it then follows that sentential meanings are truth conditions, and truth is epistemically unconstrained: not every true sentence is possibly known to be true. However, according to P_3, this consequence does not hold, and thus the initial supposition that classical logic is sound has to be rejected.

---

[1] Dummett endorses this assumption because (i) he (1977: 256) claims that logical theories have to be semantically validated if an "anti-holistic" conception of language is correct, and (ii) he (1973b: 218-20, 1977: 253-6) claims that such a conception of language is indeed correct.

The main focus of this chapter will be Dummett's argument against classical truth conditional semantics; that is, I will mainly deal with his argument in favour of P_3. This argument is known as the *manifestation argument*. It is based on the assumption that understanding is fully manifestable, while knowledge of epistemically unconstrained truth conditions cannot always be fully manifested.

In the first section, I will present an initial sketch of the manifestation argument. I will give a short introduction to truth conditional semantics in 4.2 and elucidate the main premises of the argument in 4.3 and 4.4. In the final section, I will indicate what I take to be its main weakness. In addition, I will introduce a slight alternative to DSA, which is based on a weakening of P_2 and a corresponding strengthening of P_3, and critically discuss DSA and this alternative (see 4.1.2).

4.1   A Sketch of the Manifestation Argument

A proponent of the manifestation argument tries to show that sentential meanings cannot consist in epistemically unconstrained truth conditions, because sentential understanding cannot consist in knowledge of epistemically unconstrained truth conditions. Furthermore, the proponent of the argument believes that this holds because understanding is always manifestable, while knowledge of epistemically unconstrained truth conditions sometimes is not. More precisely, the conclusion of the manifestation argument states that if sentential meanings are truth conditions, then truth is epistemically constrained: for every intelligible and true sentence S a situation in which S is known to be true is possible.[2]

First, I will present a more detailed sketch of the manifestation argument (4.1.1). Then, I will consider three variants of this argument (4.1.2): (i) a variant that does not involve the claim that *understanding* is manifestable, but rather the assumption that *meanings* have to be determined by linguistic uses, (ii) a variant whose conclusion states that if sentential meanings are truth conditions, then not every sentence is true or false, and (iii) a variant whose conclusion states that if it is *conceptually necessary* that sentential meanings are truth conditions, then it is also *conceptually necessary* that truth is epistemically constrained. The third variant is linked to a modification of DSA that is based on the assumption that classical logic depends on a truth conditional semantics in which truth is not *conceptually required* to be epistemically constrained. I will close this section by discussing this alternative anti-classical argument.

---

[2]   I take Dummett 1973a: 460-71, 1973b: 216-25, 1976: 44-6, and 1977: 257-60 as the primary sources for my reconstruction, though I will also take later writings of Dummett, in which he modifies his earlier argumentation, into account. (A rudimentary version of the manifestation argument can already be found in Dummett 1959.)

It should be noted that Dummett does not claim to have *refuted* classical truth conditional semantics but only to have presented severe difficulties for it (see Dummett 1976: 34). I will nevertheless reconstruct a truth-preserving argument to the conclusion that truth is epistemically constrained if sentential meanings are truth conditions. Dummett's cautious remark should be understood as qualifying his commitment to the premises of the reconstructed argument.

4.1.1   A Basic Version of the Argument

In this chapter, I will mostly use "sentence" as an abbreviation for "intelligible declarative sentence". In the few places in which I mention non-declarative/non-intelligible sentences (e.g. 4.2.1 & 4.5.1), it will be clear from the context that "sentence" is to be understood in the usual wide sense of that term.

I will divide the reasoning underlying the manifestation argument into five steps. The first step sets the stage. The proponent of the manifestation argument wants to establish a conditional. Consequently, she temporarily assumes that the antecedent of this conditional holds:

Meaning and Truth Condition (M-TC) For every English sentence S, the meaning of S is the truth condition of S.

This temporary assumption that sentential meanings are truth conditions will be discharged in the final step of the argumentation. (M-TC will be discussed in 4.2.) The proponent of the manifestation argument now tries to show that for every true English sentence S a situation in which someone knows that S is true is possible. Since this is a general thesis about true English sentences, she considers an arbitrary sentence S^* and, in addition, an arbitrary person a^* who understands this sentence:[3]

A$_1$ S^* is an English sentence.

A$_2$ S^* is true.

A$_3$ a^* understands S^*.

The proponent of the manifestation argument now tries to show that a situation in which a^* knows that S^* is true is possible. This would then trivially imply that a situation in which someone knows that S^* is true is possible.

In the second step, the fundamental premise of the manifestation argument is introduced. It states that understanding is fully manifestable, that is, that someone's understanding consists in a capacity which can be fully accounted for in terms of her potential linguistic behaviour. To begin with, consider the following weak version of this premise:

Weak Manifestation Thesis (WMT) For every English sentence S and for every person a who understands S, a's understanding of S is fully manifestable in a's behaviour relating to utterances of English sentences.

According to WMT, someone can fully show her understanding of some English sentence by using English sentences and by reacting to such uses.

---

[3]   Notice that a certain simplification is involved here. Strictly speaking, one can only assume that for every (intelligible) sentence there *could* be a person who understands it. But then, to obtain a valid argument, one would have to modally strengthen the following premises. This would not introduce any serious additional problems, but it would lead to a discussion of different possible versions of these premises and corresponding underlying modal assumptions. Since this discussion would not be relevant for what follows, it is here assumed that for every (intelligible) sentence there is a person who actually understands it.

The weak manifestation thesis does not specify, for every English sentence S, a restricted class $c(S)$ of English sentences such that someone's understanding of S has to be manifestable in behaviour relating to utterances of sentences in $c(S)$. It is compatible with the claim that for every English sentence S someone's understanding of S can only be manifested in (infinitely many) acts relating to *every* English sentence. This weakness of WMT is overcome in the standard version of the manifestation thesis:

Manifestation Thesis (MT) For every English sentence S and for every person a who understands S, a's understanding of S is fully manifestable in a's behaviour relating to utterances of S.

The manifestation thesis states that someone's understanding of a sentence can always be fully accounted for in terms of her potential behaviour relating to utterances of this very sentence. Instead of using the phrase "behaviour relating to utterances of S", I will mostly use the shorter "S-related behaviour" in the following.

As an example, consider the sentence "it is raining". What kind of behaviour constitutes a manifestation of its understanding? Paradigmatic examples of acts which display such understanding surely include utterances of this very sentence and reactions to such utterances. In particular, one can partly manifest one's understanding of "it is raining" by displaying one's acceptance of it while it is raining (e.g. by assertively uttering it), and also by displaying one's rejection of it while it is not raining (e.g. by shaking one's head or saying "that is not true" if someone utters it). Now, according to MT, someone's understanding of "it is raining" is *fully* manifestable by such paradigmatic kinds of manifesting behaviour. More generally, MT states that someone's understanding of *any* declarative sentence S can be fully manifested by her utterances of S and her reactions to such utterances. (MT will be the topic of 4.4.) The manifestation thesis is applied to the assumptions A_1 and A_3 to yield the following temporary conclusion:

C_4 a^*'s understanding of S^* is fully manifestable in a^*'s S^*-related behaviour.

This completes the second step of the argumentation.

The third step is based on a premise that links understanding to knowledge of truth conditions. It states that if sentential meanings are truth conditions, then sentential understanding involves knowledge of truth conditions:

Understanding and Knowledge of Truth Conditions (U-KTC)
For every English sentence S and for every person a who understands S, if the meaning of S is its truth condition, then a's understanding of S involves knowledge of the truth condition of S.

One may note that U-KTC follows from the widespread assumption that understanding involves knowledge of meaning. (U-KTC will be dealt with in 4.3.) This premise is applied to A_1 and A_3. Then, the temporary assumption M-TC is applied to A_1. Finally, both results are combined with C_4 to yield the following claim:

C_5 a^*'s knowledge of the truth condition of S^* is fully manifestable in a^*'s S^*-related behaviour.

The justification for this inference is rather obvious: if a^*'s understanding of S^* involves knowledge of the truth condition of S^*, then C_4 implies that a^*'s knowledge of the truth condition of S^* is fully manifestable in a^*'s S^*-related behaviour.

In the fourth step, manifestability of knowledge of truth conditions is linked with the possibility to recognise that obtaining truth conditions do indeed obtain. Suppose that someone is disposed to show certain S-related behaviour in certain circumstances and that this possible behaviour fully accounts for her knowledge under which condition S is true. This seems to require that there is a connection between her possible S-related behaviour and her possible attitude towards the question of whether S is true: if there were no such connection, then it would be mysterious how her possible S-related behaviour could account for her knowledge of the condition under which S is true. At a minimum, one would expect that she has a capacity to recognise in certain possible situations that S is true if it is indeed true. But if there is a possible situation in which she is in a position to recognise that S is true, then there is also a possible situation in which she knows that S is true: a possible situation in which she has exercised her recognitional capacity. Consequently, the following principle seems to hold:

Knowability Thesis (KT) For every true English sentence S and for every person a, if a has knowledge of the truth condition of S that is fully manifestable in a's S-related behaviour, then there is a possible situation in which a knows that S is true.

Having the capacity to display one's knowledge of the condition under which some true sentence S is true by S-related behaviour presupposes the possibility of a situation in which one knows that S is true.[4] (KT will be treated in 4.4.4.)

The final step of the argumentation is routine. First, one applies KT to A_1 and A_2, and then combines the relevant instance of the result with C_5 to obtain the following temporary conclusion:

C_6 There is a possible situation in which a^* knows that S^* is true.

It directly follows that there is a possible situation in which *someone* knows that S^* is true. Therefore, discharging the assumptions A_1, A_2, and A_3, it follows that every true English sentence is possibly known to be true. Finally, one discharges M-TC and obtains the conclusion of the manifestation argument:

Conclusion If the meaning of every English sentence is its truth condition, then for every true English sentence S there is a possible situation in which someone knows that S is true.

In short, if sentential meanings are truth conditions, then truth is epistemically constrained.

---

[4]   Notice that it is not claimed that for every person who can fully manifest her understanding of a true sentence S, a situation in which she recognises that S is true is possible. It is only claimed that for every person who can fully manifest her *knowledge of the truth condition* of a true sentence S *by uttering S and reacting to utterances of S*, a situation in which she recognises that S is true is possible (cp. 4.3.2).

4.1.2   Three Variants of the Argument

Before closing this introductory section, I would like to mention three alternative versions of the manifestation argument. The first alternative is a simplification of the presented argument that does not make use of the notions of
understanding and knowledge of truth conditions. By way of motivation, it may
be noted that Dummett invokes two rather different variants of the manifestation thesis: the *semantic* manifestation thesis says that *meaning* is manifestable,
while the *epistemic* manifestation thesis says that *understanding* is manifestable. In Dummett's writings, these theses are closely connected. For example,
in an early presentation, he begins his manifestation argument by stating the
semantic manifestation thesis:

> The meaning of a mathematical statement determines and is ex-
> haustively determined by its *use*. The meaning of such a statement
> cannot be, or contain as an ingredient, anything which is not mani-
> fest in the use made of it [...] (Dummett 1973b: 216).

In the next paragraph, Dummett brings considerations of knowledge of meaning
explicitly into play, and he reaches, at the end of that paragraph, a variant of
the epistemic manifestation thesis:

> Hence it follows, once more, that a grasp of the meaning of a math-
> ematical statement must, in general, consist of a capacity to use
> that statement in a certain way, or to respond in a certain way to
> its use by others (Dummett 1973b: 217).

Dummett seems to think that manifestability of meaning and manifestability
of understanding (or grasp of meaning) more or less amount to the same thing.

Given this view of Dummett's, one might ask whether the notions of understanding and knowledge of truth conditions really have to be employed in the
manifestation argument. That is, one might ask whether it is possible to present
Dummett's argument independently of such notions. This would constitute a
great simplification since difficult questions about the nature of understanding
and knowledge would not have to be discussed at all. A semantic version of the
manifestation argument might be put as follows. "Let S^* be some true English
sentence, and assume that the meaning of S^* is the condition for S^* to be true.
The semantic manifestation thesis then implies that the condition for S^* to be
true is fully determined by S^*-related behaviour. Now, a thesis corresponding
to KT implies that there is a possible situation in which it is known that S^*
is true. Generalising, for every true English sentence S whose meaning is its
truth condition, there is a possible situation in which it is known that S is true.
Thus, if sentential meanings are truth conditions, then truth is epistemically
constrained." In this argument the notions of understanding and knowledge of
truth conditions do not play any role.

Unfortunately, without further supplementation at least, such a *semantic
manifestation argument* seems to be less convincing than the original epistemic
one. In particular, one might question the transition from the claim that S^*-
related behaviour determines the truth condition of S^* to the claim that if S^*

is true, then someone can know that it is true: even if the use of S^* determines its truth condition, this determination relation might be not accessible to individual language users (cp. Williamson 1994: 206). Importantly, uses of S^* by different persons might contribute to the determination of the truth condition of S^*, and there is no reason to assume that anyone is in a position to know that socially determined truth conditions obtain.

Now, I think that Dummett would reject this criticism. But he would do so for the reason that he takes the notions of meaning and understanding to be very closely related:

> [P]hilosophical questions about meaning are best interpreted as questions about understanding: a dictum about what the meaning of an expression consists in must be construed as a thesis about what it is to know its meaning (Dummett 1976: 35).

However, given such a view (see also Dummett 1973a: 92 and 1991: 88), the semantic manifestation argument is no longer a simplification of the epistemic one. The notions of understanding and knowledge of truth conditions would be essential ingredients of the reasoning underlying the semantic manifestation argument. This justifies focussing on the epistemic manifestation argument in what follows.

The second alternative argument is based on a different understanding of classical truth conditional semantics. According to this understanding, classical truth conditional semantics is based on a *bivalent* notion of truth, that is, a notion of truth for which the following *Sentential Principle of Bivalence* holds:

PB$_{Sent}$ Every English sentence is true or false.

The resulting notion of bivalent truth conditions is employed in many discussions of the manifestation argument (see e.g. McDowell 1976 and Lievers 1998). It is therefore worth considering a modification of Dummett's argument which concerns truth conditional semantics that is based on a bivalent notion of truth.

The basic manifestation argument (from 4.1.1) has to be *extended* to ground the claim that a sound truth conditional semantics cannot be based on a bivalent notion of truth. Here is one possibility for such an extension. "Assume that sentential meanings are truth conditions, let S^* be an English sentence, and let not-S^* be its negation. The principle BP$_{Sent}$ then has the consequence that either S^* is true or not-S^* is true. (Here it is assumed that a sentence is false iff its negation is true.) Therefore, the conclusion of the basic manifestation argument implies that either S^* is possibly known to be true or not-S^* is possibly known to be true. Generalising, it follows that for every English sentence either it or its negation is possibly known to be true. Consequently, since not every English sentence is such that either it or its negation is possibly known to be true,[5] it can be inferred that not every English sentence is true or

---

[5]   Notice that this is an additional substantive assumption on which the extended argument depends.

false. Discharging the initial assumption, it follows that truth is not bivalent if sentential meanings are truth conditions."

Despite its popularity, I will set aside this extension of the basic manifestation argument. There are two reasons for this. First, since the premises of the basic manifestation argument are also relied on in this extension, my criticism of them pertains to both arguments. Second, there are reasons to doubt that even the conclusion of the extended argument suffices for Dummett's larger strategy of attacking classical logic, and that a different modification of the basic manifestation argument is called for. To see this, consider the question of whether the conclusion of the extended argument, namely that truth is not bivalent if sentential meanings are truth conditions, can be used to undermine classical logic. Here, Dummett himself presents what I take to be compelling reasons for doubting that a truth conditional semantics which validates classical logic has to be based on a bivalent notion of truth:

> The forms of reasoning employed in classical mathematics, namely those embodied in classical logic, can be justified only by the two-valued semantics, *or, at least, by a semantical valuation system whose elements form a Boolean algebra* (Dummett 1977: 257, emphasis mine).

Consider a model-theoretic semantics according to which there is a collection of 'classical' structures $\mathfrak{C}$ (e.g. the classical models of Zermelo-Fraenkel set theory; cp. Dummett ibid.) such that a sentence is true iff it is true with respect to all elements of $\mathfrak{C}$. Then, every sentence which is logically true according to classical logic is true with respect to all elements of $\mathfrak{C}$, and it follows that every such sentence is validated by the model-theoretic semantics. But this does not imply that the semantics can be conceived of as two-valued. In particular, there are many cases like the mentioned set-theoretic example, in which some sentences are neither true with respect to all elements of $\mathfrak{C}$ nor not true with respect to all elements of $\mathfrak{C}$. Consequently, classical logic does not depend on a semantic theory which is based on a bivalent notion of truth.

Similar considerations (see below) also invite the question of whether the basic manifestation argument can be used for undermining classical logic, or whether a modification that is more far-reaching than the suggested extension in terms of bivalent truth conditions is called for. This leads to contemplating another variant of the basic manifestation argument.

The third alternative argument invokes the idea of an *epistemic notion* of truth, and it leads to an important modification of Dummett's anti-classical argument DSA. Recall that the initial two premises of DSA imply that classical logic can be sound only if sentential meanings are epistemically unconstrained truth conditions (while its third premise is the conclusion of the basic manifestation argument):

P$_1$ A logical theory $\mathcal{T}$ is sound only if every argument which is logically valid according to $\mathcal{T}$ is semantically guaranteed to be truth-preserving.

P$_2$ If every argument which is logically valid according to classical logic is semantically guaranteed to be truth-preserving, then sentential meanings are truth conditions, and truth is epistemically unconstrained.

P$_3$ If sentential meanings are truth conditions, then truth is epistemically constrained.

C The theory of classical logic is not sound.

Now, generalising the foregoing considerations about bivalent truth conditions, it seems likely that the premise P$_2$ is false. Not only is classical logic compatible with non-bivalent truth conditions; it is also compatible with truth conditions that are epistemically constrained. This is because classical logic and truth conditional semantics might hold, although it is a *non-semantic* fact that truth is epistemically constrained. Setting aside arguments which purport to show that truth has to transcend knowability,[6] it is easily conceivable that every truth can in principle be recognised as such, although this is no *semantic* fact.[7]

To deal with this problem, one has to find a suitable weakening of P$_2$. And indeed, Dummett can perhaps be understood as relying only on such a weaker assumption:

> It is an essential feature of any theory of meaning that will yield a semantics validating classical logic that each sentence is conceived of as possessing a determinate truth-value, independently of whether or not we know it or have at our disposal the means to discover it (Dummett 1977: 257-8).

Here, the important point is the following: the statement that each sentence has to be *conceived of* as possessing a truth-value *independently of certain epistemic constraints* could be understood as only expressing that it must not be *conceptually necessary* for truth to be epistemically constrained. Dummett would then only claim that classicists are committed to the thesis that truth is a *non-epistemic notion*, that is, a notion which is not fundamentally characterised in epistemic terms. Instead of P$_2$, Dummett would then endorse the claim that classical logic can only be sound if sentential meanings are truth conditions, and it is not conceptually necessary that truth is epistemically constrained. Applying the notion of conceptual necessity also to the assumption

---

[6] The most famous such argument is the one usually referred to as "Fitch's paradox of knowability", an argument that was given by Alonzo Church in 1945 in a referee-report to an article of Frederic Fitch (see Salerno 2009). It was presented by Fitch in 1963, rediscovered by Hart and McGinn in 1976, and has been widely discussed since then. As has been pointed out by Williamson (1982), intuitionists can evade the argument because it depends on intuitionistically unacceptable logical principles.

[7] Thanks to Hannes Leitgeb for emphasising the importance of this possibility. To use his example: it might be possible to meet God, to recognise him to be God, and, for any given true sentence S, to hear him say that S is true. That is, for every true sentence S, it might be possible to obtain from God the testimonial knowledge that S is true.

that sentential meanings are truth conditions (for reasons that will become clear in a moment), one arrives at the claim that classical logic can only be sound if it is conceptually necessary that sentential meanings are truth conditions but not conceptually necessary that truth is epistemically constrained.

If one adopts this approach, then one has to compensate for the weakening of P$_2$ with a corresponding strengthening of P$_3$:

Dummett's Refined Semantic Argument (DRSA)

P$_1$ A logical theory $\mathcal{T}$ is sound only if every argument which is logically valid according to $\mathcal{T}$ is semantically guaranteed to be truth-preserving.

P$_2^*$ If every argument which is logically valid according to classical logic is semantically guaranteed to be truth-preserving, then it is conceptually necessary that sentential meanings are truth conditions but not conceptually necessary that truth is epistemically constrained.

P$_3^*$ If it is conceptually necessary that sentential meanings are truth conditions, then it is also conceptually necessary that truth is epistemically constrained.

C The theory of classical logic is not sound.

According to P$_3^*$, it is conceptually necessary that truth is epistemically constrained if sentential meanings are conceptually required to be truth conditions. It then follows by P$_2^*$ that not every argument logically valid according to classical logic is semantically guaranteed to be truth-preserving. Consequently, P$_1$ implies that classical logic is not sound.

The most important part of a defence of DRSA seems to be a defence of its third premise P$_3^*$.[8] To this end, it suggests itself to modify the basic manifestation argument in such a way that the modification yields the conclusion P$_3^*$. I will mention what I take to be the most promising strategy for such a modification. According to this strategy, one takes a meta-perspective on the reasoning underlying the basic manifestation argument and tries to exploit its strength. The idea is that conceptual competence alone already suffices for recognising the correctness of this reasoning. This would arguably show that the conclusion of the basic manifestation argument expresses a conceptual necessity; i.e., that it is conceptually necessary that if sentential meanings are truth conditions, then truth is epistemically constrained. Furthermore, this surely implies that it is conceptually necessary that truth is epistemically constrained if it is conceptually necessary that sentential meanings are truth conditions: for any necessity operator $\sharp$ worth of its name, the sentence "$\sharp$(if S_1, then S_2)" implies the sentence "if $\sharp S_1$, then $\sharp S_2$". (The need for such an inference was the reason for applying the notion of conceptual necessity to the assumption that sentential meanings are truth conditions.)

---

[8] It should be noted, I think, that P$_2^*$ is still no *trivial* truth. It is not *obvious* that someone who, for example, identifies the notion of truth with the notion of knowability is precluded from accepting the combination of classical logic and truth conditional semantics. But although it is no triviality, P$_2^*$ is still a plausible assumption, and I will not question it.

Now, while I tend to think that DRSA is superior to DSA, I will nevertheless concentrate on the basic manifestation argument rather than on its modification that yields P_3^*. A proper treatment of the notion of conceptual necessity would have to be rather complex and would lead far off the main theme of this book. Furthermore, it turns out to be unnecessary to deal with this intricate topic. The reason is twofold. First, my criticism of the basic manifestation argument also pertains to any modification of it that relies on the manifestation thesis, and, as far as I can see, any plausible version of the manifestation argument relies on this thesis. Second, my criticism of the basic manifestation argument is accompanied by a direct argument to the conclusion that Dummett has not shown that truth must be epistemically constrained if sentential meanings are truth conditions. If cogent, a slight modification of this argument establishes that Dummett has also not shown that truth must be epistemically constrained if it is conceptually necessary that sentential meanings are truth conditions. Consequently, the premises P_3 and P_3^* both have to be rejected, or so I will argue (see 4.5). In the following sections, I will discuss the manifestation argument in more detail.

4.2   Truth Conditional Semantics

Many philosophers and linguists subscribe to semantic theories in which the notion of truth plays a central role.[9] In this section, I will introduce some of their ideas by discussing the following thesis:

Meaning and Truth Condition (M-TC) For every English sentence S, the meaning of S is the truth condition of S.

My aim is to give a short introduction to some philosophical questions that concern meaning theories which identify the meaning of a sentence with its truth condition. This identification characterises an important family of semantic theories that are based on the notion of truth.[10] More importantly, it is the antecedent of the conclusion of Dummett's manifestation argument.

First, I will make three preliminary remarks (4.2.1) and mention a well-known problem for truth conditional semantics (4.2.2). Afterwards, I will introduce *truth-theory*, a main type of truth conditional semantics. Finally, I will discuss the semantic potential of truth-theory (4.2.4), which leads to a presentation of a refined version of it (4.2.5). Although Dummett's argument is meant to

---

[9]   I will use the expressions "semantic theory" and "meaning theory" interchangeably. My use of these phrases correspond to Dummett's use of "meaning-theory" in his 1991 book. To a rough approximation, a meaning theory for some language $\mathcal{L}$ is a theory which states what the meaningful expressions of $\mathcal{L}$ mean (see 4.2.4).

Early philosophers who ascribe a central semantic significance to the notion of truth are Frege and Wittgenstein in the Tractatus. Later influential philosophers and linguists who do so are, among many others, Davidson (1967a), Montague (1970a, 1970b), Lewis (1970), Higginbotham (1992), Larson & Segal (1995), and Heim & Kratzer (1998).

[10]   Of course, not every semantic theory in which the notion of truth is of fundamental importance implies M-TC (see, e.g., Higginbotham 1992: 4-5).

apply to all types of truth conditional semantics, I will use refined truth-theory as its primary test case.

4.2.1 Preliminaries

To begin with, it is worth stressing that according to the thesis M-TC, *sentences* have truth conditions. Suppose it were assumed that only propositions (or utterances, or judgements) could be true. Then, clearly, sentential meanings could not be truth conditions: no sentence would be true under any condition. Someone who believes that "true" applies only, say, to propositions might still adopt a thesis similar to M-TC, e.g. the thesis that the meaning of a sentence is the condition for it to express a true proposition. However, here I will stick to the simpler thesis M-TC.

Second, one has to mention that there are complications with M-TC because of the fact that expressions may be context-sensitive. Consider the following sentences:

(1) I am hungry.   (2) You are hungry.

These sentences have different meanings, and they have their meanings independently of possible contexts of use. By contrast, it is often claimed that sentences of such pairs can have the same truth condition relative to appropriately related contexts of use (see, e.g., Lepore & Loewer 1989: 182). There is no generally agreed upon solution to this problem. However, since Dummett's criticism of classical truth conditional semantics does not rest on this point, I will set aside complications arising from the phenomenon of context-sensitivity.

Third, there is a complication with M-TC that is connected to the phenomenon of grammatical mood.[11] The grammatical mood of a sentence contributes to its meaning, but it does not contribute to its truth condition. Consider the following sentences:

(3) Peter is hungry.   (4) Is Peter hungry?

These sentences are not synonymous, but their difference in meaning is not a difference in associated truth conditions.

There is a standard response to this problem which goes back at least to Frege (1918: 62); cp. Dummett 1976: 38-40 and Schiffer 2003: 104. The idea is that the meaning of a sentence consists of two parts: one part derives from its grammatical mood, while the other part derives from constituents and syntactic operations which are independent of grammatical mood. One should then interpret M-TC as relating to the mood-independent part of meaning only. I will adopt this standard response. However, I will not introduce new terminology to account for this distinction. In what follows, I will normally use the noun "meaning" and the verb "means" in the narrow sense, in which "meaning" relates to the mood-independent part of the overall meaning of a sentence and "means" stands for the relation between a sentence and this mood-independent

---

[11]   In the remainder of this subsection, I will use the word "sentence" in its ordinary wide sense relative to which there are also non-declarative sentences.

part. The context will make it clear when these expressions are to be understood in the ordinary wide sense.

4.2.2   A Problem for Truth Conditional Semantics

I will now turn to a well-known problem for truth conditional semantics, which plays a certain role in what follows. It will give rise to different versions of truth conditional semantics, one of which will be treated as the primary test case for Dummett's manifestation argument.

The problem concerns the individuation of truth conditions. It consists in the fact that truth conditions seem to be individuated less finely than meanings. Consider the question when two sentences have the same truth condition. According to the *basic option*, two sentences S and S^* have the same truth condition iff the following holds: S is true if and only if S^* is true.[12] But then the meaning of a sentence cannot be its truth condition: not all true sentences are synonymous. In response to this, one might consider a finer individuation of truth conditions. According to the *modal option*, two sentences S and S^* have the same truth condition iff the following holds for every possible world w: S is true in w if and only if S^* is true in w. But still the meaning of a sentence could not be its truth condition: not all necessarily true sentences have the same meaning.

This individuation problem is related to the issue of *compositionality*. I adopt the view that meanings are compositional: the meaning of a complex expression is determined by the meanings of its parts and the significance of the syntactic operations according to which these parts are put together. More strongly, I think it is plausible to assume that the meaning of a sentence S has a structure which corresponds to the syntactic structure of S. However, according to the previous proposals for individuating truth conditions, the truth condition of a sentence S cannot have a structure which corresponds to the syntactic structure of S: sentences of totally dissimilar structures can have the same basic and modal truth condition. More strongly, basic and modal truth conditions are not compositional: two sentences S and S^* can have the same modal truth condition, although the truth value of a complex sentence containing S may be different from the truth value of the complex sentence obtained by replacing S with S^*. For example, although the sentences "2 is smaller than 3" and "the smallest prime number is smaller than 3" are true in the same possible worlds, the sentences "Peter believes that 2 is smaller than 3" and "Peter believes that the smallest prime number is smaller than 3" might have different truth values.

At first sight, this is a severe problem for truth conditional semantics. If one is not willing to completely give up the truth conditional approach, there seem to be two possible responses. First, one can make a move which is similar to the move of separating the significance of mood from the remaining part of meaning. The idea would be that there is a central part of the meaning of a sentence which can be identified with its truth condition, where truth conditions

---

[12]   Proponents of a three-valued semantics might want to add "and S is false if and only if S^* is false". However, I will not take such views into account here.

are coarsely individuated. The thesis M-TC would then be replaced by a weaker one:

M≺TC For every English sentence S, the truth condition of S is a part of the meaning of S.

The thesis M≺TC is still a substantial thesis denied by those who do not believe in any connection between truth and meaning. However, it has to be supplemented by an account of the remaining parts of sentential meanings and by an account of the relation between truth conditions and these remaining parts.

In view of the programatic character of a version of truth conditional semantics based on M≺TC, I take it to be worth considering a different response to the individuation problem. In this alternative response, it is simply denied that truth conditions are individuated differently from meanings:

M≈TC For all English sentences S and S^*, the meaning of S is identical to the meaning of S^* iff the truth condition of S is identical to the truth condition of S^*.

A justification of M≈TC could be an important part of a justification of the thesis M-TC that sentential meanings are truth conditions.

How could the theses M≈TC and M-TC be justified? I take it to be rather unlikely that conceptual analysis of the notion of a truth condition will settle the individuation question. Rather, a suitable approach of a truth conditional semanticist should consist in the attempt to construct a viable semantic theory which yields an appropriate notion of a truth condition. Now, some followers of Davidson can be seen to have pursued this approach. Partly for this reason, but mainly for having a well described target of Dummett's manifestation argument, I will now turn to truth-theories, one of which is Davidson's version of a truth conditional semantics.

4.2.3   Truth-Theory

What I will call *truth-theory* is one of the main types of truth conditional semantics. Its fundamental assumption is that a semantic theory for some language can be based on a *materially adequate theory of truth.*[13] (A second main type is a semantics based on the assignment of semantic values to the meaningful expressions of a language. Such a semantics is called *truth conditional* iff truth values are the semantic values of the sentences of the language.)

In what follows, I will restrict attention to a minor variant of the formal arithmetical language $\mathcal{L}^{Ar}$ (cp. 2.3.1). This variant corresponds to a simple

---

[13]   The term "truth-theory" is used here to stress the close relationship to theories of truth. An alternative terminological choice would have been "Davidsonian truth conditional semantics" since Davidson was the first to claim that a meaning theory might be based on a (materially adequate) theory of truth (see Davidson 1967a). But apart from being quite long, this term has the disadvantage of suggesting a meaning theory which incorporates Davidson's views about *radical interpretation* (see Davidson 1973), and here I prefer a wider concept which does not incorporate such additional views of Davidson. For an elaborate discussion of truth-theory, see Hoeltje 2012.

fragment of English, and I will also refer to it by $\mathcal{L}^{Ar}$. The basic signs of this language are over-lined symbols: brackets, variables, the logical constants $\overline{\neg}$, $\overline{\vee}$, $\overline{\wedge}$, $\overline{\Rightarrow}$, $\overline{\exists}$, $\overline{\forall}$, and the arithmetical predicates $\overline{N}$, $\overline{ID}$, $\overline{SC}$, $\overline{SM}$, $\overline{PR}$.[14] The reason for using over-lined symbols here is to prevent confusion with expressions of the meta-language (see below).

I will assume that the reader is familiar with post-Tarskian theories of truth and will only give a very brief introduction to a simplified version of one such theory. The simplification consists in the fact that I will use *quasi-sentences* of $\mathcal{L}^{Ar}$ instead of open sentences of $\mathcal{L}^{Ar}$ so that I do not have to take variable assignments into account:

QS₁ A sequence consisting of an n-place relation symbol followed by n objects which are not variables of $\mathcal{L}^{Ar}$ is a quasi-sentence of $\mathcal{L}^{Ar}$; each of these objects is then said to saturate the relation symbol.

QS₂ If S_1 and S_2 are quasi-sentences of $\mathcal{L}^{Ar}$, then so are $\overline{\neg}S_1$, $S_1\overline{\vee}S_2$, $S_1\overline{\wedge}S_2$, $S_1\overline{\Rightarrow}S_2$.[15]

QS₃ If S is a quasi-sentence of $\mathcal{L}^{Ar}$ which contains a non-variable o that saturates one or more relation symbols, if S^* results from S by replacing o wherever it saturates some relation symbol with a token of the variable ν, and if no newly introduced token of ν is bound in S^*, then $\overline{\exists}\nu S^*$ and $\overline{\forall}\nu S^*$ are also quasi-sentences of $\mathcal{L}^{Ar}$.

This is meant to be a recursive definition so that nothing else is a quasi-sentence of $\mathcal{L}^{Ar}$. A quasi-sentence is like a sentence except that it may contain arbitrary objects in places in which an ordinary sentence contains variables (or other terms).[16]

A theory of truth $\mathcal{T}$ for $\mathcal{L}^{Ar}$ is presented in a formal language of first-order predicate logic which is disjoint from $\mathcal{L}^{Ar}$ and corresponds to a richer fragment of English. In particular, the language of $\mathcal{T}$ contains canonical terms for expressions of $\mathcal{L}^{Ar}$ and an additional unary predicate symbol T corresponding to the predicate "is a true sentence of $\mathcal{L}^{Ar}$". Materially adequate theories of truth are defined by their consequences. To this end, a sentence of the form

$$T(t) \leftrightarrow S^*$$

is called a *T-sentence for S* iff t is a canonical term for a sentence S of $\mathcal{L}^{Ar}$. Such a T-sentence is called *interpretive* iff S and S^* have the same meaning. Using the notion of an interpretive T-sentence, one can present the relevant definition:

---

[14]   Recall that $\mathcal{L}^{Ar}$ contains no function symbols or individual constants. This has the advantage that I do not have to take the relation of denotation into account. In addition, it is assumed here that $\mathcal{L}^{Ar}$ contains no parameters so that only (bound) variables occupy the positions of singular terms in sentences of $\mathcal{L}^{Ar}$.

[15]   Here and in what follows, I use concatenation to indicate sequences, and I omit outer brackets.

[16]   The notion of a quasi-sentence is thus similar to the notion of an M-sentence (as defined in 1.2.2). However, it is not assumed here that there is any one set to which all the additional objects belong.

Material Adequacy A theory of truth $\mathcal{T}$ is called *materially adequate for* $\mathcal{L}^{Ar}$ iff it implies an interpretive T-sentence for every sentence of $\mathcal{L}^{Ar}$.

For every sentence S of $\mathcal{L}^{Ar}$, a materially adequate theory of truth implies a biconditional in which S is matched with a synonymous sentence of the language of $\mathcal{T}$.

I will sketch here an example $\mathcal{T}_+$ of such a theory. It will be called the *standard theory of truth for $\mathcal{L}^{Ar}$*. The syntactic part of $\mathcal{T}_+$ contains axioms which express basic assumptions about the quasi-sentences of $\mathcal{L}^{Ar}$. In particular, it comprises individual constants for the symbols of $\mathcal{L}^{Ar}$, a two-place function symbol for building sequences, unary predicate symbols Q and S for quasi-sentences and sentences of $\mathcal{L}^{Ar}$, and a ternary function symbol $sb(x, y, z)$ for the result of substitution of z for the free occurrences of the variable y in x. I will use the over-lined symbols of $\mathcal{L}^{Ar}$ as constants for themselves. I will also use concatenations to indicate sequences and write x_y^z instead of $sb(x, y, z)$.

The semantic part of $\mathcal{T}_+$ comprises a unary truth-predicate T and synonymous counterparts of all the basic symbols of $\mathcal{L}^{Ar}$; these counterparts resemble the symbols of $\mathcal{L}^{Ar}$ except that they do not have a horizontal line on top of them. In addition, it contains a unary predicate T^Q ("is quasi-true"), which is meant to apply to a quasi-sentence q in which the non-variables $o_1, \ldots, o_n$ saturate relation symbols iff the sentential fragment resulting from q by deleting $o_1, \ldots, o_n$ in these places is *true of* the n-tuple $(o_1, \ldots, o_n)$.

The semantic part contains the axiom $\forall a\, (Ta \leftrightarrow (Sa \wedge T^Q a))$,[17] which states that something is a true sentence of $\mathcal{L}^{Ar}$ iff it is a sentence of $\mathcal{L}^{Ar}$ and quasi-true. Furthermore, it contains axioms for the arithmetical and logical symbols of $\mathcal{L}^{Ar}$. These axioms involve the predicate T^Q. First, there are the following axioms for the relation symbols of $\mathcal{L}^{Ar}$:

$$\forall a\, \left(T^Q(\overline{N}a) \leftrightarrow Na\right);$$

$$\forall a \forall b\, \left(T^Q(\overline{ID}\, ab) \leftrightarrow ID\, ab\right);$$

$$\forall a \forall b\, \left(T^Q(\overline{SC}\, ab) \leftrightarrow SC\, ab\right);$$

$$\forall a \forall b \forall c\, \left(T^Q(\overline{SM}\, abc) \leftrightarrow SM\, abc\right);$$

$$\forall a \forall b \forall c\, \left(T^Q(\overline{PR}\, abc) \leftrightarrow PR\, abc\right).$$

As an example, consider the axiom for $\overline{N}$. It states that a quasi-sentence in which the unary predicate symbol $\overline{N}$ is applied to an object is quasi-true iff this object is a natural number; that is, it states that $\overline{N}$ is true of some object iff this object is a natural number. Second, there are the axioms for the sentential operators:

$$\forall a\, (Q(a) \rightarrow (T^Q(\overline{\neg}a) \leftrightarrow \neg T^Q(a)));$$

$$\forall a \forall b\, (Q(a \overline{\vee} b) \rightarrow (T^Q(a \overline{\vee} b) \leftrightarrow (T^Q(a) \vee T^Q(b))));$$

---

[17] In this and the following axioms, the quantifiers should be conceived of as restricted: the variables a, b, c do *not* range over variables of $\mathcal{L}^{Ar}$.

$$\forall a \forall b \left(Q(a \overline{\wedge} b) \rightarrow (T^Q(a \overline{\wedge} b) \leftrightarrow (T^Q(a) \wedge T^Q(b))) \right);$$

$$\forall a \forall b \left(Q(a \overline{\Rightarrow} b) \rightarrow (T^Q(a \overline{\Rightarrow} b) \leftrightarrow (T^Q(a) \rightarrow T^Q(b))) \right).$$

As an example, consider the axiom for $\overline{\neg}$. It states that the negation of a quasi-sentence is quasi-true iff the quasi-sentence is not quasi-true; that is, it states that the negation of a sentential fragment is true of a tuple of objects iff the sentential fragment is not true of this tuple of objects. Third, there are the axioms for the quantifiers:

$$\forall a \forall \nu \left(Q(\overline{\forall} \nu a) \rightarrow (T^Q(\overline{\forall} \nu a) \leftrightarrow \forall c\, T^Q(a_\nu^c)) \right);$$

$$\forall a \forall \nu \left(Q(\overline{\exists} \nu a) \rightarrow (T^Q(\overline{\exists} \nu a) \leftrightarrow \exists c\, T^Q(a_\nu^c)) \right).$$

As an example, consider the axiom for $\overline{\exists}$. It states that a quasi-sentence in which $\overline{\exists}$ is followed by a variable ν and a sequence a is quasi-true iff there is some object such that the quasi-sentence resulting from a by replacing every unbound occurrence of ν with this object is quasi-true; that is, it states that a sentential fragment in which $\overline{\exists}$ is followed by a variable ν and a sentential fragment a^* is true of a tuple of objects $(o_1, \ldots, o_n)$ iff there is some object o such that the sentential fragment resulting from a^* by deleting the unbound occurrences of ν is true of the tuple $(o_1, \ldots, o_n, o)$.

Note that these axioms are not only true with respect to the intended meanings of the used symbols; they are also *interpretive*: the over-lined symbols of $\mathcal{L}^{Ar}$ mentioned in these axioms have the same meanings as their counterparts in the language of $\mathcal{T}_+$ which are used in these axioms. Consequently, the theory $\mathcal{T}_+$ is materially adequate: for every sentence S of $\mathcal{L}^{Ar}$ it implies an interpretive T-sentence for S.

In recent developments of truth conditional semantics based on materially adequate theories of truth, a special class of calculi has played an important role (see Davies 1981: 33, Lepore & Ludwig 2007: 36, 95-7, and Hoeltje 2012: sec. 2.3.1):

Canonical Calculi A Hilbert-style calculus $\mathcal{R}$ is called *canonical for $\mathcal{T}$* iff

(i) for every sentence S of $\mathcal{L}^{Ar}$ some T-sentence for S is derivable in $\mathcal{R}$ from the axioms of $\mathcal{T}$;

(ii) every T-sentence which is derivable in $\mathcal{R}$ from the axioms of $\mathcal{T}$ is interpretive.

A canonical calculus for a materially adequate theory of truth yields interpretive T-sentences for all sentences of $\mathcal{L}^{Ar}$, and it yields no non-interpretive T-sentences.

Now, the usual Hilbert-style calculi that yield the classical or the intuitionistic logical truths are not canonical. For example, if one can derive an interpretive T-sentence $T(t) \leftrightarrow S_1^*$ in such a calculus $\mathcal{R}$ from the axioms of $\mathcal{T}$, and if S_2^* is any other sentence which can be derived in $\mathcal{R}$ from the axioms of $\mathcal{T}$, then the non-interpretive T-sentence $T(t) \leftrightarrow (S_1^* \wedge S_2^*)$ can also be derived in $\mathcal{R}$ from the axioms of $\mathcal{T}$. Fortunately, for theories like the standard theory

$\mathcal{T}_+$ canonical calculi are known (see Lepore & Ludwig 2007: 95-8 and Hoeltje 2012: appx. B). I will use $\mathcal{R}_+$ to refer to some canonical calculus for $\mathcal{T}_+$.

4.2.4   The Semantic Significance of Theories of Truth

I will now turn to Davidson's claim that materially adequate theories of truth may serve as semantic theories and to the corresponding version of Dummett's thesis that the notion of truth figuring in a theory of truth $\mathcal{T}$ has to be epistemically constrained if $\mathcal{T}$ may serve as a semantic theory.[18]

To this end, I would like to contrast two notions of a semantic theory. A *partially semantic* theory is any theory which makes essential use of semantic vocabulary. Uncontroversially, theories of truth are partially semantic theories in this sense. The predicate "is a true sentence of $\mathcal{L}^{Ar}$" is a semantic predicate (cp. Künne 2003: 179); it applies only to expressions of $\mathcal{L}^{Ar}$, and it applies to the sentences to which it applies in virtue of their meanings. But Davidson's claim cannot be understood in terms of the notion of a partially semantic theory. This would result in a thesis that is simply too weak to characterise his views about the semantic significance of theories of truth. Davidson's claim that a materially adequate theory of truth is semantically relevant is not based on the observation that it depends on the meaning of a sentence whether a truth predicate applies to it. Rather, his idea is that a materially adequate theory of truth can be the central part of a semantic theory which is in some sense *complete*. Similarly, Dummett is concerned with *complete* semantic theories in which truth is the central notion.

What is a complete semantic theory? I will adopt an answer which speaks directly about what the expressions of the object language $\mathcal{L}$ *mean*. That is, restricting attention to the sentences of $\mathcal{L}$, a theory $\mathcal{T}$ is called a *complete semantic theory for* $\mathcal{L}$ iff its theorems state what the sentences of $\mathcal{L}$ mean.

In what follows, I am primarily interested in complete semantic theories for a specific fragment of English. More precisely, I am interested in complete semantic theories for a formal language which serves as a model for a fragment of English, namely the first-order arithmetical language $\mathcal{L}^{Ar}$. Now, it is useful to present the notion of a complete semantic theory for $\mathcal{L}^{Ar}$ in analogy to the notion of a materially adequate theory of truth for $\mathcal{L}^{Ar}$. Following Dummett (1975: 7) and using the expression $M(t, S^*)$ for "t means that S^*", I will call a sentence of the form

$$M(t, S^*)$$

an *M-sentence for S* iff t is a canonical name for a sentence S of $\mathcal{L}^{Ar}$. Employing this expression, one can characterise the notion of a complete semantic theory for $\mathcal{L}^{Ar}$ as follows:

---

[18]   A large amount of work by truth-theorists is independent of these claims by Davidson and Dummett. In this work, truth-theorists try to extend simple theories like $\mathcal{T}_+$ so that they become applicable to more complex formal languages, which can serve as models for larger fragments of natural languages. However, this additional work does not play a role in what follows.

Complete Semantics A theory $\mathcal{T}$ is called a *complete semantic theory for* $\mathcal{L}^{Ar}$ iff it implies a true M-sentence for every sentence of $\mathcal{L}^{Ar}$.

The theorems of a complete semantic theory for $\mathcal{L}^{Ar}$ explicitly state what all the sentences of $\mathcal{L}^{Ar}$ mean.

According to Davidson, a semantic theory should yield an explanation of how a finite being can acquire a certain infinite amount of semantic knowledge. One can use the notion of a complete semantic theory to make this project of Davidson's more precise. The idea would be to identify the infinite amount of semantic knowledge which is to be explained with the information expressed by the sentences of a set containing a true M-sentence for every sentence of $\mathcal{L}^{Ar}$. A theory which implies a true M-sentence for every sentence of $\mathcal{L}^{Ar}$, i.e. a complete semantic theory for $\mathcal{L}^{Ar}$, is surely a plausible candidate for a theory which meets Davidson's desiderata.[19]

4.2.5   Refined Truth-Theory

How might a theory of truth be of any help if one aims at a complete semantic theory? A materially adequate theory of truth yields T-sentences, not M-sentences. Its axioms do not express knowledge whose possession puts one in a position to know what every sentence of the object language means.

Here, we face a variant of the individuation problem that was mentioned in 4.2.2, namely the problem that truth conditions seem to be individuated more coarsely than meanings. For example, to know that "Peter ist hungrig" is true iff Peter is hungry does not seem to be a strong enough basis for knowing that "Peter ist hungrig" means that Peter is hungry. Knowledge of what is expressed by T-sentences does not suffice for knowledge of what is expressed by M-sentences.

Consider the following *truth-meaning rule* (see Larson & Segal 1995: 40 n. 15, Kölbel 2001: 619-20, Ludwig 2002: 159, Hoeltje 2012: sec. 2.4.4):

$$\frac{T(t) \leftrightarrow \varphi}{M(t, \varphi)}$$

where t and φ are schematic letters for terms and sentences. To some extent, this rule measures the distance between the standard theory $\mathcal{T}_+$ for $\mathcal{L}^{Ar}$ and a complete semantic theory for $\mathcal{L}^{Ar}$: if one adds this rule to the canonical calculus $\mathcal{R}_+$, then one obtains a calculus with which one can derive, from the axioms of $\mathcal{T}_+$, a true M-sentence for every sentence of $\mathcal{L}^{Ar}$ (while one cannot derive any false sentences). However, this rule is not truth-preserving. For example, while "snow is white" is true iff grass is green, "snow is white" does not mean

---

[19]   It could be objected that one might have the knowledge expressed by the axioms of a complete semantic theory without being in a position to have the knowledge expressed by its theorems. To deal with this problem, one has to consider whether the pertinent implicational relations between axioms and theorems have the potential to transfer knowledge. But I will not dwell on this here (cp. Hoeltje 2012: sec. 1.2.2 and 2.4.2).

that grass is green. The problem is that it seems to be rather unlikely that our semantic knowledge might be based on a rule which is not truth-preserving.[20]

How can this gap between the information expressed by interpretive T-sentences and the information expressed by true M-sentences be closed? I have mentioned two strategies for dealing with the individuation problem. According to the first strategy, truth conditions are coarsely individuated and only parts of sentential meanings. Transferred to the present context, the strategy suggests that a materially adequate theory of truth should be supplemented by axioms which induce additional information about the meanings of the sentences of the object language. There would be two pieces of information about a sentence denoted by t: the piece of information given by the derivable T-sentence $T(t) \leftrightarrow S_1^*$ and some further piece of information given by another sentence $S_2^*(t)$. The idea would be that the two sentences $T(t) \leftrightarrow S_1^*$ and $S_2^*(t)$ together imply the true M-sentence $M(t, S_1^*)$. One might then say that a coarsely individuated truth condition, expressed by an interpretive T-theorem, constitutes a part of the meaning that is ascribed by a true M-sentence. However, as far as I know, no proposal along these lines has been worked out.

According to the second strategy, truth conditions are individuated as finely as meanings. There should then be a plausible semantic theory containing the resources to introduce an appropriate notion of a truth condition. Now, although the idea that a materially adequate theory of truth already constitutes a complete semantic theory is not viable, there have been proposals to modify a materially adequate theory of truth into a theory which fares better in this regard. The idea is to strengthen the axioms of an appropriate materially adequate theory of truth by embedding them into a suitable operator. This idea has its root in Foster's (1976: 20) proposal that one should strengthen a materially adequate theory of truth by prefixing to the conjunction of its axioms a sentential operator like the following: "there is a materially adequate theory of truth such that", a proposal adopted by Davidson (1976: 174) (cp. Wiggins 1997: 19). The idea is that if we know not only the interpretive T-sentences but also that they are implied by an appropriate theory, then we can infer that they are interpretive.

In contrast to Foster's proposal, I adopt a version in which each axiom is embedded in an appropriate sentential operator. A proposal along these lines has been made by Higginbotham (1992).[21] The version presented below is obtained by transferring a proposal of Fine (2010) to the framework proposed by Davidson.[22]

---

[20]   On the other hand, this rule might be relevant for providing a *causal* explanation of our semantic beliefs. Consider the following challenge: specify a possible mechanism which produces, for every sentence S of $\mathcal{L}^{Ar}$, a belief that is expressed by a true M-sentence for S. Such a mechanism might be given by the standard theory of truth $\mathcal{T}_+$ and the canonical calculus $\mathcal{R}_+$ supplemented by the truth-meaning rule. But I will not discuss this alternative project here. As will be explained in 4.4.1, such causal accounts of understanding do not seem to be targets of Dummett's manifestation argument.

[21]   See especially p. 9. For a critique, see Soames 2008.

[22]   Fine sharply contrasts his theory with a Davidsonian theory and accuses the latter of speaking about semantic theories rather than about semantic facts. I will not discuss here whether this criticism is justified.

Applied to the simple object language $\mathcal{L}^{Ar}$, the idea is to strengthen all the axioms of the standard theory $\mathcal{T}_+$ by prefixing a sentential operator which expresses semantic necessity: the operator $\Box_s$, corresponding to the phrase "it is a semantic requirement that". The resulting theory will be referred to by $\mathcal{T}_+^{\Box_s}$. For example, the new axiom for $\overline{N}$,

$$\Box_s(\forall a\,(T^Q(\overline{N}a) \leftrightarrow Na)),$$

states that it is a semantic requirement that $\overline{N}$ is true of some object iff this object is a natural number. Analogously, the new axiom for negation,

$$\Box_s(\forall a\,(Qa \rightarrow (T^Q(\neg a) \leftrightarrow \neg T^Q(a)))),$$

states that it is a semantic requirement that the negation of a sentential fragment is true of a tuple of objects iff the sentential fragment is not true of this tuple of objects. The idea is to account for our semantic knowledge by invoking a finite amount of knowledge to the effect that certain universally quantified biconditionals about being true (of something) express semantic requirements.

If one were to develop this proposal into a serious semantic theory, then one would have to do two things: first, one would have to say more about the notion of a semantic requirement, and, second, one would have to indicate how the knowledge expressed by the true M-sentences might be obtained from knowledge of the axiomatically given semantic requirements. Here I only want to indicate a possibility for dealing with the second of these tasks. (For the first one, see Fine 2010: 65-8.) The idea is that the rules of $\mathcal{R}_+$ *respect* semantic requirements: whenever a rule from $\mathcal{R}_+$ allows one to move from sentences $S_1, \ldots, S_n$ to a sentence S, then the sentences $\Box_s S_1, \ldots, \Box_s S_n$ imply the sentence $\Box_s S$. Suppose this were true. Then, for every derivation in $\mathcal{R}_+$ of a T-sentence $T(t) \leftrightarrow S^*$ from some axioms A_1, ..., A_n of $\mathcal{T}_+$, the axioms $\Box_s A_1, \ldots, \Box_s A_n$ of $\mathcal{T}_+^{\Box_s}$ would imply the sentence $\Box_s(T(t) \leftrightarrow S^*)$. That is, for every sentence S of $\mathcal{L}^{Ar}$, the theory $\mathcal{T}_+^{\Box_s}$ would imply a sentence of the following form:

$$\Box_s(T(t) \leftrightarrow S^*),$$

where t is a canonical term for S, and where S and S^* have the same meaning.

Now, the idea is that such $\Box_s$-T-sentences directly imply the corresponding M-sentences: if it is a semantic requirement that t is true iff S^*, then t means that S^*. This might be put by saying that the *refined truth-meaning rule* is valid:

$$\frac{\Box_s(T(t) \leftrightarrow \varphi)}{M(t, \varphi)}$$

But then $\mathcal{T}_+^{\Box_s}$ would be a complete semantic theory for $\mathcal{L}^{Ar}$: for every sentence S of $\mathcal{L}^{Ar}$, the theory $\mathcal{T}_+^{\Box_s}$ would imply a true M-sentence for S.

I will not try to substantiate this result here. To this end, I would have to defend the claim that the rules of $\mathcal{R}_+$ respect semantic requirements, and this task is beyond the scope of this book. I only wanted to indicate what I take to be the most promising version of a truth conditional semantics for a simple formal language like $\mathcal{L}^{Ar}$, namely a theory resulting from a suitable materially adequate theory of truth for $\mathcal{L}^{Ar}$ by embedding each of its axioms into an appropriate intensional operator. This will suffice for the following discussion of the manifestation argument.

Before closing this section about truth conditional semantics, I would like to return to the individuation problem mentioned above. Suppose that t_1 and t_2 are terms for sentences S_1 and S_2, and suppose that S_1 means that S^*; i.e., suppose that the formal sentence $M(t_1, S^*)$ is true. Under which condition do the sentences S_1 and S_2 have the same meaning? Obviously, they have the same meaning iff S_2 also means that S^*; i.e., they have the same meaning iff the formal sentence $M(t_2, S^*)$ is also true. But then, given that a sentence means that S^* iff it is a semantic requirement that it is true iff S^*, it follows that the formal sentence $\Box_s(T(t_1) \leftrightarrow S^*)$ is true iff the formal sentence $\Box_s(T(t_2) \leftrightarrow S^*)$ is true. This, in turn, suggests that there is a suitable notion of a truth condition according to which two sentences have the same meaning iff they have the same truth condition: to have the meaning that S^* is nothing but to satisfy the semantic requirement to be true iff S^*. This can be taken as a suitable explication of the claim that sentential meanings are truth conditions.

4.3   Understanding and Knowledge

My critical discussion of the manifestation argument will centre around the manifestation thesis, which states that understanding is capable of showing itself in suitable linguistic behaviour (4.4 & 4.5). In this section, I will provide crucial background information about the pertinent notion of understanding. In particular, I will discuss the thesis U-KTC that sentential understanding involves knowledge of a truth condition if sentential meanings are truth conditions (4.3.1). In addition, I will clarify Dummett's view about the claim that sentential meanings are truth conditions and the claim that sentential understanding involves knowledge of a truth condition (4.3.2). Questionable assumptions about the relation between these claims and the manifestation argument have resulted in a widespread underestimation of the power of this argument.

4.3.1   Ideal Understanding

How is the concept of *understanding* to be understood? Dummett claims that sentential understanding involves knowledge of a truth condition if sentential meanings are truth conditions, and he claims understanding to be manifestable:

Understanding and Knowledge of Truth Conditions (U-KTC)
> For every English sentence S and for every person a who understands S, if the meaning of S is its truth condition, then a's understanding of S involves knowledge of the truth condition of S.

Manifestation Thesis (MT) For every English sentence S and for every person a who understands S, a's understanding of S is fully manifestable in a's behaviour relating to utterances of S.

I will consider two related problems for these premises and propose a solution to the problems that elucidates the central concept of understanding as it figures in Dummett's argument.

Given the assumption M-TC, the premise U-KTC provides a connection between sentential understanding and knowledge of truth conditions. Furthermore, this relation is important for the argumentation, since the manifestation thesis MT is about understanding while the knowability thesis KT is about knowledge of truth conditions. If understanding and knowledge of truth conditions were not linked, an opponent of Dummett's argument could grant that someone's *understanding* has to be fully manifestable, and also that someone has to be able to recognise the truth of a true sentence if her *knowledge of its truth condition* is fully manifestable. If sentential understanding did not involve knowledge of a truth condition, then it would neither follow that someone's understanding a true sentence S implies the possibility to recognise S as true, nor would it follow that her knowledge of the truth condition of S implies this possibility.

Dummett accepts U-KTC because he believes that understanding involves knowledge of meaning.[23] However, the manifestation argument does not depend on such a view, and there are meaning theories which do not incorporate it. What the manifestation argument does depend on is the premise that *truth conditional semanticists* are committed to the claim that sentential understanding involves knowledge of a truth condition. Now, on the face of it, this premise is reasonable. Semanticists that identify sentential meanings with truth conditions seem to be committed to the claim that someone who understands a sentence S is in a suitable propositional mental state to the effect that S satisfies some truth condition. Furthermore, the most obvious candidate for such a state seems to be a state of knowledge. For example, a proponent of refined truth-theory (see 4.2.5) seems to be committed to claiming that understanding the sentence "$2 + 2 = 4$" involves knowledge that this sentence is semantically required to be true iff $2 + 2 = 4$. Consequently, the thesis U-KTC appears to be plausible.

On closer inspection, however, there might be reason to doubt that sentential understanding always involves knowledge of a truth condition, even if sentential meanings are truth conditions. According to Pettit (2002), there are Gettier cases in which knowledge of meanings and knowledge of truth conditions fail, while understanding does not. He gives the following example (p. 519-20): A tourist, travelling in Germany and moderately competent with respect to German, does not understand the German word "Krankenschwester". A person that appears to be a reliable informant answers the tourist's question about its meaning by (correctly) saying: "it means *nurse*". Unbeknownst to the tourist, however, the person does not understand the sentence he has uttered. This person once overheard this sentence and utters it whenever a tourist asks him anything. In this situation, Pettit argues, the tourist understands the word

---

[23]   In fact, he (1973b: 216-7) *identifies* understanding with knowledge of meaning. However, as Williamson (2000: 110 n. 4) notes, this identification is questionable in certain cases since knowledge of meaning, unlike understanding, does not always seem to be compositional (cp. Dummett 1976: 36).

"Krankenschwester" but does not know its meaning. It would then follow that the tourist understands a sentence like "sie ist eine Krankenschwester" without knowing its truth condition. Such cases seem to be problematic for Dummett's argument. If understanding but not knowledge of truth conditions is fully manifestable, and if the claim that someone has the ability to recognise the truth of a true sentence only follows from the claim that his knowledge of its truth condition is fully manifestable, then the argument breaks down.

If one accepts Pettit's description of his example,[24] there seem to be two possible responses that an adherent of the manifestation argument can give. The first option is to replace the references to *knowledge* of truth conditions in the argument by references to propositional mental states that are less elusive than knowledge. (This fits to the moral that Pettit wants his thought-experiment to have; cf. p. 548). The second possibility is to reformulate the argument in terms of certain *ideal cases of understanding* in the hope that these *do* involve *knowledge* of a truth condition. In my view, the second option is to be preferred. On the one hand, the first option would require a reformulation of the thesis KT that links manifestable knowledge of truth conditions with the possibility to recognise true sentences as such, which is significantly less plausible than KT itself. On the other hand, there is also an objection to the *manifestation thesis* which suggests that one should concentrate on certain ideal cases of understanding. I will now present this objection to MT and indicate what a proponent of the manifestation argument has to do if she wants to work out the second possibility.

Suppose that Peter is competent in English but does not understand the word "gorse". Now, suppose that he overhears a few sentences like

(5) Everyone likes gorse.

in which the word "gorse" is used, although none of these sentences puts Peter in a position to know what gorse is. It is possible, I assume, that Peter then understands the sentence (5), because he is suitably related to competent uses of sentences containing "gorse" (cp. 2.4.2). His understanding of "gorse" is then based other persons' understanding of this word. However, this creates a problem for MT: it is seems questionable that Peter has to have the capacity to fully manifest his understanding of (5) solely by his uses of (5) and his reactions to such uses. Since his understanding obtains in virtue of a suitable (causal) relation between his uses of "gorse" and competent uses of others, it seems plausible to assume that a manifestation of his understanding would have to display this (causal) relation.

How should a proponent of the manifestation argument react to Pettit's counterexample to U-KTC and this counterexample to MT? What I would like to suggest is that the best way to develop the manifestation argument might be based on a notion of *ideal understanding*. The idea would be that by relying on a suitable notion of ideal understanding one can present variants of the

---

[24]   Dummett could, of course, also respond by questioning Pettit's characterisation of the mental state of the tourist either by claiming that the tourist does not understand "Krankenschwester", or by claiming that he knows its meaning.

theses U-KTC and MT which are not undermined by the above mentioned examples (because Peter and the tourist in Pettit's example do not have such ideal understanding).

In more detail, I propose the following modification of the manifestation argument. One assumes that for every (intelligible) English sentence S there is some person who has ideal understanding of S, and one modifies the premises U-KTC and MT by replacing the notion of understanding with the narrower notion of ideal understanding. Then, one still reaches the desired conclusion that if sentential meanings are truth conditions, then every true English sentence is possibly known to be true. The benefit of introducing a notion of ideal understanding would be twofold. First, it makes the resulting version of the manifestation thesis more plausible: although the above example seems to show that not every kind of understanding of a sentence S is manifestable in S-related behaviour, *ideal* understanding of a sentence S is manifestable in such a way. Second, it also makes the resulting version of the thesis U-KTC more plausible: although not all states of understanding involve knowledge of a truth condition, states of *ideal* linguistic understanding do involve such knowledge (if sentential meanings are truth conditions).[25]

The crucial questions for a proponent of the manifestation argument then are, of course, what exactly ideal understanding amounts to, and whether it can be assumed plausibly that every (intelligible) sentence can be understood in the relevant ideal way. I will largely set aside these questions, but I would like to note that for Dummett's semantic argument against classical logic DSA (see 4.1.2), it suffices to answer these questions with respect to a suitable fragment of English, such as the fragment of first-order arithmetical sentences. (The fragment has to be such that the argument resulting from DSA by replacing references to English with references to this fragment is no less plausible than DSA itself. It seems plausible to assume that this is true with respect to the fragment of first-order arithmetical sentences.) If classical logic does not hold for arithmetical arguments (because classical truth conditional semantics is inappropriate for arithmetical sentences), then classical logic fails because it is a theory which incorporates commitments about arithmetical arguments (see 2.1.1).

---

[25] If I understand Dummett correctly, then he does not want to develop his argument in precisely this way. He also invokes the idea that a meaning theory only has to yield an *idealised* account of understanding and knowledge of meaning:

> [A] meaning-theory aims at providing, not a faithful representation of a speaker's linguistic knowledge, but a systematisation of it (Dummett 1991: 103).

However, he does not require that every sentence can be understood in an ideal way. (A possible reason for this is Dummett's thesis (1991: 85-6) that there are some words of natural languages which *no one* fully understands.) But the question is how such an approach can be combined with suitable counterparts of U-KTC, MT, and KT such that the conclusion of the manifestation argument is still forthcoming. (In his early presentations, Dummett (1973b: 217, 1976: 36, 1977: 259) employs the notion of *implicit knowledge of meaning*, which is related to the notion of *ideal understanding*. In later publications, Dummett (1991: 95-7, 1993: x-xii, 2007: 562) expresses dissatisfaction with his use of the notion of implicit knowledge of meaning because it seems to suggest that it is a kind of knowledge which can be turned into explicit verbalisable knowledge given suitable prompting.)

Now, arguably, with respect to first-order arithmetical sentences, it is less problematic to assume that they can be understood in a way such that Gettier cases and cases of dependent understanding can be set aside. One might claim that an *ordinary* understanding of such sentences, an understanding that most persons have acquired as children by observing and practicing arithmetic and by observing and practicing the use of the standard logical constants, can be identified with *ideal* understanding.

It has to be admitted that even with respect to such restricted linguistic resources, it is a substantial task to work out an account of ideal understanding and to defend the claim that the relevant sentences can be understood in the explicated ideal way. Nevertheless, I think that an approach along these lines constitutes a promising possibility to improve the manifestation argument. In what follows, I will assume that such an account works. However, I will leave the restrictions to ideal understanding implicit: henceforth, understanding is always meant to be ideal understanding.

4.3.2   Understanding and Knowledge of Truth Conditions

I will now discuss the claim that sentential meanings are truth conditions and the claim that sentential understanding involves knowledge of a truth condition. Dummett's views about the relation between meanings and understanding on the one hand and truth conditions and knowledge of truth conditions on the other hand are rather complex and underwent some changes throughout his career. I would like to comment on one aspect of these views in the hope of showing that two common criticisms of the manifestation argument are based on two (complementary) misinterpretations of Dummett's views. In each case Dummett's opponents claim that the manifestation argument depends on a strong thesis about the relation between understanding and knowledge of truth conditions. But since Dummett's argument does not depend on such a thesis, these interpretations make the argument appear less powerful than it is.[26]

According to the first interpretation, Dummett's argument is based on the assumption that someone's understanding of a true sentence S is manifestable only if it is possible for her to know that S is true. As an example, consider the following assertion by Boulter:

> [A]ny form [the manifestation argument] is given will have to rely on the defining assumption of semantic anti-realism, namely, that understanding cannot outstrip recognitional capacities (Boulter 2001: 325).

Here, a recognitional capacity is a capacity to recognise whether the truth condition of the relevant sentence obtains (p. 326). According to Boulter, Dummett

---

[26]   For what follows, one should keep in mind that Dummett *identifies* understanding and knowledge of meaning (see 4.3.1 n. 23). He therefore treats the assumption that sentential meanings are truth conditions as equivalent to the assumption that sentential understanding amounts to knowledge of a truth condition, while my reconstruction of the manifestation argument only makes use of the weaker assumption that sentential understanding *involves* knowledge of a truth condition if sentential meanings are truth conditions.

is committed to the thesis that someone's understanding of a true sentence S necessitates her ability to recognise that S is true. The background for such an interpretation seems to be the assumption that Dummett's argument should be interpreted as being based on the *premise* that sentential meanings are truth conditions so that sentential understanding involves knowledge of a truth condition. This interpretation is also adopted, for example, by Byrne (2005: 103), who thinks that proponents of "the least radical version" of the argument "hold that speakers' understanding of declarative sentences incorporate [...] *evidence-constrained* truth conditions".

According to the second interpretation, Dummett rejects the thesis that sentential understanding involves knowledge of a truth condition and, correspondingly, the thesis that sentential meanings are truth conditions. More precisely, it is claimed that the conclusion of Dummett's argument states that sentential meanings are not truth conditions:

> What that argument shows, if cogent, is that meaning cannot be taken to consist in truth conditions, on any conception of truth according to which truth is undecidable (Pagin 1998: 167).

Note that truth does not have to be decidable if it is epistemically constrained: the fact that every truth is knowable does not imply that there is a procedure which can be applied to an arbitrary sentence S and whose application puts one in a position to know the truth value of S. According to Pagin, Dummett tries to establish the thesis that sentential meanings cannot be truth conditions irrespective of whether truth is epistemically constrained or not.[27]

Thus, there are two contradictory interpretations of Dummett's views about the relation between sentential meanings and truth conditions and, correspondingly, about the relation between sentential understanding and knowledge of a truth condition. To obtain a plausible interpretation of the manifestation argument, this dispute about the correct interpretation of Dummett's views has to be settled.

Dummett explicates his views about the relation between truth and meaning in 1978a: xxii-xxiii. In particular, he makes it plain that he accepts the claim that sentential meanings are truth conditions under a certain assumption:

> [U]nder any theory of meaning whatever - at least, any theory of meaning which admits a distinction like that Frege drew between *sense* and *force* - we can represent the meaning (sense) of a sentence as given by the condition for it to be true, on some appropriate notion of construing 'true': the problem is not whether meaning is to be explained in terms of truth-conditions, but of what notion of truth is admissible (Dummett 1978a: xxii).

---

[27]   See Pagin 1998: 164-9, and cp. Pagin 2009a: 223 and 2009b: 716-19. A look at early discussions of the manifestation argument might create the impression that Pagin's interpretation is widely shared because Dummett's argument is then often presented as having the conclusion that sentential meanings are not truth conditions (see, e.g., Appiah 1985 and Edgington 1985). However, this is only due to the fact that in these discussions it was an unpronounced presupposition that the relevant notion of truth is epistemically unconstrained (cp. Miller 2002: 355-6 and Byrne 2005: 107-8).

Similarly, Dummett (1977: 258) claims that according to those meaning theories which admit a distinction between sense and force (meaning theories which he (1977: 256) also calls "anti-holistic") sentential meanings can be identified with truth conditions.[28] Since Dummett (1977: 253-6) furthermore takes anti-holistic meaning theories to be correct, it follows that he accepts the thesis that sentential meanings can be identified with truth conditions.[29] Thus, Pagin's interpretation has to be rejected.

However, the fact that Dummett accepts the claim that sentential meanings can be identified with truth conditions does not imply that his manifestation argument depends on this claim. This becomes clear once one notices that Dummett reckons with holistic meaning theories according to which sentential understanding cannot be identified with knowledge of a truth condition, while in his presentations of the manifestation argument he assumes that he is not dealing with such holistic meaning theories:

> A distinction between sense and force is implicit in any thesis such
> as the one which we are considering, that to know the meaning of a
> sentence is to know its truth-condition (Dummett 1976: 39).

In particular, the argument is not based on the premise that *knowledge of the truth condition* of a sentence S is fully manifestable in S-related behaviour; instead, it is only based on the premise that an *understanding* of S is thus manifestable. According to Dummett, the manifestation thesis is compatible with holistic meaning theories:

> [O]ne very radical way of interpreting [Wittgenstein's celebrated
> slogan 'Meaning is use'] is as repudiating altogether any distinction
> between sense and force [...] (Dummett 1976: 38-9).

Consequently, his argument does not depend on the assumption that full manifestability of understanding a true sentence presupposes the ability to recognise its truth; instead, it only depends on the assumption KT which states that full manifestability of knowledge of the truth condition of a true sentence presupposes the ability to recognise its truth. Therefore, the interpretation put forward, for example, by Boulter and Byrne has to be rejected.

The best version of the argument, and a version which seems to be most faithful to Dummett's texts, has a conditional conclusion: *if* sentential meanings are truth conditions, then truth is epistemically constrained. One cannot

---

[28]   In an early presentation of a variant of his manifestation argument, Dummett (1959) did not conceive of the possibility of adopting an epistemically constrained notion of truth in a meaning theory and, therefore, suggested that the meaning of a sentence should not be identified with its truth condition but with its assertability condition (cp. Dummett 1978a: xxi-xxii).

[29]   Dummett does not seem to have a constant opinion about the precise role of the notion of truth in such meaning theories. For example, in his 1978 book (1978a: xxiii), he claims that the notion of truth "will play an essential rôle in the account of the connection between the way in which the meaning of a sentence is given and the use that is made of it", whereas in 1994 (p. 294-7) he expresses his belief that a notion of truth will not play an essential role in (intuitionistic) meaning theories for mathematical sentences, although it will play an important role in meaning theories for empirical sentences.

refute it by refuting the claim that manifestability of understanding does not presuppose the ability to recognise the truth of a true sentence. To refute it, one has to refute one of its premises: MT, U-KTC, or KT, and these do not depend on the assumption that sentential meanings are truth conditions or on the assumption that sentential understanding involves knowledge of a truth condition.

4.4   Understanding and Use

In this section, I will discuss the central thesis of Dummett's argument:

Manifestation Thesis (MT) For every English sentence S and for every person a who understands S, a's understanding of S is fully manifestable in a's behaviour relating to utterances of S.

My primary aim will be to elucidate MT. First, I will discuss the nature of the manifestation relation which might hold between the understanding of a sentence S and potential S-related behaviour (4.4.1). Then, I will deal with possible kinds of S-related behaviour (4.4.2) and legitimate characterisations of it (4.4.3). Finally, I will consider the consequences of combining the manifestation thesis with the temporary assumption M-TC and the premise U-KTC (4.4.4): Does the manifestability of someone's knowledge of the truth condition of a true sentence S by means of her S-related behaviour imply that there is a possible situation in which she knows that S is true (as it is stated in the knowability thesis KT)?

As a preparation, I have to include one clarifying observation about the relation between the manifestation thesis and general questions about mental phenomena and their behavioural manifestations, namely that Dummett's claim that understanding is manifestable is *not* a corollary of a general thesis to the effect that any mental state has to be manifestable:

> Philosophical behaviourism is, I think, the view that there is nothing to any mental state or process but the behaviour that manifests it. There is nothing to pain but pain-behaviour; there is nothing to fear but frightened actions. I have certainly never maintained a general doctrine of this form (Dummett 2007: 558-9).

Rather, Dummett's manifestation thesis is specifically about linguistic understanding.

This view of the exceptional nature of linguistic understanding is related to Dummett's conception of languages. According to Dummett, the sentences of a language are essentially tools for performing acts of various kinds. They are types of meaningful sounds uttered for all kinds of reasons. A theory of meaning, according to Dummett, has to explain these performances; it has the goal of making the complex of behaviour related to these speech acts intelligible (see 4.4.1). This conception of languages should not be misinterpreted as involving the claim that a language is primarily an instrument of communication as opposed to a vehicle for thought (cp. Dummett 1991: 103). The present view only states that the sentences of a language are essentially things which are *used* (in all sorts of speech acts). It does not presuppose that language use is a "purely practical" activity (see Dummett 1991: 93-5), and it does not presuppose that sentences are only used for one special purpose, such as communication.

Dummett's conception of languages has to be contrasted with strict cognitivist conceptions such as the one put forward by Chomsky (1986) (cp. Boeckx 2010: 5-6). According to Chomsky, a language is a part of a human mind/brain, which develops in (nearly) every child according to an innate mechanism given a modest amount of triggering events. According to such a view, activities of language *use* are only indirectly related to the mental/neurophysiological entities called "languages". Accordingly, there seems to be as little reason for believing that these mental entities are tied to certain forms of behaviour as there is with respect to pain or fear. As one of Chomsky's followers puts it:

> The entire question of the external 'function' of language may be of unclear significance and does not seem to provide a useful heuristic for understanding it (Hinzen 2006: 138).

I will not try to resolve this dispute here. For the sake of the argument, I will simply assume that Dummett's conception of languages is correct. Criticisms of the manifestation argument which are based on strict mentalist conceptions of languages are beyond the scope of this chapter.[30]

4.4.1 Dummettian Accounts of Understanding

What does it mean to say that someone's understanding of a sentence S can be fully manifested in her S-related behaviour? The purpose of this subsection is to suggest the direction in which an answer to this question is to be sought.

Dummett interprets the manifestation thesis as stating that it is possible to give a certain *account* of someone's understanding of a sentence S by describing her possible S-related behaviour (see below). These accounts will be called *Dummettian accounts of understanding*. I will assume that such accounts are given by lists of open sentences in which possible S-related behaviour is described. The entries of these lists will be called *Dummettian descriptions of S-related behaviour*. As an example, consider a person who understands the following sentence:

(6) 28 is a perfect number.

This sentence is true because the following sentences are true:

(7) The proper divisors of 28 are 1, 2, 4, 7, 14;

(8) $1 + 2 + 4 + 7 + 14 = 28$.

(A *perfect number* is a natural number which is the sum of its proper divisors.) Here is an example of a Dummettian description of (6)-related behaviour:

(9) x establishes (7) and (8) by calculation and then assertively utters (6).

---

[30]   For a criticism of Dummett's argument from a Chomskian viewpoint, see Burgess 1984.

The idea is that if someone acts in the way described in (9), then this description might be part of an account of her understanding of (6) in terms of her (6)-related behaviour. Of course, the fact that someone assertively utters (6) after having established (7) and (8) by calculation does not *imply* that she understands (6), and the fact that someone understands (6) does not *imply* that she assertively utters it after having established (7) and (8) by calculation. But this does not contradict the proposal that someone's understanding of (6) might partly be *accounted for* by using (9). Note that not only the act of assertively uttering (6) is described in (9); there are also descriptions of previous acts, which can be seen as justifying the assertive utterance.

The kind of (6)-related behaviour that is described in (9) might be called *assertion based on conclusive verification* (see Dummett 1976: 41). (More specifically, one might speak about assertion based on mathematical proof.) Now, there are numerous further kinds of S-related behaviour and, accordingly, countless types of Dummettian descriptions of such behaviour (see 4.4.2 and 4.4.3). The general idea behind the manifestation thesis is that for every English sentence S and for every person a who understands S, there are a number of Dummettian descriptions of S-related behaviour which yield an account of a's understanding of S. Four inter-related questions arise in this context:

Q_1 What is the nature of a Dummettian account of understanding?

Q_2 What kinds of behaviour may be treated in such an account?

Q_3 What are legitimate descriptions of such behaviour?

Q_4 Is there a Dummettian account for every state of understanding?

To begin with, I will deal with Q_1. The question Q_2 will be the topic of 4.4.2, and the question Q_3 will be discussed in 4.4.3. I will not pursue Q_4 in this section; the plausibility of the manifestation thesis will concern me in 4.5.

What is the nature of a Dummettian account of sentential understanding? According to what Shieh (1998: 37-42) has called the *standard interpretation*, the manifestation thesis is a claim about epistemological presuppositions of successful communication. According to this reading, the manifestation thesis is meant to express the assumption that linguistic communication can only be successful if each participant can come to know how the other participants understand the sentences they use on the basis of observation of how they use them. It would then be an epistemological thesis about the basis of knowledge of how speakers understand the sentences they use, i.e. a thesis which states that knowledge of understanding is based on observation of use.[31] (The standard interpretation has a second component which I will mention in 4.4.3.)

Shieh (1998: 42) rejects the standard interpretation because it involves the assumption that "Dummett's account of language depends on epistemological

---

[31] Shieh (1998: 34 n. 1) cites the following proponents of this interpretation: Appiah (1985, 1986), Burgess (1984), Campell (1982), Devitt (1983), Edgington (1985), George (1984), McDowell (1981), McGinn (1980), Tennant (1987), Weir (1985), Wright (1993). Another adherent of this interpretation is Pagin (see 2009b: 717, and cp. 2008).

'results'," while Dummett takes "the theory of meaning as the only part of philosophy whose results do not depend upon those of any other part." I am not entirely sure that this is a convincing response since Dummett's account of language surely has an epistemological character (see 4.1.2 & 4.3.1). Nevertheless, I will also not rely on the standard interpretation. As far as I can see, it only introduces a superfluous complication into Dummett's argumentation. (Note that the pertinent question is *not* whether Dummett believes that successful communication requires such knowledge of understanding. The pertinent question is whether Dummett's manifestation argument is based on such a thesis.) Dummett explains the kind of account of linguistic understanding he has in mind as follows:

> We can, in general, make some unfamiliar human activity - say, a social function or ceremony - intelligible without either circularity or anything resembling a causal theory [...]. To do so, we describe the practice and the institutions that surround the practice, and then it becomes intelligible as an activity of rational agents. And that is all the understanding that we seek of language (Dummett 1991: 92).

A Dummettian account of linguistic understanding is an account which makes the practice of performing and reacting to speech acts intelligible as an activity of rational agents (cp. Dummett 1973a: 92 and 1976: 37). It is meant to be an account of what a given state of understanding (or knowledge of meaning) *consists in*, or an account of *what it is* to be in such a state (see Dummett 1976: 70-1 and 1978b: 101).

This suggests, I think, that the manifestation thesis is not about an actual or possible basis of someone's knowledge of someone else's understanding. Rather, it only states that someone's understanding of a sentence can be made intelligible in terms of how this person is prepared to use this sentence and to react to uses of this sentence by others. Whether or not such a description of possible behaviour also delivers an account of how one may come to know about someone's understanding of a sentence is then a separate question.

It might be helpful to consider the analogy between languages and games, which has been invoked by Dummett in his presentation of the manifestation argument. As a justification for the semantic manifestation thesis that meaning is exhaustively determined by use (see 4.1.2), Dummett presents the following consideration:

> The reason is that the meaning of a statement consists solely in its rôle as an instrument of communication between individuals, just as the powers of a chess-piece consist solely in its rôle in the game according to the rules (Dummett 1973b: 216).

Dummett compares the meaning of a sentence with the significance of a chess-piece, i.e. with the rules which state how the chess-piece can be moved.[32] Similarly, one might compare sentential understanding with grasping the significance of a chess-piece.

---

[32]   There is one problematic aspect of this formulation (cp. the introduction to 4.4):

Transferred to the case of chess, the manifestation thesis would amount to the claim that someone's grasp of the significance of a bishop, say, is fully manifestable in how he uses the bishop and reacts to such uses. And here it seems indeed quite plausible to say that one can account for someone's grasp of the significance of the bishop by describing him as moving the bishop only diagonally. Furthermore, such an account may be part of a larger account (which also speaks about the other chess-pieces and specifies what counts as winning) that makes the practice of playing chess intelligible. Now, to say that there is an account of someone's grasp of the significance of chess-pieces by which the activity of playing chess can be made intelligible is independent of the claim that someone can obtain knowledge about someone else's grasp of the significance of a chess-piece by observing how he uses this chess-piece. The way that Dummett invokes the analogy between language and chess thus also suggests that Dummettian accounts of understanding are not accounts of knowledge of understanding but accounts of the nature of understanding itself.[33]

When Davidson asks about a possible basis of someone's semantic knowledge (see 4.2.4), he emphasises that such a basis has to be finite. A related point is made by Dummett with respect to accounts of understanding. This topic has been brought up early in the discussion of Dummett's manifestation argument. Prawitz addresses this issue in relation to the semantic manifestation thesis: might it be that infinitely many possible uses are needed to fix the meaning of a sentence?

> In the formulation above ["the meaning of a sentence in a language is determined (at least) by the total possible use in this language"] the principle leaves open the possibility of sentences whose meaning is never exhausted by any finite portion of the actual use in the language. It is possible that although each ingredient of the meaning of a sentence is capable of showing itself in some use in the language, there are sentences such that no finite use in the language can fully determine their meaning (Prawitz 1977: 10).

Suppose that there are infinitely many independent aspects of the meaning of a sentence; i.e., suppose that there are countably many yes-no questions concerning the meaning of a sentence S such that no answers to any finite number of them determines an answer to any other. It may be assumed that the meaning of S is determined by an infinite sequence of answers to these questions. However, even if the answer to each question is determined by a finite number of possible uses, there is no finite number of possible uses that determines the answers to all the questions together.

> Dummett seems to assume here that linguistic behaviour always consists in *communicative* speech acts, which is a rather implausible assumption. Fortunately, the analogy between languages and games does not depend on this assumption. The basic idea of the analogy is only that the meaning of a sentence, like the significance of a pawn in a game, can be exhaustively given by saying *what can be done with it.*

[33] Heck (2007: 533) reaches the corresponding verdict about the semantic manifestation thesis when he calls it a metaphysical rather than an epistemological doctrine. Similarly, Shieh (1998: 33) speaks about a *conceptual* as opposed to an epistemological motivation for the manifestation thesis.

Prawitz (1994: 84-5) repeats this issue with respect to understanding and the (epistemic) manifestation thesis, and, in his reply, Dummett (1994: 293) makes it clear that he believes that someone's understanding of a sentence S can be fully manifested by finitely many uses and reactions to uses of S (see also Dummett 1993: xii-xv). According to Dummett, there has to be a 'finite account' of someone's understanding of a sentence S, i.e. an account in which a finite portion of her possible S-related behaviour is described.

One may note that Davidson's finiteness requirement is, in a sense, much stronger than that attributed to Dummett here: while Davidson demands a finite basis of someone's complete semantic knowledge, I have presented Dummett here only as demanding a finite account of each item of someone's semantic knowledge, i.e. a finite account of each instance of someone's sentential understanding. Now, I think that Dummett also accepts the stronger thesis that there is a finite account for the totality of someone's linguistic understanding (see Dummett 1994: 292). But since this stronger thesis does not play a role in what follows, I will set it aside here.

4.4.2   Kinds of Manifesting Behaviour

Which kinds of S-related behaviour have to be considered if one wants to give an account of someone's understanding of a sentence S? I have already mentioned one example of S-related behaviour: one can partly manifest one's understanding of S by displaying that one accepts S (e.g. by assertively uttering it) on the basis of a conclusive verification of S. Somewhat complementary, one should also mention behaviour which displays that one rejects S on the basis of a conclusive falsification of S. For example, someone might partly show her understanding of the sentence "8 is a perfect number" by reacting in the following way to the corresponding yes-no question: she establishes the sentences "the proper divisors of 8 are 1, 2, 4" and "$1 + 2 + 4 = 7$" by calculation, and then she shakes her head. Signs of acceptance on the basis of conclusive verification and signs of rejection on the basis of conclusive falsification seem to be equally important examples of behaviour which manifest someone's understanding of sentences.

In addition, there are instances of S-related behaviour which come into mind particularly when one considers manifestations of understanding of *mathematical* sentences: linguistic acts which someone performs when she presents a mathematical argumentation. Clearly, not all such acts are outright assertions. One may cite at least two additional kinds of such S-related behaviour: (i) S might be inferred from some conditionally asserted sentences (and then itself be conditionally asserted); (ii) some sentence might be inferred from conditionally asserted sentences including S (and then itself be conditionally asserted). Uses of a mathematical sentence S in presentations of mathematical argumentations are surely important instances of S-related behaviour which display one's understanding of S.

Now, there are all further kinds of S-related behaviour which might be mentioned in an account of someone's understanding of S. In particular, there are numerous kinds of S-related behaviour in which the relevant person does not utter S but in which she responds to an utterance of S in a certain way. (Think about a person who stops eating an apple in reaction to an assertive utterance of the sentence "the apple is poisonous". This seems to be a prime example of behaviour which manifests her understanding of the uttered sentence.) Dummett (1976: 41) explicitly endorses the thesis that S-related behaviour of the mentioned kinds and of several further kinds may figure in an account of someone's understanding of S. In particular, he does *not* claim that there is a small class of types of S-related behaviour such that every person can manifest her understanding of S by performing behaviour of one of these types. I conjecture that some commentators of Dummett's argument have thought otherwise because they conflated the question of which kinds of behaviour can account for someone's *sentential understanding* with the question of which kinds of behaviour can account for someone's *knowledge of a truth condition* (cp. 4.3.2). I will return to the possibilities of manifesting one's knowledge of a truth condition in 4.4.4.

4.4.3   Descriptions of Manifesting Behaviour

What are legitimate descriptions of S-related behaviour? By way of motivation, consider the proposal that all descriptions of possible S-related behaviour are legitimate and that one can account for someone's understanding of a sentence S by presenting a list L of descriptions of S-related behaviour such that the following holds: the fact that a person satisfies the entries of L implies that she understands S. According to this proposal, the manifestation thesis would be implied by the following thesis:

Manifestation Thesis$^?$ (MT$^?$) For every English sentence S and every person a who understands S, there is a list L of descriptions of S-related behaviour such that the following holds: the fact that a satisfies the entries of L implies that a understands S.

According to this proposal, *every* description of possible S-related behaviour is allowed to be an entry of the list L.

The problem with this proposal is that it turns the manifestation thesis into a triviality. To see this, consider a description of S-related behaviour like the following:

(10) x uses the sentence "28 is a perfect number" with understanding.

The fact that someone satisfies (10) obviously implies that she understands the sentence "28 is a perfect number". Consequently, the proposal fails. Dummett does not try to refute classical truth conditional semantics on the basis of such trivialities (cp. Heck 2007: 539-41). Adequate accounts of understanding have to be *substantive*, while nothing substantive is said about someone's understanding of the sentence "28 is a perfect number" if it is only said that she uses this sentence with understanding.

The question therefore arises as to which descriptions of possible behaviour have the potential to yield a substantive account of sentential understanding.

According to the second part of the *standard interpretation* of the manifestation thesis (see 4.4.1 for the first part), these descriptions are physicalistic descriptions of dispositions to assent to or dissent from sentences under suitable promptings (see Shieh 1998: 38). According to this interpretation, the manifestation thesis requires that for every person a and every sentence S there is a list of physicalistic descriptions of dispositions to assent to or dissent from S which yield an account of a's understanding of S. Now, I have already expressed my reservations about the idea that Dummett is committed to the view that only S-related behaviour of a small number of types (like behaviour which expresses assent to S or dissent from S) may figure in an account of someone's understanding of S. Furthermore, Dummett is not committed to the thesis that the relevant behaviour has to be described in physicalistic terms. As Shieh (1998: 42-3) notes, Dummett (1973a: 614) explicitly denies that such descriptions can yield a satisfactory theory of language.

I will not try to present a proposal about which types of descriptions of S-related behaviour have the potential to yield a substantive account of someone's understanding of S. (For a suggestion, see Shieh 1998: 47-52.) I assume that substantiveness is one of the desiderata for legitimate Dummettian descriptions of S-related behaviour. However, there are further desiderata. I will mention one such. Suppose that unbeknownst to Peter, there is exactly one thing which makes Mary angry: establishing the sentences "the proper divisors of 28 are 1, 2, 4, 7, 14" and "$1 + 2 + 4 + 7 + 14 = 28$" by calculation. Now, consider the following two sentences:

(11) x establishes the sentences "the proper divisors of 28 are 1, 2, 4, 7, 14" and "$1 + 2 + 4 + 7 + 14 = 28$" by calculation, and then x assertively utters "28 is a perfect number".

(12) x does something which makes Mary angry, and then x assertively utters "28 is a perfect number".

I would like to suggest that (11) has the greater chance of being part of a plausible account of Peter's understanding of "28 is a perfect number" than (12). According to this idea, a Dummettian account of Peter's understanding does not only have to be substantive; it also has to be framed in terms accessible to Peter. The significance of the descriptive material used in a Dummettian description of S-related behaviour is not exhausted by the fact that it is used to refer to possible S-related behaviour; for a Dummettian account of understanding it is also important *how* the relevant possible S-related behaviour is singled out.

4.4.4   Manifestability of Knowledge of Truth Conditions

I will now explore the consequences of combining the manifestation thesis with the claim that sentential understanding involves knowledge of a truth condition, a claim that results from the premise U-KTC and the temporary assumption M-TC, according to which sentential meanings are truth conditions. Does manifestability of knowledge of the condition for some true sentence S to be true lead to the possibility to know that S is true? The fact that is does is stated in the remaining premise KT of Dummett's manifestation argument:

Knowability Thesis (KT) For every true English sentence S and for every person a, if a has knowledge of the truth condition of S that is fully manifestable in a's S-related behaviour, then there is a possible situation in which a knows that S is true.

If someone can display her knowledge under which condition some true sentence S is true by S-related behaviour, then, according to KT, there is a possible situation in which she knows that S is true.

For what follows, suppose that S^* is a true English sentence, and that a^* can fully manifest her knowledge of the condition for S^* to be true by performing certain S^*-related behaviour. What might such S^*-related behaviour be? Dummett mentions two possibilities. First, he (1976: 45) observes that if someone can immediately recognise whether the truth condition of a sentence S obtains, then she can display her knowledge of the truth condition of S by directly displaying her acceptance or rejection of S (by S-related behaviour). Second, he (1976: 45-6) observes that if someone has a decision procedure for some sentence S, then she can display her knowledge under which condition S is true by carrying out this procedure and then displaying acceptance or rejection of S (by S-related behaviour) on the basis of the outcome of the procedure.

Since Dummett does not mention any further possibilities for how someone might manifest her knowledge of the truth condition of a sentence S by S-related behaviour, one might speculate that he believes that there are no further possibilities. However, this would lead to rather implausible results. Consider the following sentence:

(13) It is raining in London.

First, suppose that Mary, a native speaker of English, is in London and that she can go to a window from which she has a clear view of the sky over London. She may then manifest her knowledge of the condition for (13) to be true by applying the following procedure: go to the window, look at the sky, and then assertively utter (13) iff you see rain! Now, however, suppose that Mary is alone at the South Pole and that she does not know whether she will ever be in a position to obtain any information about the weather in London. In this case, she does not know of a method whose execution will put her into a position to know whether (13) is true. Is Dummett then committed to saying that Mary does not possess knowledge of the condition for (13) to be true that is fully manifestable in (13)-related behaviour? As a second example, consider the following sentence from Appiah 1985: 26 (cp. 4.5.2):

(14) It rained here a million years ago.

It is quite likely that no one knows of a method whose application puts her in a position to know whether this sentence is true. Does it follow that no one has knowledge of the truth condition of (14) that is fully manifestable in (14)-related behaviour? Finally, consider the following true mathematical sentence:

(15) There are infinitely many prime numbers.

Suppose that Peter is a mathematically gifted person who knows many arithmetical proofs but who does not know that (15) is true and who does not have any method to find out. Does one have to say that Peter does not have knowledge of the truth condition of (15) that is fully manifestable in (15)-related behaviour?

One might contemplate attributing to Dummett the radical claim that if someone can fully manifest her knowledge of the truth condition of some true sentence S by S-related behaviour, then this person must know of a decision procedure whose application puts her in a position to know that S is true. And there are certainly traces in some of Dummett's discussions of the manifestation argument which point in this direction.[34] However, there are also some traces which favour a weaker and more plausible claim about the consequences of manifestability of knowledge of truth conditions, e.g. those places in which Dummett stresses the modal component of the notion of manifestability (see Dummett 1993: xii-xiii and 1994: 293). Here, the idea is that someone has *manifestable* knowledge iff there are *possible* situations in which she can manifest her knowledge. Applied to the previous examples: even if Mary is alone at the South Pole, her knowledge of the truth condition of (13) may count as manifestable because if she were in London, then she could manifest it; similarly, if a person had lived a million years ago, then she could manifest her knowledge of the truth condition of (14); and if Peter were presented with a proof of (15), then he could manifest his knowledge of the truth condition of (15) by recognising this proof as such and then assertively uttering (15).

For what follows, I will rely on this interpretation of the notion of manifestability. It is reflected in my formulation of the knowability thesis KT, in which it is only stated that there is a *possible* situation in which the relevant person knows the sentence to be true. I do not claim that this is the version that Dummett always had in mind; however, I think, it is the most plausible one, and it coheres with many of Dummett's discussions.

At this point, one may mention Williamson's (2000: 111) objection to the manifestation argument. Briefly put, the objection is that if the argument were cogent, then it could also be applied to a semantic theory according to which sentential meanings are assertability conditions. It would then imply that someone who understands an assertable sentence S is in a position to know that S is assertable. Williamson argues, however, a sentence may be assertable although someone who knows its assertability condition is not in a position to know that it is assertable. Consequently, he claims, Dummett's argument cannot be cogent.

However, this objection is only applicable to such versions of the argument which are based on the premise that if someone has fully manifestable knowledge of the truth condition of some true sentence S, then she is in a position to know that S is true. That is, it only applies to such versions in which it is not allowed that someone's knowledge of a truth condition is manifestable because there is some *possible* situation in which she performs certain S-related

---

[34]   Dummett (1976: 46), for example, claims that sentences about inaccessible regions of space-time are problematic for classical truth conditional semantics.

behaviour. Thus, even if Williamson's claim is correct, and a sentence may be assertable although someone who knows its assertability condition is not in a position to know that it is assertable, it does not follow that a sentence may be assertable although there is no possible situation in which someone who knows its assertability condition is in a position to know that it is assertable.[35] It therefore follows that Williamson's objection is not applicable to the version of the argument which it relied on here.

Now, the fact that the premise KT is weaker than related assumptions relied on in alternative versions of the manifestation argument does not show, of course, that it is true. It has to be acknowledged, however, that KT seems to have some plausibility. Suppose that a person a^* knows under which condition S^* is true, and that the truth condition of S^* indeed obtains. How could a^*'s possible uses of S^* and her possible reactions to uses of S^* fully display her knowledge of the truth condition of S^* other than by relating to possible situations in which a^* can recognise that the truth condition of S^* obtains?

Consider the suggestion that descriptions of possible uses of a mathematical sentence S^* account for a^*'s knowledge of the truth condition of S^*, namely possible uses in which S^* is inferred from other sentences and possible uses in which another sentence is inferred from S^* and further sentences (cp. 4.4.2). On the face of it, descriptions of such inferential acts can provide a full account $\mathcal{A}$ of a^*'s knowledge of the truth condition of S^* only if they include accounts of a^*'s knowledge of the truth conditions of the additional sentences mentioned in $\mathcal{A}$. One cannot fully account for someone's knowledge of the truth condition of a sentence S_1 by saying that she is disposed to infer S_1 from S_2 (in certain circumstances) unless one includes an account of her knowledge of the truth condition of S_2. Furthermore, one has to avoid circular accounts of knowledge of truth conditions. For example, one does not provide a full account of someone's knowledge of the truth conditions of S_1 and S_2 if one only states that she is disposed to infer S_1 from S_2 and that she is disposed to infer S_2 from S_1.

This observation can be generalised. Suppose $(\mathcal{A}_S)_{S \in \mathcal{L}}$ is a *complete* collection of Dummettian accounts of a^*'s knowledge of the truth conditions of the sentences of some language $\mathcal{L}$, i.e. a collection that contains a relevant account $\mathcal{A}_S$ for *every* sentence S of $\mathcal{L}$. Then, one would expect that there is some *quasi-finite strict partial ordering* $<$ of the sentences of $\mathcal{L}$,[36] such that the account $\mathcal{A}_S$ only presupposes a^*'s knowledge of truth conditions of sentences $<$-related to S. (See 4.4.1 for the finiteness requirement for Dummettian accounts of understanding.) In this way, a circular overall account of the totality of knowledge of truth conditions would be avoided: an account $\mathcal{A}_S$ would only presuppose 'earlier' accounts $\mathcal{A}_{S^*}$, i.e. accounts $\mathcal{A}_{S^*}$ for which $S^* < S$.

---

[35] Williamson (ibid.) cites Dummett (1973a: 586) as having claimed something which is equivalent to the claim that if a sentence is assertable, then those who understand it are in a position to know that it is assertable. However, the fact that Dummett had this belief does not show that his manifestation argument depends on it.

[36] That is, $<$ is an irreflexive and transitive (binary) relation on the set of sentences of $\mathcal{L}$ such that for every sentence S of $\mathcal{L}$ there are only finitely many sentences of $\mathcal{L}$ that are $<$-related to S.

The crucial question is whether such a collection of inferential accounts of someone's knowledge of truth conditions lends itself to undermining the premise KT. And one might worry that it does not, because of the restrictions placed on the accounts that are allowed to figure in this collection. The accounts of $(\mathcal{A}_S)_{S \in \mathcal{L}}$ can only display inferential relations between *sentences* of the object language $\mathcal{L}$; they cannot pertain to the inferential properties of sub-sentential expressions.[37] Then, given that an inferential account $\mathcal{A}_S$ of someone's knowledge of the truth condition of S can only describe inferential acts in which S is inferred from other sentences and inferential acts in which some sentence is inferred from S and additional sentences, one may suspect that $\mathcal{A}_S$ has to provide the means for an evaluation of S: if such S-related inferential acts *fully* account for someone's knowledge of the *truth* condition of S, then, one might think, there should be either a sequence of such acts which constitutes a *proof* of S, or one that constitutes a *refutation* of S. Then, however, assuming that S is true, it would follow that there is a proof of S, and this would suggest that the relevant person can, by following this proof, come to know that S is indeed true.

Clearly, these sketchy remarks do not provide a sufficient justification of KT, and I do not want to commit myself to this thesis. However, I think it is worth considering whether there might not be a possibility for undermining Dummett's argument that is more convincing than the objection that no one has presented a complete justification of KT. Now, it seems to me that the manifestation thesis MT, according to which there must be a full account of someone's understanding of a sentence in terms of how this person is disposed to use *this very sentence*, is quite strong, and that one should take a closer look at this assumption. Thus, without endorsing KT myself, I think it is worth considering whether the manifestation thesis in the presented form is really plausible. I will turn to this task in the following (final) section of this chapter.

4.5   The Simplicity of Manifestability

I will now argue that Dummett's manifestation argument is not convincing. My initial aim is to show that the manifestation thesis MT has to be rejected, and that a plausible weakening of it does not support Dummett's argument. In addition, I will spell out how the conclusion of the manifestation argument can be rejected from a classical perspective that incorporates feasible assumptions about manifestability. Developing a proposal of McDowell (1976) and Appiah (1985), I will show that the compositionality of understanding suggests a compositional account of manifestability that does not support the manifestation argument, and which can coherently be combined with classical logic and truth conditional semantics.[38]

---

[37]   In the following section, I will indicate an account of knowledge of truth conditions that relates to the inferential properties of sub-sentential expressions. However, such an account is not sanctioned by the manifestation thesis (see 4.5.1) and can therefore not be cited in a refutation of KT.

[38]   I would like to emphasise that I will not defend classical logic or the identification of sentential meanings with truth conditions. My goal is the more limited one of presenting

In the first subsection, I will present my skepticism about the manifestation thesis MT and present a compositional account of someone's knowledge of the truth conditions of arithmetical sentences. In the second subsection, I will then defend the claim that this account coheres with classicism and, in particular, with the claim that truth is epistemically unconstrained.

4.5.1   Compositional Manifestation

According to Dummett's manifestation thesis, someone's understanding of a sentence has to be fully manifestable in her uses and reactions to uses of *this very sentence*:

> [A] grasp of the meaning of a mathematical statement must, in general, consist of a capacity to use *that statement* in a certain way, or to respond in a certain way to *its* use by others (Dummett 1973b: 217, emphasis mine).

> An argument of this kind is based upon a fundamental principle, which may be stated briefly, in Wittgensteinian terms, as the principle that a grasp of the meaning of an expression must be exhaustively manifested by the use of *that expression* (Dummett 1977: 260, emphasis mine).

> The question is, therefore, whether, for any intelligible meaning that someone may attach to an expression, it is possible that he should make sufficiently many and, sufficiently various uses of *it* [...] (Dummett 1994: 293, emphasis mine).

By way of motivation, recall Dummett's analogy between language and chess (see 4.4.1).[39] Someone's grasp of the significance of a chess-piece can plausibly be taken to be fully manifestable in her behaviour that relates to uses of this very piece. Thus, if a grasp of the significance of a chess-piece is comparable to sentential understanding, then someone's understanding of a sentence must also be fully manifestable in her behaviour that relates to uses of this very sentence.

The problem is that the analogy between language and chess breaks down at a crucial point. Sentential meanings and sentential understanding are *compositional*. The significance of chess-pieces and our grasp of their significance, however, are not compositional (in any comparable sense). No chess-piece is built up of components which have an individual significance in the game, and no two chess-pieces of different types (such as a knight and a bishop) have corresponding parts which make the same contribution to their overall significance. This stands in sharp contrast to the linguistic case: sentences *are* built up of meaningful components, and, in many cases, two sentences *do* have such meaningful components in common.

---

a coherent picture in which these assumptions hold, and which is not undermined by any of Dummett's considerations about manifestability that survive closer inspection.

[39]   The importance of this analogy for Dummett's understanding of the manifestation thesis has been stressed by Shieh (1998: 45-7).

This undermines the motivation for MT. The compositionality of understanding strongly suggests that in an account of someone's understanding of a sentence S, one may speak about linguistic behaviour that does not relate to utterances of S but to utterances of other sentences that contain constituents of S. Just as understanding is compositional, so is its manifestation. Since an understanding of a complex expression derives from an understanding of its parts, one can at least partly manifest one's understanding of a complex expression by manifesting one's understanding of its parts.

I am not sure to what extent Dummett is under an illusion here. Perhaps he has simply not taken enough care in his formulations of the manifestation thesis. Consider the following statement:

> What *is* it to know the truth-condition of a sentence? We can make no progress with this question without taking account of the fact that we want that knowledge of the truth-condition of a sentence which is to constitute an understanding of that sentence to be derived from an understanding of the words which compose the sentence and the way they are put together (Dummett 1976: 35-6).

If someone's understanding of a sentence S is derived from his understanding of the parts of S and the way in which they are put together, then, I assume, he can manifest his understanding of S by manifesting his understanding of the parts of S and the way in which they are put together. On balance, I therefore think that the indicated possibility of compositional manifestability violates the letter but not the spirit of Dummett's manifestation thesis.

Still, there is at least one author who considers and rejects the claim that someone can manifest her understanding of a sentence by manifesting her understanding of its parts:

> [T]his suggestion [that we can manifest the understanding of sentences by manifesting understanding of their component words] delivers some counterintuitive results. Consider the following sentence: 'I rocked a slice above the quality.' While we understand the component words of this seemingly well-formed sentence, and we can manifest understanding of them by correctly using them in a wide range of cases, we still do not, and indeed cannot, understand [it] (Murzi 2012: 22).

I am not entirely sure that competent speakers of English do not understand a sentence like the one mentioned by Murzi. (Normally, they will not have much use for such a sentence, but this is a different point.) But even if there are sentences that cannot be understood although their parts and the involved modes of combination can be understood,[40] this would not contradict the thesis that the understanding of a sentence is based on the understanding of its parts and

---

[40] Sentences which involve nested central embeddings might be alternative examples. Most, if not all, actual speakers of English fail to understand simple sentences of this kind, e.g.: "The mouse the cat the dog the man hit chased bit ran away". Thus, possibly, more complicated sentences of this kind *cannot* be understood.

the involved modes of combination, or with the thesis that the understanding of a sentence can be manifested by manifesting the understanding of its parts and the involved modes of combination. Instead of presenting a sentence which cannot be understood, Murzi would have to present a sentence which can be understood but whose understanding is not based on the understanding of its parts and the involved modes of combination. In short, what Murzi would have to present is a counterexample to the assumption that understanding is compositional, and this he has not done.

In what follows, I will assume that meaning and understanding are compositional.[41] But then it seems plausible to assume that manifestability is also compositional, that is, that someone can manifest her understanding of a complex expression, at least partly, by manifesting her understanding of its parts.

For the remainder of this section, I will assume that sentential understanding is nothing but knowledge of truth conditions. My aim is to provide a coherent picture which combines even a strong form of classical truth conditional semantics with a plausible weakening of Dummett's manifestation thesis.

I will now present a compositional account of someone's understanding of arithmetical sentences (i.e. someone's knowledge of truth conditions of arithmetical sentences), including those like Goldbach's Conjecture (cp. 3.3.1), which, according to classicists, might be true although no one can know that they are true. More precisely, I will speak about the sentences of a formal language that corresponds to a simple arithmetical fragment of English. Since it simplifies the following presentation, I will here use a formal language $\mathcal{L}$ that contains numerals $0, S0, SS0, \ldots$ in addition to arithmetical predicates. Specifically, I will assume that $\mathcal{L}$ contains brackets, parameters, variables, and the following logical and arithmetical vocabulary:

$$\neg \quad \vee \quad \wedge \quad \rightarrow \quad \forall \quad \exists \quad 0 \quad S \quad N \quad ID \quad SM \quad PR$$

where the arithmetical relation symbols N, ID, SM, PR are to be understood in the same way as those of $\mathcal{L}^{Ar}$ (see 2.3.1 & 4.2.3).

According to the manifestation thesis MT, there has to be a full account of someone's understanding of Goldbach's Conjecture G in terms of her G-related behaviour. Someone's possible utterances of G and her possible reactions to such utterances must fully manifest her understanding of this complex arithmetical sentence. As was seen, however, this assumption is highly questionable: someone's understanding of G only has to be fully manifestable in behaviour that also relates to sentences containing the logical and arithmetical expressions of G. This leads to the following central hypothesis:

---

[41] For a discussion of various precisifications of the notion of compositionality and for a discussion of arguments for and against corresponding compositionality theses, see Pagin & Westerståhl 2010a and 2010b.

Compositional Manifestation (CM) Someone can manifest her knowledge of the truth condition of a sentence S of $\mathcal{L}$ by doing three things:

(i)   manifesting her knowledge of the syntactic modes of combination that are used to build S,

(ii) manifesting her knowledge of the contribution of the logical and arithmetical vocabulary contained in S to truth conditions of arbitrary sentences which contain this vocabulary,

(iii) manifesting her knowledge of the identity of S: knowledge that states which logical and arithmetical symbols with which truth conditional properties are combined in which way to form S.

In the remainder of this subsection, I will indicate how CM can be spelled out, and I will mention that the presented account of knowledge of truth conditions is intuitionistically acceptable. In 4.5.2, I will defend the claim that this account also coheres with classicism and, in particular, with the claim that truth is epistemically unconstrained.

As regards (i), I will simply assume that a person can display her syntactic knowledge of $\mathcal{L}$ by using a suitable number of sentences of $\mathcal{L}$ in a syntactically correct manner.

The main motivation for including (iii) is to account for the relevant mental difference between a person who *would* understand a sentence S if she *encountered* it, a capacity displayed in behaviour of the kind referred to in (i) and (ii), and a person who *actually understands* S.[42] I assume that a person can display her knowledge of the identity of S by uses of S which fit with her other uses of the components of S and with her other applications of the syntactic rules that are involved in the construction of S. However, I will not try to develop a precise description of this linguistic behaviour here since it does not seem to constitute a critical difference between possible classical and intuitionistic accounts of knowledge of arithmetical truth conditions.

As regards (ii), I will consider arithmetical and logical vocabulary in turn. Simplifying slightly, I conjecture that a person can manifest her knowledge of the contribution of the arithmetical vocabulary to truth conditions of sentences which contain this vocabulary, by presenting a suitable collection of correct *calculations* which only involve *atomic* sentences of $\mathcal{L}$.[43] I will treat the arithmetical predicate *SM* as a representative example. Here, the idea is that by

---

[42]   Thanks to Timothy Williamson for pointing out the importance of this distinction for a full account of someone's sentential understanding.

[43]   Strictly speaking, I doubt that this will provide a *full* account of the relevant knowledge because it is not ensured that the numerals of $\mathcal{L}$ denote *all* natural numbers. Arguably, a full account of the relevant knowledge has to describe potential inferential behaviour which displays acceptance of the rule of induction, or some such principle. And this seems to require linguistic resources beyond those of *atomic* sentences of $\mathcal{L}$. Thus, I suspect that a completely satisfying account would acknowledge a certain inter-dependency of arithmetical and logical vocabulary. Such an account would extend the following account of linguistic behaviour that expresses commitment to *logical* rules by also mentioning linguistic behaviour that expresses commitment to the rule of induction, or a similar principle.

presenting a sufficiently diverse collection of computations that involve atomic sentences in which numerals satisfy the predicate SM, a person can display the following knowledge (cp. 4.2.5):

(16) It is a semantic requirement that SM is true of a triple of numbers (n_1, n_2, n_3) iff n_3 is the sum of n_1 and n_2.

More precisely, she can express her commitment to these rules by presenting sufficiently many computations in which she performs inferences in accordance with the fundamental recursive rules for SM (corresponding to the recursive equations $x + 0 = x$ and $x + Sy = S(x + y)$; see Appx. A.1). In this way, she can display her knowledge that it is a semantic requirement that SM is true of every triple of numbers (n_1, n_2, n_3) such that n_3 is the sum of n_1 and n_2. Furthermore, by displaying rejection of sufficiently many incorrect inferences that involve sentences in which numerals satisfy the predicate SM,[44] the person can express that her commitment to the fundamental recursive rules for SM is, in a sense, complete: every true atomic sentence in which numerals satisfy the predicate SM can be proven on the basis of these rules. In this way, she can display her knowledge that it is a semantic requirement that SM is true of a triple of numbers (n_1, n_2, n_3) only if n_3 is the sum of n_1 and n_2. Taken together, by performing this linguistic behaviour, the person can display her semantic knowledge of the contribution of SM to truth conditions of arbitrary sentences of $\mathcal{L}$ which contain this predicate, as it is reported in (16).[45]

It might be objected that due to the finiteness requirement for Dummettian accounts of understanding, it is impossible to preclude the possibility that the relevant person understands SM in a 'quus'-like manner (see Kripke 1982: 9):

(17) It is a semantic requirement that SM is true of a triple of numbers (n_1, n_2, n_3) iff

(a) $\max(n_1, n_2) < 57$, and n_3 is the sum of n_1 and n_2, or

(b) $\max(n_1, n_2) \geq 58$, and $n_3 = 5$.

Now, it has to be agreed that no finite number of uses of arithmetical sentences can *guarantee* that a person has a normal understanding of arithmetical vocabulary. But here I follow Dummett (1993: xii-xv & 1994: 293; cp. 4.4.1) in assuming that this does not preclude the possibility of a (finite) Dummettian account of arithmetical understanding. If this possibility was precluded, then Dummett's project of undermining classical truth conditional semantics on the basis of assumptions about (finite) Dummettian accounts of sentential understanding would be doomed anyway.

Turning to the remaining vocabulary, my hypothesis is that a person can manifest her knowledge of the contribution of the logical vocabulary to truth

---

[44]   Notice that classicists and intuitionists agree about which such inferences are correct/incorrect.

[45]   Obviously, this is only a sketch of a precise account of the relevant semantic knowledge. In particular, it is not specified *how many* computations the relevant person has to present. But since the different possibilities to complete this sketch do not seem to be relevant for the conflict between classicists and intuitionists, I will not try to be more precise here.

conditions of sentences containing this vocabulary by presenting a suitable collection of *simple arithmetical proofs*. In particular, I assume that a person can
manifest this knowledge by presenting proofs that display her commitment to
fundamental logical rules that govern the logical vocabulary. I will mention three
examples: ∧, ∀, and ¬. As regards ∧, the idea is that by presenting mathematical proofs in which conjunctions are inferred from both conjuncts and in which
conjuncts are inferred from conjunctions, a person can express her commitment
to the rules $(\wedge_E)$ and $(\wedge_I)$, thereby displaying the following knowledge:

(18) It is a semantic requirement that a conjunction is true iff both conjuncts
      are true.[46]

As regards ∀, the idea is that by presenting mathematical proofs in which
instances are inferred from universal quantifications and in which a universal
quantification is inferred from an arbitrary instance (formulated by means of a
suitable parameter), the person can express her commitment to the rules $(\forall_E)$
and $(\forall_I)$, thereby displaying the following knowledge:

(19) It is a semantic requirement that a universal quantification $\forall n\varphi(n)$ is true
      iff the open sentence $\varphi(x)$ is true of every number.

As regards ¬, the idea is that by presenting mathematical proofs in which
arbitrary sentences are inferred from a sentence and its negation and in which
sentences are refuted by showing that they imply a sentence and its negation,
the person can express her commitment to the rules $(\neg_E)$ and $(\neg_I)$, thereby
displaying the following knowledge:

(20) It is a semantic requirement that a negation is true iff the negated sentence
      is not true.

In this way, linguistic behaviour that relates to simple arithmetical proofs accounts for knowledge of the contribution of logical vocabulary to truth conditions of arbitrary sentences of $\mathcal{L}$ which contain this vocabulary.

Thus, in sum, the idea is that by presenting calculations and simple arithmetical proofs, a person can show her commitment to fundamental arithmetical
and logical rules, and that, from a semantic perspective, this behaviour can be
interpreted as displaying her knowledge of the contribution of arithmetical and
logical vocabulary to truth conditions of arbitrary sentences of $\mathcal{L}$ which contain
this vocabulary.

I would like to emphasise that intuitionists who accept the claim that linguistic behaviour can fully account for knowledge of truth conditions, and who
also accept the possibility of compositional manifestation, can hardly object to
an account along these lines. First, the linguistic behaviour referred to in this
account is intuitionistically acceptable. In particular, the relevant logical and
arithmetical commitments correspond to those of standard calculi of intuitionistic arithmetic (cp. the calculus $\mathcal{R}^{Ar}$ from 2.3.1). Intuitionists thus accept the

---

[46] Strictly speaking, one should here and in (19) and (20) refer to semantic requirements
      about the notion of being *true of* a tuple of objects to allow for the case of sentences
      with parameters. See 4.2.5 for adequate formulations.

claim that a person who acts in a way that conforms to this account performs legitimate linguistic behaviour. Consequently, if an intuitionist who accepts the claim that linguistic behaviour can fully account for knowledge of truth conditions wanted to reject the presented account, then she would have to deny that the presented description of linguistic behaviour is *strong enough* to account for knowledge of arithmetical truth conditions. However, this seems to be rather implausible. Recall that the *logical* rules of a calculus like $\mathcal{R}^{Ar}$ suffice to uniquely characterise the involved logical operators (see 2.2.3) and that the *arithmetical* rules of a calculus like $\mathcal{R}^{Ar}$ present a maximal limitation of possible distributions of truth values for arithmetical sentences (in the sense of 2.3.1). This suggests that if any arithmetical behaviour suffices to account for someone's knowledge of arithmetical truth conditions, then the mentioned behaviour is strong enough to do so as well. Consequently, intuitionists who accept the claim that linguistic behaviour can fully account for knowledge of truth conditions cannot sensibly object to an account along these lines. This leads to the question of what classicists should say to such an account.

4.5.2   Manifestation and Epistemic Constraints

I will now defend the claim that the presented compositional account of arithmetical understanding coheres with classicism and, in particular, with the claim that truth is epistemically unconstrained. I will use "(non-)classical semantics" as an abbreviation of "(non-)classical truth conditional semantics" here.

To begin with, I would like to mention that I am not the first to suggest that considerations about the compositionality of understanding can be used to undermine Dummett's manifestation argument. In particular, such considerations have been put forward by Appiah:[47]

> Suppose the sentence is: 'it rained here a million years ago'. Why is it not evidence that someone assigns this sentence the correct truth conditions, that they use 'rain' properly in sentences about present rain, that they can count to a million, know how long a year is, and display a grasp of the past tense in relation to the recent past (Appiah 1985: 26)?

Appiah's basic idea is that someone can manifest her understanding of sentences which are not (known to be) decidable by using sentences for which decision procedures are known and which contain all the constituents of the original sentence.

Unfortunately, there is one salient problem with Appiah's proposal for countering Dummett's argument, and there is one crucial open question that he does not answer. First, Appiah interprets the manifestation thesis as an epistemological thesis: a thesis about possible evidence that someone understands the relevant sentence. However, as was argued in 4.4.1, this interpretation is wrong. The manifestation thesis is about the nature of understanding itself and not about the basis of our knowledge of understanding.

---

[47]   An early suggestion along these lines was made by McDowell (1976: 63).

Second, Appiah does not justify the claim that a *classical* semanticist has the right to say that someone can manifest her knowledge of the truth condition of some undecidable sentence by manifesting her knowledge of the truth conditions of decidable sentences. This has been pointed out, for example, by Byrne:[48]

> [T]he anti-realist may concede with equanimity that a speaker can, by appropriate use of other sentences, manifest (indirectly) that she understands S (or at any rate, would understand it if she considered it), and given compatibilism, it follows that she thereby manifests grasp of S's truth condition. But the anti-realist should insist that such a performance does not, for all that has been said so far, support any particular constitutive account of that understanding – i.e., any particular construal of that truth condition, such as the realist one (Byrne 2005: 107).

This objection can be rephrased as follows: "A *classical* semanticist has to give accounts of knowledge of epistemically *unconstrained* truth conditions. But then it is insufficient for her to give accounts of knowledge of truth conditions in terms of linguistic behaviour which could equally be understood as providing accounts of knowledge of epistemically *constrained* truth conditions."

The point can be brought out more concretely by comparing two competing truth conditional semanticists. One of them believes that the meaning of a sentence is its knowability condition; i.e., she identifies truth with knowability. The other one believes that the meaning of a sentence is its truth condition but rejects the claim that every truth is knowable. Now, there is nothing in the presented compositional account of someone's knowledge of arithmetical truth conditions that suggests that it concerns truth conditions which can obtain although no one can know that they obtain. All the linguistic behaviour that is referred to is compatible with the assumption that the knowledge that has been accounted for is knowledge of knowability conditions. But then, it is objected, this behaviour cannot amount to a *full* manifestation of knowledge of truth conditions for a notion of truth that *transcends* knowability.

However, this objection can be met. Consider Goldbach's Conjecture (G). A classical semanticist does not have to claim that the *linguistic behaviour* which accounts for knowledge of the truth condition of G *shows* that G can be un-knowably true. She can say that the mentioned linguistic behaviour accounts for knowledge of arithmetical truth conditions without indicating whether truth is or is not epistemically constrained. The proponent of the manifestation argument tries to show that truth conditions whose knowledge is accounted for by linguistic behaviour have to be based on an epistemically constrained notion of truth. But this does not imply that a semanticist who subscribes to classical logic is under the obligation to show that truth conditions whose knowledge is accounted for by linguistic behaviour have to be based on an epistemically unconstrained notion of truth. Put otherwise, if someone wants to show that classical semantics is wrong, then it is not enough for her to point out that

48   See Wright (1980: 101) for the corresponding objection to McDowell's (1976: 63) suggestion.

epistemically constrained truth conditions are compatible with the linguistic data. Rather, she seems to be under the obligation to show that epistemically unconstrained truth conditions are incompatible with the linguistic data.

Recall the two semantic arguments against classical logic DSA and DRSA (4.1.2). The argument DSA contains the premise that truth is epistemically constrained if sentential meanings are truth conditions, while the argument DRSA contains the premise that it is conceptually necessary that truth is epistemically constrained if it is conceptually necessary that sentential meanings are truth conditions. Now, suppose that it is indeed conceptually necessary that sentential meanings are truth conditions. For neither of the two arguments it suffices to show that truth *could* be epistemically constrained, in the sense that this assumption is compatible with certain linguistic data. Rather, it has to be shown that truth *is* epistemically constrained, or even that it is conceptually necessary that truth is epistemically constrained. However, this has not been done.

If the manifestation thesis MT were true, so that compositional accounts of understanding were precluded, then there might be some reason to suppose that truth has to be epistemically constrained, given the assumption that someone's knowledge of truth conditions has to be fully manifestable (see 4.4.4). But as soon as one allows for compositional manifestation, as one has to, a behavioural account of someone's knowledge of truth conditions can leave it open whether truth is epistemically constrained or not.

There are different possible responses that adherents of the Dummettian strategy of undermining classical semantics by means of manifestability considerations might suggest. One possibility is to grant the claim that the relevant linguistic data is compatible with classical semantics, but to insist that the linguistic data fits *better* to a non-classical semantics. Prawitz spells out this idea in terms of simplicity considerations:

> [T]here is another and simpler theory that also does this [accounting for knowledge of truth condition in terms of less disputed uses of quantifiers], viz. a verificationist theory, which takes knowledge of meaning of a statement to consist in knowledge of what counts as a direct verification of the statement. The problem for the classical meaning theorist is that it is he who explicitly insists on the conceptual difference between truth and verifiability. He maintains [...] that knowledge of the truth-condition contains an ingredient over and above what is contained in knowledge of what counts as a verification. The problem for him is thus what empirical import that extra ingredient has [...] (Prawitz 1994: 87).

Prawitz claims that a semantic theory according to which understanding a sentence S is identified with knowledge of the condition for something to be a direct verification of S is simpler than classical truth conditional semantics because it does not insist on a conceptual difference between truth and verifiability, and because its adherents do not claim that knowledge of a truth condition contains an ingredient over and above what is contained in knowledge of a verification condition.

In my view, however, Prawitz' two superiority claims are not convincing. Consider the first one. He claims that classical semanticists make the assumption that truth and verifiability/knowability are different concepts, although this assumption is not justified by the relevant data about uses of (arithmetical) sentences. However, the assumption that truth and verifiability are different concepts is not in more tension with the mentioned linguistic data than the assumption that they are the same concepts. The mentioned arithmetical behaviour simply leaves it open whether truth coincides with verifiability or not. Furthermore, while non-classical semanticists usually conceive of the claim that every truth is knowable as a fundamental ingredient of their semantic theory, classical semanticists might adopt the thesis that not every truth is knowable for a *non-semantic* reason such as Fitch's paradox of knowability (cp. 4.1.2).

Consider now the second point. Prawitz claims that someone's knowledge of the truth condition of some sentence is *stronger* according to classical semantics than it is according to non-classical semantics. He says that classical semanticists claim that such knowledge contains an ingredient which is not assumed to be present by non-classical semanticists. This claim can be related to Dummett's view that truth conditional semanticists have to explain how competent language users may come to obtain the notion of truth. On the one hand, Dummett seems to think that *classical* semanticists attribute semantic knowledge to competent language users that is not easy to come by because it involves a notion that is difficult to be obtained: a notion of truth which might apply to a sentence although no one can know that it does. On the other hand, Dummett seems to think that *non-classical* semanticists attribute semantic knowledge to competent language users that is easy to come by because it is based on a notion that is not difficult to be obtained: a notion of truth which could not apply to a sentence although no one can know that it does.

However, neither of these views seems to be warranted. *Pace* Prawitz, a classical semanticist does not attribute stronger semantic knowledge to competent language users than a non-classical semanticist does. In particular, a classical semanticist does not claim that everyone who knows the truth conditions of arithmetical sentences also knows that some arithmetical sentences might be unknowably true. The belief that not all truths are knowable is not part of the semantic knowledge attributed to competent language users. A classical semanticist can coherently combine the following claims about Goldbach's Conjecture (G) and some person a: (i) G might be true although no one could possibly know that G is true, (ii) a knows under which condition G is true, and (iii) a cannot even grasp the thought that G might be true although no one could possibly know that G is true.

Similarly, *pace* Dummett, the semantic knowledge attributed to competent language users by classical semanticists is not based on a notion which is more difficult to obtain than the notion on which the semantic knowledge is based that is attributed to competent language users by non-classical semanticists. To begin with, I would deny that the semantic knowledge referred to by classical semanticists differs from the semantic knowledge referred to by non-classical semantics. In particular, they are based on the same notion, the notion of truth to wit, about which they have contradictory opinions. Just as intuitionists and classicists understand logical and mathematical vocabulary in the same way (see Ch. 2), it is most plausible to assume that they also understand the predicate "is a true sentence of $\mathcal{L}$" in the same way. Furthermore, even if there were two different notions of truth, it would be questionable that a notion of truth according to which a sentence might be true although no one could possibly know it to be true has to be a notion that is more difficult to be obtained than a notion of truth according to which every true sentence can be known to be true. For example, a classical semanticist can claim that competent speakers obtain the relevant notion of truth by coming to know the truth conditions of the sentences of their language, and she can also claim that this latter semantic knowledge is 'compositional knowledge', which is easily obtained and can be accounted for in precisely that way in which a *non-classical* semanticist tries to account for knowledge of truth conditions.

Thus, the claim that the presented linguistic data fits better to non-classical semantics than to classical semantics misrepresents classical semantics and its relation to the claim that truth is not epistemically constrained. Classical semanticists allow for a conceptual difference between truth and knowability. But this does not induce a gap between their preferred theory and the relevant linguistic data, because classical semanticists do not have to claim that this conceptual difference exists *in virtue of* the relevant linguistic data. Furthermore, classical semanticists do not specify semantic knowledge which is *stronger* or *more difficult to grasp* than the semantic knowledge specified by non-classical semanticists. The different views about truth (that it is or is not epistemically constrained) are not part of the semantic knowledge attributed to competent speakers, and there is no reason to claim that the task of explaining how competent speakers obtain the notion of truth is more difficult to carry out for classical semanticists than it is for non-classical semanticists.

The present findings conform to those of Chapter 2. It was then argued that there could not be two different ways of understanding logical and arithmetical vocabulary, an intuitionistic way and a classical way. Part of the reason for this fact is that logical constants are governed by (intuitionistically and classically acceptable) rules which uniquely characterise them and that arithmetical vocabulary is governed by (intuitionistically and classically acceptable) rules which preclude that this vocabulary is associated with different kinds of meanings that sanction different arithmetical theses. Accordingly, one should not expect that there have to be different accounts of understanding: one account which must be put forward by classical semanticists and another one which must be put forward by non-classical semanticists. Rather, one should expect that if some account of arithmetical understanding is such that an *intuitionist* can claim that it implies that the relevant arithmetical meanings sanction basic intuitionistic rules, then a *classicist* can claim that the same account implies that these meanings sanction basic classical rules. Classicists and intuitionists attribute inferential properties to the logical and arithmetical constants which are sufficiently similar to be compatible with the claim that they understand these constants in the same way. Consequently, it should not be surprising that clas-

sical and non-classical semanticists can give the same accounts of understanding
these constants. They do so although they have contradictory opinions about
the semantic knowledge attributed to competent speakers in these accounts.

I would like to close this chapter by indicating what the presented view suggests for the relation between logical theories and semantic theories. Recall that
the two anti-classical arguments DSA and DRSA (4.1.2) proceed from the assumption that a sound logical theory has to be semantically validated: every
argument that is logically valid according to the theory has to be semantically
guaranteed to be truth-preserving. What I would like to suggest is that a truth
conditional semanticist possesses the means to substantiate this claim, and, depending on whether she has classical or intuitionistic inclinations, that she will
end up with claiming that classical logic or intuitionistic logic is semantically
validated. Focussing on logical *truth* rather than on logical validity,[49] a truth
conditional semanticist can claim that a sentence S is *semantically* guaranteed
to be true if the semantic requirements for the logical constants of S suffice to
guarantee that S is true.[50]

Consider Goldbach's Disjunction $(G \vee \neg G)$, which is a classical logical truth
that intuitionists claim to be a weak counterexample to classical logic (3.3.1).
Is this sentence semantically guaranteed to be true? The truth conditional semantics that underlies the presented compositional account of arithmetical understanding explicitly states that it is a semantic requirement that $G \vee \neg G$ is
true iff every or not every even number larger than 2 is the sum of two primes
(see 4.2.5). However, classicists and intuitionists will react differently to this
claim. A classicist will say that this semantic requirement provides an immediate guarantee that $G \vee \neg G$ is true because she takes it to be a *logical triviality*
that every or not every even number larger than 2 is the sum of two primes. An
intuitionist, by contrast, will deny that this semantic requirement provides a
guarantee that $G \vee \neg G$ is true. She will say that it is still an open arithmetical
question whether $G \vee \neg G$ is true because she takes it to be an unproven arithmetical claim that every or not every even number larger than 2 is the sum of
two primes.

From a perspective that attributes an ultimate foundational significance
to semantic theories, the presented view is probably somewhat dissatisfying.
Roughly speaking, it seems to allow for a semantic validation of a logical theory that presupposes the soundness of this logical theory. However, instead of
inferring that the presented view is inadequate, one might question the ultimate
foundational significance of semantics. On an alternative view, any serious intellectual achievement, including the validation of ones preferred logical theory,
presupposes basic logical principles. Then, it is not surprising that there cannot

---

[49]   The treatment of logical validity is similar: an argument α is *semantically* guaranteed to
be truth-preserving if the semantic requirements for the logical constants of α guarantee
that S is truth-preserving. However, since I want to use Goldbach's Conjecture as my
illustrating example, I will stick to logical truth here.

[50]   Note that the possible alternative suggestion that S is semantically guaranteed to be true
iff it is a semantic requirement that S is true is clearly inadequate. No sensible logical
theory would result from such an approach, which treats logic as a *part of* semantics.

be a certain kind of non-circular justification of fundamental logical principles. To apply this to the case at hand, recall that Dummettian accounts of understanding are meant to be accounts that make a certain practice (namely the use of language) intelligible as an activity of rational agents (see 4.4.1). But then, since classicists and intuitionists have contradictory commitments even with respect to the most fundamental (logical and mathematical) level, one should expect that they have different views about how some practice can intelligibly be explained. They simply have different standards of intelligibility.

For all that has been said, an adherent of Dummett's manifestability considerations can formulate a legitimate challenge to someone who wants to combine classical logic, truth conditional semantics, and the presented compositional account of arithmetical understanding: "If the combination of classical logic and truth conditional semantics is really compatible with the presented account of knowledge of arithmetical truth conditions, then it must also be compatible with an extension of this account which also mentions distinctly classical inferential behaviour. It would be utterly mysterious if one could only account for knowledge of truth conditions which substantiate *classical* logic by referring to intuitionistically acceptable inferential acts and commitments. Concretely, the presented account could only be classically acceptable if it can be extended by describing distinctly classical uses of the negation operator $\neg$. That is, the extended account would also speak about presentations of proofs in which a sentence is, for example, inferred from its double negation, thereby displaying commitment to $(\neg_{\neg E})$."

  I think that a classicist has to accept this challenge. She has to agree that an *explicitly classical* extension is no less convincing than the presented neutral account itself. But then, intuitionists can try to strike back, because they can claim that an explicitly classical extension of the presented account of knowledge of arithmetical truth conditions is demonstrably *incoherent*. This claim, which I take to lie at the heart of Dummett's attempted semantic attack on classical logic, will be the topic of the following final chapter of this book. In this chapter, I will consider so-called *proof-theoretic* arguments, which are meant to show that classical logic is not sound, because it is not based on an adequate system of inferential rules. Furthermore, the most convincing proof-theoretic arguments will be precisely arguments that try to spell out that classical logic has to rely on an underlying incoherent inferential practice.

Chapter 5

Proof-Theoretic Arguments

Systems of rules, also referred to as *calculi*, are not only of mathematical interest. They are also relevant for various non-mathematical theories, especially for logical theories. In this chapter, I will discuss arguments against the theory of classical logic which are based on considerations about calculi. These arguments will be called *proof-theoretic arguments (against classical logic)*.

Proof-theoretic arguments are usually based on certain *mathematical* results on calculi. This might be taken to suggest that the prospects of classical logic depend on future mathematical research, that is, one might think that there are open mathematical questions about calculi on whose answers the soundness of classical logic depends. However, as far as I can see, this is not the case. The relevant mathematical landscape is well explored, and the relevant facts seem to be known. The philosophical disputes about proof-theoretic arguments against classical logic are rather disputes about non-mathematical consequences of the known mathematical truths. For an evaluation of these arguments, it will be helpful to clearly separate the relevant mathematical results from their alleged philosophical implications.

In most philosophical discussions, proof-theoretic arguments against classical logic are only partly specified. A main goal in what follows will be the development of precise versions of these arguments. Against the background of these clarifications, it will then be possible to single out their most contentious assumptions.

To begin with, I will introduce central concepts employed in proof-theoretic arguments (5.1) and present the general form of these arguments (5.2). The remaining sections then focus on two proof-theoretic arguments. First, I will deal with an argument that has been suggested by Neil Tennant (5.3). It will be argued that it is based on an unwarranted assumption about the kinds of rules that may figure in adequate calculi. A suggestion will be made as to how Tennant's argument can be improved by weakening its crucial premise. Second, I will introduce the notion of harmonious collections of rules which lies at the centre of most proof-theoretic arguments (5.4) and present Dummett's version of such an argument (5.5). I will conclude by criticising Dummett's proof-theoretic argument (5.6).

5.1   Proof-Theoretic Preliminaries

This section contains definitions of the most general proof-theoretic concepts that are relied on in the following discussion.[1] The pertinent proof-theoretic arguments are already applicable to fragments of natural languages which can be modelled by formal languages of *sentential logic*. Accordingly, I will restrict attention to such simpler languages. In 5.1.1, fundamental notions that relate to rules and calculi are defined, and in 5.1.2, the class of calculi of natural deduction is introduced. In 5.1.3, an important type of calculi of natural deduction will be singled out, and the presentation of calculi by means of schemata will be discussed.

5.1.1   General Proof-Theoretic Concepts

Most of the formal languages considered in this chapter are based on the same countable supply of sentential constants: p, q, r, p_1, p_2, *etc.* However, they are based on different sets of sentential operators. Of special importance will be the language whose operators are the unary operator $\neg$ and the binary operators $\wedge$, $\vee$, $\rightarrow$. This language is denoted by $\mathcal{L}^*$.

Several kinds of *expressions*, i.e. one- or two-dimensional complexes of symbols, will be important.[2] Some of these kinds comprise expressions which are built up from sentential constants. Others comprise expressions which are built up from schematic letters. I will use φ, ψ, χ (with sub- and superscripts) as schematic letters for sentences and Γ, Δ, Σ (with sub- and superscripts) as schematic letters for *fs-sets*, i.e. for finite sets of sentences.

First, the notion of a logical framework and the notion of a sequent will be introduced. These notions induce a central classification of the different calculi.

A *schematic sentence* is like a sentence except that it contains schematic letters for sentences instead of sentential constants; I will use A, B, C, D for schematic sentences. A *sentential list* is a finite (possibly empty) and alternating sequence of sentences and commas which begins and ends with a sentence, for example, $\neg p, q, q \vee r, q$. A *schematic list* is like a sentential list except that it contains schematic sentences and schematic letters for fs-sets instead of sentences, for example, $\Gamma, \varphi, \Delta, \neg \varphi$. A sentence in a sentential list should be thought of as representing its singleton and a comma should be thought of as representing the operation of set-union. Thus, for example, the sentential list $\neg p, q, q$ should be interpreted as $\{\neg p\} \cup \{q\} \cup \{q\}$, i.e. as $\{\neg p, q\}$. Similarly, a schematic list like $\Gamma, \varphi, \neg \varphi$ should be interpreted as $\Gamma \cup \{\varphi, \neg \varphi\}$. I will use X, Y, Z for sentential and schematic lists.

One important respect in which calculi differ concerns the category of linguistic objects they deal with, e.g. sentences or arguments. In general, these

---

[1]   Most of these concepts are taken from Humberstone 2011.

[2]   In a precise presentation of the following concepts, one distinguishes between expressions and tokens of expressions which occur in more complex expressions (see Troelstra & Schwichtenberg 2000: 5-6). For increased readability, I will be less precise about this distinction.

objects are called *sequents*. By choosing a type of sequent, one chooses a *logical framework*. In what follows, I will only deal with logical frameworks whose sequents can be identified with pairs (x, y) of fs-sets. The members of x are called *antecedents*, and the members of y are called *succedents* (cp. Dummett 1991: 185). By placing constraints on the set of antecedents and the set of succedents, one can single out various logical frameworks. There are three relevant examples:

(i)   If the set of succedents is required to be an fs-set with exactly one member, one obtains the logical framework ARG of (finite) *arguments*.[3]

(ii) If the set of succedents is required to be an fs-set with at most one element, one obtains the logical framework ARG_0 of (possibly) *improper arguments*.

(iii) If no restriction is placed on the sets of antecedents and succedents, one obtains the logical framework G-ARG of *generalised arguments*.[4]

Sequents are usually represented by expressions of the form $t_1 \succ t_2$, in which t_1 stands for the set of antecedents and t_2 stands for the set of succedents. A *schematic sequent* is obtained from such an expression by replacing t_1 and t_2 by appropriate schematic lists. For example, *schematic sequents of* G-ARG are expressions of the form $X \succ Y$, and *schematic sequents of* ARG are expressions of the form $X \succ A$.

Logical frameworks are not tied to a specific language. I will call the restriction of a logical framework to the sequents of some language a *specialised logical framework*. A logical framework can thus be identified with a function from languages to specialised logical frameworks (see Humberstone 2011: 103). A term for a logical framework with a subscript for a language denotes such a specialised logical framework; e.g., the term $\text{ARG}_{\mathcal{L}^*}$ stands for the specialised framework of arguments of $\mathcal{L}^*$.

Now, the notion of a calculus and the notion of a derivation in a calculus will be introduced. These notions are of fundamental significance for the proof-theoretic arguments.

A *transition* of some framework $\mathcal{F}$ is a pair (x, s) consisting of a finite set x of sequents of $\mathcal{F}$, called the *initial sequents*, and another sequent s of $\mathcal{F}$, called the *final sequent*. The antecedents of initial/final sequents are called *initial/final antecedents*, and the succedents of initial/final sequents are called *initial/final succedents*. A *rule* of some framework $\mathcal{F}$ is a set of transitions of $\mathcal{F}$,[5] and a *calculus* of $\mathcal{F}$ is a set of rules of $\mathcal{F}$. A transition that is an element

---

[3]   I here deviate from the notion of an argument introduced in 1.1.1, where it was assumed that an argument consists of a *list* of premises rather than of a *set* of premises. Since I do not consider logical theories according to which the order or multiplicity of the premises of an argument is relevant for the question of whether it is logically valid, the difference between these two notions of an argument is not important here.

[4]   Humberstone (2011: 108) uses the terms SET-FMLA, SET-FMLA$_0$, and SET-SET for the logical frameworks ARG, ARG$_0$, G-ARG.

[5]   I will adopt a simple conception of rules according to which they are tied to specific languages. In 2.2.3 and 2.2.4, it was argued that a revised conception of rules according to which they are functions from languages to sets of transitions has important philosophical applications. For the purposes of this chapter, however, the simpler conception suffices.

of a rule is also called an *application* of it. Transitions are usually represented by expressions of the form

$$\frac{t_1 \ldots t_n}{t}$$

in which the terms $t_1 \ldots t_n$ stand for the initial sequents and the term t stands for the final sequent. (Transitions do not have to have any initial sequents. Those which do not are usually represented by expressions of the form $\bar{t}$.) *Schematic transitions* are obtained from such expressions by replacing the terms $t, t_1, \ldots, t_n$ by schematic sequents of the relevant framework. (Note that schematic transitions are expressions, although transitions are not.)

In 1.2.3, a notion of derivability for finite arguments was introduced. This notion has an immediate counterpart in the present setting. A collection $\mathcal{R}$ of rules of some logical framework induces the following property of sequents of this framework:

$\mathcal{R}$**-Derivability** A sequent s is called $\mathcal{R}$-*derivable* iff there is a finite sequence $s_0, \ldots, s_n$ of sequents such that $s_n = s$ and such that for every $i \leq n$ there is a subset J_i of $\{0, 1, \ldots i-1\}$ such that $(\{s_j : j \in J_i\}, s_i)$ is an application of a rule in $\mathcal{R}$.

That is, a sequent s is $\mathcal{R}$-derivable iff there is a *derivation sequence* for it: a finite sequence ending with s such that every member can be obtained from earlier members or from the empty set by means of the rules in $\mathcal{R}$.

Instead of using derivation sequences in this chapter, I will follow the more usual practice of using *derivation trees*. (This is at least the choice of the proponents of the proof-theoretic arguments discussed here.) Part of the reason for preferring derivation trees over derivation sequences is that the former embody more information than the latter. If one is given a derivation sequence $(s_0, \ldots, s_n)$, then one does not thereby know from which earlier sequents each sequent has been obtained; i.e., one does not know the identity of the sets J_i for the different members s_i of the sequence. In a derivation tree, by contrast, this information is made explicit; i.e., one always knows the identity of the involved transitions.

In the present context, a *tree* is a partially ordered set $(T, <)$ with a least element, the *root*, such that the set $\{y; y \leq x\}$ is a chain for every $x \in T$. The elements of T are the *nodes* of the tree, the maximal chains are the *branches* of the tree, and the top elements are the *leaves* of the tree. (Trees grow upwards: "x is *below* y" amounts to $x < y$.) Two nodes $x < y$ are called *neighbours* iff there is no node z such that $x < z < y$.

A finite tree together with a function that maps every node v of the tree to a sequent s^v of $\mathcal{F}$ is called a *sequent tree* in the framework $\mathcal{F}$. A calculus $\mathcal{R}$ then determines a set of sequent trees:

$\mathcal{R}$**-Derivations** An $\mathcal{R}$-*derivation* is a sequent tree whose transitions are applications of rules of $\mathcal{R}$: for every node v with upper neighbours $v_1, \ldots, v_n$, the transition with initial sequents $s^{v_1} \ldots s^{v_n}$ and final sequent s^v is an element of a rule of $\mathcal{R}$.[6]

Note that transitions may belong to different rules. That is, a derivation does not necessarily determine for each transition a unique rule which contains it. The roots of derivations then correspond to final elements of derivation sequences: a sequent may be called $\mathcal{R}$-*derivable* iff there is an $\mathcal{R}$-derivation whose root it is. It is easily seen that this agrees with the sequential definition of $\mathcal{R}$-derivability.

Sequent trees have simple representations. They are obtained from representations of finite trees by replacing (representations of) the nodes v of the tree by (representations of) its images s^v. For example, the representation of the tree with the nodes $v_1, \ldots, v_7$ displayed to the left yields a representation of the sequent tree displayed to the right

$$
\begin{array}{cc}
v_1 \quad v_2 & \\
\hline
v_3 \quad\quad v_6 & \\
\hline
v_4 \quad\quad v_5 & \\
\hline
v_7 &
\end{array}
\qquad\qquad
\begin{array}{cc}
s_1 \quad s_2 & \\
\hline
s_3 \quad\quad s_6 & \\
\hline
s_4 \quad\quad s_5 & \\
\hline
s_7 &
\end{array}
$$

if the function of the sequent tree maps a node v_i to the sequent s_i for every $i \in \{1, \ldots, 7\}$. A horizontal line indicates that the object below the line is a neighbour of each object above the line.

Finally, let me recall the sentential fragments of the calculi from 1.2.3. More precisely, I will consider calculi of the logical framework ARG (whose sequents are pairs (x, y) of fs-sets such that y contains exactly one member).[7] The rules of the two calculi are introduced by presenting schematic transitions. The intuitionistic calculus $\mathcal{R}^I$ comprises the following rules. First, there is one so-called *structural* rule, *Assumption*, which is represented by a schema in which no logical operator occurs:

$$\frac{}{\varphi \succ \varphi} \text{ (A)}$$

Second, there are so-called *introduction* and *elimination rules* (see 5.4.2). For every logical operator there are one or two introduction rules and one or two elimination rules in whose schemata the operator figures:

$$\frac{\Gamma \succ \varphi \quad \Delta \succ \psi}{\Gamma, \Delta \succ \varphi \wedge \psi} \ (\wedge_I) \qquad\qquad \frac{\Gamma \succ \varphi \wedge \psi}{\Gamma \succ \varphi} \ (^1\!\wedge_E) \quad \frac{\Gamma \succ \varphi \wedge \psi}{\Gamma \succ \psi} \ (^2\!\wedge_E)$$

$$\frac{\Gamma \succ \varphi}{\Gamma \succ \varphi \vee \psi} \ (^1\!\vee_I) \quad \frac{\Gamma \succ \psi}{\Gamma \succ \varphi \vee \psi} \ (^2\!\vee_I) \qquad \frac{\Gamma \succ \varphi \vee \psi \quad \Delta, \varphi \succ \chi \quad \Sigma, \psi \succ \chi}{\Gamma, \Delta, \Sigma \succ \chi} \ (\vee_E)$$

$$\frac{\Gamma, \varphi \succ \psi}{\Gamma \succ \varphi \rightarrow \psi} \ (\rightarrow_I) \qquad\qquad \frac{\Gamma \succ \varphi \rightarrow \psi \quad \Delta \succ \varphi}{\Gamma, \Delta \succ \psi} \ (\rightarrow_E)$$

$$\frac{\Gamma, \varphi \succ \psi \quad \Delta, \varphi \succ \neg\psi}{\Gamma, \Delta \succ \neg\varphi} \ (\neg_I) \qquad\qquad \frac{\Gamma \succ \varphi \quad \Delta \succ \neg\varphi}{\Gamma, \Delta \succ \psi} \ (\neg_E)$$

---

[6]   If the node v is a leaf, then this amounts to the requirement that the transition without initial sequents and with final sequent s^v is an element of a rule of $\mathcal{R}$.

[7]   Since it is assumed that premises assemble into sets rather than into lists (see n. 3), it is not necessary here to include the rules (E) and (C) from 1.2.3.

The classical calculus $\mathcal{R}^C$ comprises the above rules plus the rule of *double negation elimination*:

$$\frac{\Gamma \succ \neg\neg\varphi}{\Gamma \succ \varphi} \; (\neg_E)$$

5.1.2   Calculi of Natural Deduction

Proof-theoretic arguments presuppose a tight connection between (representations of) derivations and informal argumentations (cp. 2.2.3). As a consequence, the quality of a calculus is not simply a function of its set of derivable sequents: derivations have to be plausible models of argumentations. Arguably, certain kinds of calculi of natural deduction are closest to systems of informal rules which can plausibly be taken to be applied in informal argumentations. I will introduce these calculi in this subsection.

Initially, one might think that if there is to be a close connection between derivations and informal argumentations, then the derivations should belong to the logical framework whose sequents are sentences. In simple argumentative steps, one moves from some sentences which one accepts, the *premises*, to another sentence which one then also accepts, the *conclusion*. If transitions are to correspond to such simple acts, then they should be identifiable with the corresponding pair of the set of premises and the conclusion.

On closer inspection, though, it becomes apparent that there are other types of argumentative moves. One not only draws inferences from sentences which one accepts but also from temporary assumptions that one does not accept. This might suggest the adoption of the logical framework ARG with a strong restriction as to the acceptable rules. A rule is called *basic* iff the set of initial antecedents in its applications equals the set of final antecedents. Transitions of basic rules can be understood as corresponding to inferences whose premises depend on certain assumptions: the antecedents of a sequent are interpreted as (temporary) assumptions, and its succedent is interpreted as a sentence which is accepted conditional on these assumptions. In an application of a basic rule, one infers the final succedent from the initial succedents, which then depends on all those assumptions on which at least one initial succedent depended.

However, even this larger set of argumentative moves does not capture all ingredients from which ordinary argumentations can be thought of as being built. In informal reasoning, one often introduces a temporary assumption, which is discharged again at a later step in the argumentation. As an example, consider a schematic fragment of an argumentation whose final step corresponds to an application of the rule of *conditional proof*:

> Suppose that A; ...; consequently, B. Thus, on the supposition that A, B. Therefore, if A, then B.

As a second example, consider a schematic fragment of an argumentation whose final step corresponds to an application of the rule of *argument by cases*:

> A or B. First, suppose that A; ...; therefore, C. Thus, on the supposition that A, C. Second, suppose B; ...; therefore, C. Thus, on the supposition that B, C. Therefore, C.

If such pieces of reasoning are to be captured by derivations of some calculus, then there has to be a means of registering the shifting dependencies of the sentences which are conditionally asserted at various points of the argumentation. A natural way to capture this is by means of rules in which not every initial antecedent is also a final antecedent. An initial antecedent that is not also a final antecedent in an application of such a rule is then interpreted as an assumption that is discharged in this application.[8] It is therefore necessary to introduce a concept that is wider than the concept of a basic rule:

N-Rules A rule is called an *N-rule* iff, for every application of it, the set of final antecedents is a (possibly improper) subset of the set of initial antecedents.

Basic rules are those N-rules in whose applications these sets are identical. Now, N-rules do not yet suffice to provide a model for ordinary argumentations. If a calculus were to contain only N-rules, then there would be no derivable sequent with a non-empty set of antecedents. Minimally, there has to be one rule which introduces new antecedents. As is usual, I will assume that in a calculus of natural deduction there is exactly one such rule: the rule (A) from the calculi $\mathcal{R}^I$ and $\mathcal{R}^C$. This gives a broad class of calculi:

Calculi of Natural Deduction A calculus $\mathcal{R}$ is called a *calculus of natural deduction* iff $\mathcal{R}$ contains only N-rules and the rule (A).

An application of a rule of natural deduction either corresponds to the introduction of a new assumption, to a simple inference in which a sentence is inferred from other sentences, inheriting their dependencies, or to a complex argumentative move in which a sentence is inferred from other sentences while, at the same time, a previously made assumption is discharged. With respect to calculi of natural deduction, antecedents are called *assumptions*, initial succedents are called *premises*, and final succedents are called *conclusions*. In an application of an N-rule, the initial assumptions which are not also final assumptions are called *discharged* assumptions.

One should note that not only N-rules and the rule (A) correspond to informal rules of inference (cp. Dummett 1991: 186-7). Consider the following schematic fragment of an argumentation:

> First, suppose A; ...; therefore, C. Thus, on the supposition that A, C. Second, suppose B; ...; therefore, C. Thus, on the supposition that B, C. Therefore, on the supposition that A or B, C.

---

[8]   Pieces of reasoning which correspond to such rules are sometimes described as inferences whose premises are derivations. One would then say, for example, that in an application of the rule of conditional proof, a conditional is inferred from a derivation of its consequent from its antecedent. I find this description not entirely satisfactory. But what it rightly emphasises is that in the application of these rules a certain ascent of perspectives is involved: one is only justified in applying such a rule at a certain time if one is aware of what one did before.

The informal rule of inference applied in the final step of this argumentation corresponds to the following (formal) rule:

$$\frac{\Gamma, \varphi \succ \chi \quad \Delta, \psi \succ \chi}{\Gamma, \Delta, \varphi \vee \psi \succ \chi} \ (\vee_?)$$

However, the rule $(\vee_?)$ is not an N-rule. In its applications, disjunctions that are not initial antecedents are final antecedents.

Nevertheless, it is legitimate to restrict attention to calculi of natural deduction in what follows. The reason is that there is a simple recipe for replacing a rule distinct from (A) which introduces new antecedents by an N-rule, where the new rule shares all features with the original rule that are relevant for the proof-theoretic arguments considered here. Suppose that $\mathcal{R}$ is a calculus which contains the rule (A) and the following N-rule, which will belong to the most important calculi discussed in this chapter (see 5.3.1):

$$\frac{\Gamma \succ \varphi \quad \Delta \succ \psi}{\Gamma, \Delta \succ \varphi} \ (\mathrm{W})$$

Now, suppose that $\mathcal{R}$ contains a rule represented by a schema of the following form

$$\frac{Y_1 \succ B_1 \quad \ldots \quad Y_m \succ B_m}{X, C_1, \ldots, C_n \succ A}$$

in which the final antecedents which are not also initial antecedents are precisely $C_1, \ldots, C_n$. Then, let $\Gamma_1, \ldots, \Gamma_n$ be distinct schematic letters for fs-sets which do not occur in the above schema and replace this rule by the following one:

$$\frac{Y_1 \succ B_1 \quad \ldots \quad Y_m \succ B_m \quad \Gamma_1 \succ C_1 \quad \ldots \quad \Gamma_n \succ C_n}{X, \Gamma_1, \ldots, \Gamma_n \succ A}$$

For example, the application of this procedure to the rule $(\vee_?)$ which introduces a new disjunction into the antecedent yields the rule $(\vee_E)$. It is easy to see that if one applies this recipe to every rule of $\mathcal{R}$ which introduces new antecedents, except (A), then one obtains a calculus of natural deduction which yields the same derivable sequents as $\mathcal{R}$.[9] Furthermore, it will be seen that the new calculus has all those features of $\mathcal{R}$ that are relevant for the proof-theoretic arguments considered here.[10] In conclusion, it is therefore no loss of generality if one restricts attention to calculi of natural deduction.

---

[9]   On the one hand, every derivation of the original calculus can be turned into a derivation of the new calculus: in an application of $\frac{Y_1 \succ B_1 \quad \ldots \quad Y_m \succ B_m}{X, C_1, \ldots, C_n \succ A}$ one adds additional leaves of the form $\frac{}{C_i \succ C_i}$ and applies the new rule. On the other hand, every derivation of the new calculus can be turned into a derivation of the original calculus: in an application of $\frac{Y_1 \succ B_1 \quad \ldots \quad Y_m \succ B_m \quad \Gamma_1 \succ C_1 \quad \ldots \quad \Gamma_n \succ C_n}{X, \Gamma_1, \ldots, \Gamma_n \succ A}$ one applies the original rule to the first m initial sequents and uses (W) to introduce new antecedents (if necessary).

[10]   This is not to say, of course, that there could not be considerations for preferring the original calculus. In general, for example, its rules will have fewer premises and contain fewer schematic letters for fs-sets. However, such properties are irrelevant for the pertinent proof-theoretic arguments.

A very useful feature of calculi of natural deduction is that their derivations can be captured by trees of sentences. Suppose that a sequent tree of a calculus of natural deduction, that is, a tree together with a function f which maps every node of the tree v to a sequent s^v, is given. Then, to a first approximation (see below), one obtains *the corresponding sentential tree* if one replaces f by the function f^* which maps a node v of the tree to the set of succedents of the sequent s^v. This can be visualised with respect to representations of sequent trees. Consider the following representation of a derivation of the calculus $\mathcal{R}^I$:

$$\frac{\dfrac{p \wedge q \succ p \wedge q}{p \wedge q \succ q} \quad \dfrac{p \wedge q \succ p \wedge q}{p \wedge q \succ p}}{p \wedge q \succ q \wedge p}$$

Its corresponding sentential tree then has the following representation:

$$\frac{\dfrac{p \wedge q}{q} \quad \dfrac{p \wedge q}{p}}{q \wedge p}$$

As it stands, however, the definition of the concept of a corresponding sentential tree is not yet adequate since it has not been taken into account that assumptions are discharged in certain transitions. To this end, one has to alter the definition of the function f^* of the corresponding sentential tree, which replaces the function f of the underlying sequent tree. Suppose that some fs-set x has been discharged in a transition with final sequent s^v. Then, the function f^* of the corresponding sentential tree maps v to the pair consisting of the set of conclusions of s^v and x. This is most easily understood with an example of representations of a sequent tree:

$$\frac{\dfrac{p \to (q \to p) \succ p \to (q \to p) \quad \dfrac{p \wedge q \succ p \wedge q}{p \wedge q \succ p}}{p \to (q \to p), p \wedge q \succ q \to p} \quad \dfrac{p \wedge q \succ p \wedge q}{p \wedge q \succ q}}{\dfrac{p \to (q \to p), p \wedge q \succ p}{p \to (q \to p) \succ (p \wedge q) \to p}}$$

and of its corresponding sentential tree:

$$\frac{\dfrac{p \to (q \to p) \quad \dfrac{p \wedge q^{(1)}}{p} \quad \dfrac{p \wedge q^{(1)}}{q}}{q \to p}}{\dfrac{p}{(p \wedge q) \to p^{(1)}}}$$

In the last transition, the assumption $p \wedge q$ has been discharged. The sentential list which denotes the set of antecedents of the sequent that labels the root in the sequent tree does not contain the sentence $p \wedge q$ any more. At the root of the corresponding sentential tree one therefore finds a pair of items: a term for the conclusion $(p \wedge q) \to p$ and an expression which represents the discharged assumption $p \wedge q$.

As in the given example, I will always use bracketed numerals as superscripts to assumptions discharged at some place in the derivation. These assumptions always label leaves of sentential trees, and it is required that only tokens of the same type are labelled with the same numeral. Correspondingly, I will use lists of bracketed numerals as superscripts to conclusions of transitions to indicate the set of assumptions which is discharged in these transitions.[11]

5.1.3 Schemata for Rules and Standard Calculi

In this subsection, I will deal with the relation between a rule and schematic transitions that represent it, and I will introduce the notion of a standard N-rule and the corresponding notion of a standard calculus. For simplicity, I will introduce these notions with respect to the logical framework ARG, whose sequents are arguments. It poses no problem, however, to transfer them to other logical frameworks.

Recall the definition of a schematic transition: a schematic transition is a two-dimensional expression consisting of a horizontal line, a finite number of schematic sequents above the line, and another schematic sequent below the line. A schematic sequent (of the logical framework ARG), in turn, consists of a schematic list X and a schematic sentence A, which are separated by the symbol $\succ$. Recall furthermore that in a schematic list a schematic sentence should be thought of as representing its singleton and that a comma should be thought of as representing the operation of set-union. One may now single out a special class of rules and a corresponding class of calculi:

Schematic Rules A rule is called *schematic* iff it can be represented by some schematic transition.

Schematic Calculi A calculus $\mathcal{R}$ is called *schematic* iff $\mathcal{R}$ is finite and contains only schematic rules.

In this chapter, I will only deal with schematic calculi.

The notion of a schematic transition is rather narrow. First of all, schematic transitions are not allowed to be qualified with *side conditions*, as, for example, representations of quantificational rules usually are. Moreover, the only set-theoretic operation permitted is the operation of set-union. Thus, for example, the term $\Gamma - \varphi$, which contains a sign for set-subtraction, is not considered as a schematic list, and, therefore, an expression such as $\frac{\Gamma, \varphi \succ \psi}{\Gamma - \varphi \succ \varphi \rightarrow \psi}$ is not considered as a schematic transition. Consequently, I would like to stress that I will not rely on the assumption that a calculus has to be schematic myself. Rather, I will try to show that even if it is granted that a calculus has to be schematic, there is no convincing argument to the conclusion that an adequate calculus cannot be classical.

I will now introduce the central notion of a standard N-rule and the corresponding notion of a standard calculus. Recall that apart from the rule (A), calculi of natural deduction contain only N-rules, i.e. rules in whose applications the set of final antecedents is a subset of the set of initial antecedents. Now, there is an important class of schemata for N-rules:

---

[11]   For a precise (recursive) definition, see Troelstra & Schwichtenberg 2000: 35-8.

Standard Schemata A schema for an N-rule is called a *standard* schema iff it has the following form:

$$\frac{\Gamma_1, C_1^1, \ldots, C_{n_1}^1 \succ B_1 \quad \ldots \quad \Gamma_m, C_1^m, \ldots, C_{n_m}^m \succ B_m}{\Gamma_1, \ldots, \Gamma_m \succ A}$$

where $m, n_1, \ldots, n_m$ are natural numbers, $\Gamma_1, \ldots, \Gamma_m$ are distinct schematic letters for fs-sets, and C_j^i, B_i, A are schematic sentences.

(The special case for $m = 0$ is worth noting: $\frac{}{\succ A}$.) In standard schemata, non-discharged assumptions have to be represented by distinct schematic letters for fs-sets. The notion of a standard schema directly yields corresponding notions of a standard rule and a standard calculus:

Standard Rules An N-rule is called *standard* iff it has a standard schema.

Standard Calculi A schematic calculus of natural deduction $\mathcal{R}$ is called *standard* iff the N-rules of $\mathcal{R}$ are standard.

In what follows, it will be convenient to restrict attention to standard calculi.[12]
    Requiring a schema for an N-rule with m initial sequents to be standard can be seen as a requirement consisting of four sub-requirements. First, it is required that the schematic lists for the discharged assumptions only contain schematic sentences; they are not allowed to contain schematic letters for fs-sets. (This would be violated by a schema like $\frac{\Gamma, \Delta \succ \varphi \quad \Sigma \succ \neg \varphi}{\Gamma, \Sigma \succ \psi}$, in which an fs-set represented by the schematic letter Δ is discharged.) Second, it is required that the schematic list for the final assumptions is the concatenation of m non-empty schematic lists such that each list occurs in exactly one initial schematic sequent. (This would be violated by a schema like $\frac{\Gamma, \Delta \succ \varphi \quad \Delta, \Sigma \succ \varphi}{\Gamma, \Delta, \Sigma \succ \varphi}$, in which Δ occurs in both initial sequents.) Third, it is required that each of the m non-empty schematic lists consists of a single schematic letter for fs-sets. (This would be violated by a schema like $\frac{\varphi \succ \psi}{\varphi \succ \psi \vee \chi}$, in which the schematic list consists of a schematic sentence.)
    I am not aware of views on logical theories which have to be based on a calculus of natural deduction whose rules violate one of these three sub-requirements. This is different with respect to the fourth sub-requirement, in which it is demanded that the m schematic letters for fs-sets are distinct. This requirement is violated by a schema like the following

$$\frac{\Gamma \succ \varphi \vee \psi \quad \Delta, \varphi \succ \chi \quad \Delta, \psi \succ \chi}{\Gamma, \Delta, \Delta \succ \chi}$$

which is a representation of the rule of argument by cases as defended by adherents of quantum logic (see Humberstone 2011: 298-302). However, I will not deal with such logical theories here. The debate about the respective merits of intuitionistic and classical calculi can be understood more easily if it is taken for granted that N-rules are standard.

---

[12] The property of being a standard rule is closely related to what Humberstone (2011: 521) calls the property of being a rule which is *general in respect to side formulas*.

A helpful feature of standard schemata is that they can be displayed by simplified expressions. Instead of the expression above, one may use the following:

$$
\begin{array}{ccc}
[C_1^1 \ldots C_{n_1}^1] & & [C_1^m \ldots C_{n_m}^m] \\
\vdots & & \vdots \\
B_1 & \ldots & B_m \\
\hline
& A &
\end{array}
$$

where A stands for the conclusion, $B_1, \ldots, B_m$ stand for the premises, and $C_1^1, \ldots, C_{n_m}^m$ stand for the discharged assumptions. In a standard schema it is not necessary to display the undischarged assumptions since they are always required to be represented by distinct schematic letters for fs-sets, one for each initial sequent.

In the following, I will often speak about 'the' schema corresponding to some schematic rule although, strictly speaking, every schematic rule is represented by infinitely many schemata. This loose way of speaking is justified by the fact that different schemata representing the same rule differ only in inessential ways. I will call two schemata *equivalent* iff they represent the same rule. Then, whenever I characterise 'the' schema of some schematic rule, I will take care that the characterisation applies to all schemata representing the rule.

I will conclude this subsection by indicating under which conditions two standard schemata are equivalent:

$$
\begin{array}{ccc}
[C_1^1 \ldots C_{n_1}^1] & & [C_1^m \ldots C_{n_m}^m] \\
\vdots & & \vdots \\
B_1 & \ldots & B_m \\
\hline
& A &
\end{array}
\qquad
\begin{array}{ccc}
[{}^*C_1^1 \ldots {}^*C_{j_1}^1] & & [{}^*C_1^i \ldots {}^*C_{j_i}^i] \\
\vdots & & \vdots \\
{}^*B_1 & \ldots & {}^*B_i \\
\hline
& {}^*A &
\end{array}
$$

There are the following five reasons for such schemata to be equivalent:

(i)   Two schemata are equivalent if their only difference consists in the order of the items of some schematic list of discharged assumptions.[13]

(ii) Two schemata are equivalent if they differ only in the order of the initial schematic sequents.

(iii) Two schemata are equivalent if their only difference consists in the multiplicity of the items of some schematic list.[14]

(iv) Two schemata are equivalent if their only difference consists in the multiplicity of some initial schematic sequent.

(v)   Two schemata are equivalent if one of them can be obtained from the other one by a systematic replacement of schematic letters for sentences

---

[13]   For example, the two displayed schemata are equivalent if their only difference consists in the fact that C_1^1, C_2^1 equals $\varphi, \neg\varphi$ while ${}^*C_1^1, {}^*C_2^1$ equals $\neg\varphi, \varphi$.

[14]   For example, the two displayed schemata are equivalent if their only difference consists in the fact that $C_1^1 \ldots C_{n_1}^1$ equals φ while ${}^*C_1^1 \ldots {}^*C_{j_1}^1$ equals φ, φ.

(Here, such a replacement is called *systematic* iff identity and distinctness of schematic letters is preserved.) I conjecture that the relation of equivalence between standard schemata is the transitive and reflexive closure of these conditions. However, nothing of what follows depends on this conjecture.

5.2 Accounts of Meaning and Logical Frameworks

I will now turn to the discussion of proof-theoretic arguments against classical logic. According to the proponents of these arguments, the theory of intuitionistic logic is to be preferred over the theory of classical logic because intuitionistic calculi are to be preferred over classical calculi. In 5.2.1, I will introduce the general form of the proof-theoretic arguments discussed in this chapter. In 5.2.2, I will then deal with the question of which logical frameworks are relevant for these arguments.

5.2.1 The General Form of Proof-Theoretic Arguments

In this subsection, I will first introduce the notion of a classical and of an intuitionistic calculus. Then, I will indicate the mathematical and the non-mathematical component of a proof-theoretic argument. Finally, I will give a preliminary explanation of how the non-mathematical component is to be understood.

Recall that $\mathcal{L}^*$ is the formal language whose logical operators are precisely the standard sentential operators: $\neg$, $\vee$, $\wedge$, $\rightarrow$. In general, a calculus is called classical iff it yields precisely the arguments of the relevant language which are logically valid according to classical logic. Since I restrict attention to sentential logic in this chapter, this leads to the following definition:

Classical Calculi A calculus $\mathcal{R}$ is called $\text{ARG}_{\mathcal{L}^*}$-*classical* iff the following holds: a sequent of $\text{ARG}_{\mathcal{L}^*}$ is $\mathcal{R}$-derivable iff it is $\mathcal{R}^C$-derivable.

Note that an $\text{ARG}_{\mathcal{L}^*}$-classical calculus is neither required to be a calculus of the logical framework ARG nor to be a calculus whose transitions are sequents of $\mathcal{L}^*$. It is only required to yield every sequent that is $\mathcal{R}^C$-derivable and to yield a sequent of $\text{ARG}_{\mathcal{L}^*}$ only if it is $\mathcal{R}^C$-derivable. In particular, an $\text{ARG}_{\mathcal{L}^*}$-classical calculus is allowed to be based on a language which properly extends $\mathcal{L}^*$, and it is allowed to belong to a logical framework which properly extends ARG.[15] Corresponding to the notion of an $\text{ARG}_{\mathcal{L}^*}$-classical calculus, there is the notion of an $\text{ARG}_{\mathcal{L}^*}$-intuitionistic calculus:

Intuitionistic Calculi A calculus $\mathcal{R}$ is called $\text{ARG}_{\mathcal{L}^*}$-*intuitionistic* iff the following holds: a sequent of $\text{ARG}_{\mathcal{L}^*}$ is $\mathcal{R}$-derivable iff it is $\mathcal{R}^I$-derivable.

---

[15]   The definition precludes $\text{ARG}_{\mathcal{L}^*}$-classical calculi based on some language $\mathcal{L}^+$ that does not contain $\mathcal{L}^*$. In 5.6.1, it will be seen that some defenders of classical logic proceed by presenting a calculus based on some such language $\mathcal{L}^+$ and by invoking a translation function between $\mathcal{L}^+$ and $\mathcal{L}^*$. To begin with, I will not take such complications into account.

An $\text{ARG}_{\mathcal{L}^*}$-intuitionistic calculus yields every sequent that is $\mathcal{R}^I$-derivable, and it yields a sequent of $\text{ARG}_{\mathcal{L}^*}$ only if it is $\mathcal{R}^I$-derivable.

By relying on these notions, it is possible to indicate the general form of a proof-theoretic argument against classical logic. Such an argument is based on a mathematical result about calculi and about certain non-mathematical theses about the significance of the mathematical result. The relevant mathematical result has the following form:

Anti-Cla$_F$ No $\text{ARG}_{\mathcal{L}^*}$-classical calculus is F.

Note that this is a result which concerns *every* $\text{ARG}_{\mathcal{L}^*}$-classical calculus. Furthermore, since the proof-theoretic arguments are put forward by intuitionists, the property of being F is exemplified by at least one intuitionistic calculus:

Pro-Int$_F$ At least one $\text{ARG}_{\mathcal{L}^*}$-intuitionistic calculus is F.

The combination of Anti-Cla$_F$ and Pro-Int$_F$ states that the property of being F is a distinguishing mark of some $\text{ARG}_{\mathcal{L}^*}$-intuitionistic calculi; i.e., it is a property that no $\text{ARG}_{\mathcal{L}^*}$-classical calculus shares with them.

Now, if a result of the form Anti-Cla$_F$ can be used to undermine the theory of classical logic, then the soundness of classical logic must depend on the existence of an $\text{ARG}_{\mathcal{L}^*}$-classical calculus which is F. This conditional constitutes the non-mathematical ingredient of a proof-theoretic argument: if the theory of classical logic is sound, then there is an $\text{ARG}_{\mathcal{L}^*}$-classical calculus which is F. Adherents of a proof-theoretic argument try to establish a link between the mathematical property of being F and a property with a certain non-mathematical force: a property which some calculus that corresponds to a sound logical theory must have.

It turns out to be helpful to split up the non-mathematical component of a proof-theoretic argument into two parts:

Existence of Adequate Calculi (EAC) If the theory of classical sentential logic is sound, then there is an adequate $\text{ARG}_{\mathcal{L}^*}$-classical calculus.

Adequateness Implies F-ness (AIF) An adequate calculus is F.

Here, adequacy is the non-mathematical property of calculi which constitutes the link between the mathematical result Anti-Cla$_F$ and the non-mathematical thesis that classical logic fails: the combination of Anti-Cla$_F$, EAC, and AIF implies that the theory of classical sentential logic is not sound. A proponent of a proof-theoretic argument tries to present a replacement of "F" and an interpretation of "adequate" such that the mathematical result Anti-Cla$_F$ can be proven and such that EAC and AIF can be made plausible. (The result Pro-Int$_F$ then ensures that it is not possible to present an analogous argument against the theory of intuitionistic logic.)

In the remainder of this subsection, I will indicate how the non-mathematical property of being adequate is to be understood. It will be a property that a calculus has iff it has a certain semantic potential. To begin with, however, I would like to make one remark about the corresponding mathematical property of being F.

Given that adequacy is a semantic property of calculi, the most obvious candidate for the corresponding mathematical property of being F is probably a model-theoretic one: the property of producing precisely those arguments of $\mathcal{L}^*$ that are truth-preserving with respect to all models of some independently specified class. The idea would be that calculi with the required semantic potential have to respect an independently specified model-theoretic notion of logical validity. Interestingly, however, proof-theoretic arguments are semantic arguments which do not concern such a model-theoretic property of calculi. And it is not difficult to see why this is so: if they were based on the thesis that classical logic is not sound because it does not capture a certain model-theoretic notion of logical validity, then considerations about calculi would constitute an unnecessary detour. Thus, the mathematical property of being F will be related to a property with semantic significance, but it will not relate (in any obvious sense) to the property of being true in a model.

What then is the relevant semantic property of being adequate? To a first approximation, a calculus is adequate iff it yields an account of the meanings of the standard sentential operators, i.e. an account of the meanings of the operators $\neg, \vee, \wedge, \rightarrow$ corresponding to "not", "or", "and", "if".[16] An adherent of a proof-theoretic argument claims that classical sentential logic would only be sound if some $\text{ARG}_{\mathcal{L}^*}$-classical calculus yielded an account of the meanings of the standard sentential operators. On the basis of this claim, she then infers that classical sentential logic is not sound because there is no $\text{ARG}_{\mathcal{L}^*}$-classical calculus which is F, while, according to her, only calculi that are F can yield an account of the relevant meanings.

It may be noted that the underlying thesis of this argument, according to which there has to be a calculus that yields an account of the meanings of the standard sentential operators, is related to the semantic manifestation thesis (4.1.2), according to which the use of a sentence determines its meaning. However, there are differences between these two assumptions, the most important of which is that the semantic manifestation thesis concerns *sentences* whose meanings are determined by their use, while the corresponding assumption of the proof-theoretic arguments concerns *logical constants* whose meanings are accounted for by inferential rules which govern them.

What does it mean to say that some mathematical structure, namely a set of rules, yields an account of the meanings of certain words? To be clear about the significance of this thesis, it is helpful to consider a specific suggestion for how the meanings of sentential operators could be accounted for. To this end, take a look at the following statement by Rumfitt, in which he recapitulates a proposal by Peacocke (1987):

> [A] connective will possess the sense that it has by virtue of its competent users' finding certain rules of inference involving it to be primitively obvious. A speaker will apprehend or grasp the connective's sense just in case he knows to go by those rules when engaged

---

[16]   Obviously, there are related notions which concern the meanings of other expressions. In this chapter, however, only the meanings of the standard sentential operators are relevant.

in deductive reasoning; and a theorist may specify the connective's sense by formulating these rules (Rumfitt 2000: 787-8).

The idea is that certain rules account for the meanings of the operators they govern because competent users of the operators have a certain attitude towards applications of these rules: they find them primitively obvious. What I would like to draw attention to is that Peacocke's conception of a rule has to be different from the present one if his proposal is to have any chance of being convincing.[17] In this chapter, a rule is conceived of as a mathematical object: it is a set of pairs of sets of sequents and sequents, where, depending on the logical framework under consideration, a sequent is a pair of finite sets of sentences which satisfies a certain condition (e.g. the condition that the second member of the pair contains precisely one element). Now, of course, Peacocke does not want to claim that a sentential operator possesses the meaning that it does because competent users of it find certain of such sets of pairs primitively obvious. Systems of rules, in the present sense of that term, can only yield an account of the meanings of the involved operators if they are interpreted as corresponding to informal rules of inference. Unlike the mathematical objects that are called rules here, informal rules of inference can sensibly be declared to be primitively obvious.

A calculus can only yield an account of the relevant meanings *relative to a correspondence between its elements and informal rules of inference* (cp. Steinberger 2011: 335). More precisely, there has to be a general correlation between the transitions of the rules of the calculus and informal argumentative moves, i.e. a correlation that maps a transition, in accordance with its structural features, to an argumentative act. As an example, recall from 5.1.2 how transitions of calculi of natural deduction of the framework ARG are taken to correspond to informal argumentative moves: (i) an application of the rule (A) is interpreted as an argumentative move in which an assumption is introduced; (ii) a transition in which the set of final assumptions equals the set of initial assumptions is interpreted as a simple inference, in which some sentence is inferred from other sentences while inheriting their dependencies; (iii) a transition in which the set of final assumptions is a proper subset of the set of initial assumptions is interpreted as a complex argumentative move, in which an inference is combined with an act in which a previously made assumption is discharged.

The claim that a certain calculus does not yield an account of the relevant meanings has to be understood as the claim that there is no permissible correlation between its elements and informal rules of inference, relative to which it yields an account of the relevant meanings. (In the following subsection, it will

---

[17]   In Chapter 2, I questioned the idea that competent users of a logical constant must accept the basic rules that govern it. Similarly, I do not think that competent users of a logical constant must find the basic rules that govern it primitively obvious. As a consequence, a plausible suggestion about how a system of rules may account for the meanings of the operators it governs differs from the one that is given by Peacocke (cp. Rumfitt 2000: 788-9). Here, however, I only want to use his proposal to illustrate that there are different conceptions of rules which have to be distinguished.

be seen that it is controversial which correlations between formal and informal rules are permissible.)

It must be emphasised, however, that this does not mean that a calculus can only yield an account of the relevant meanings if there is a permissible correlation between its elements and informal rules of inference such that the corresponding informal rules are precisely those that are generally accepted as the basic rules for the relevant vocabulary. On the contrary, adherents of proof-theoretic arguments invoke the thesis that formal rules have to be suitably related to informal rules to substantiate the claim that certain standardly used informal rules do *not* yield an account of the relevant meanings. The idea is not that an adequate calculus must precisely capture the totality of standardly accepted informal rules. The idea is rather that an adequate calculus must correspond to a legitimate *systematisation* of the basic informal rules that govern the standard sentential operators. The applications of the rules of an adequate calculus have to constitute an *idealised* model of the class of basic informal argumentative moves which involve these operators. This idea will be further clarified in 5.4.1.

By way of conclusion, recall the form of a proof-theoretic argument against classical logic:

P$_1$　There is no ARG$_{\mathcal{L}^*}$-classical calculus which is F.

P$_2$　If the theory of classical sentential logic is sound, then there is an adequate ARG$_{\mathcal{L}^*}$-classical calculus.

P$_3$　An adequate calculus is F.

C　The theory of classical sentential logic is not sound.

A proof-theoretic argument is obtained from this form if "F" is replaced by a suitable predicate. In each of the following proof-theoretic arguments, the property of being F will be a conjunctive property: to be F is to be F_1, F_2, $\ldots$, and F_n (for some F_1, $\ldots$, F_n). The third premise of a proof-theoretic argument then amounts to the claim that an adequate calculus exemplifies each of these conjunctive components of the property of being F. I will refer to the claims that an adequate calculus exemplifies such components of the property of being F as *desiderata*. Adherents of proof-theoretic arguments claim that a calculus has to satisfy a number of such desiderata if it is to yield an account of the relevant meanings (relative to a permissible correlation between its elements and informal rules of inference.)

5.2.2　Permissible Logical Frameworks

In the following sections, mathematical properties of calculi which no ARG$_{\mathcal{L}^*}$-classical calculus has are discussed. Adherents of proof-theoretic arguments claim that classical sentential logic would only be sound if there were adequate ARG$_{\mathcal{L}^*}$-classical calculi, and that an adequate calculus has to have one or the other of these properties. It would then follow that classical sentential logic is not sound. As will be seen, the relevant mathematical properties will all be exemplified by slight variants of the $\text{ARG}_{\mathcal{L}^*}$-intuitionistic calculus $\mathcal{R}^I$. Consequently, an adherent of a proof-theoretic argument has to show that for every $\text{ARG}_{\mathcal{L}^*}$-classical calculus $\mathcal{R}$ there is a semantically relevant property which (the pertinent variant of) $\mathcal{R}^I$ exemplifies but $\mathcal{R}$ does not, i.e. a property which an adequate calculus has to have. In this subsection, I will introduce an $\text{ARG}_{\mathcal{L}^*}$-classical calculus which closely resembles $\mathcal{R}^I$, thereby motivating a first desideratum on adequate calculi.

The relevant $\text{ARG}_{\mathcal{L}^*}$-classical calculus is a counterpart of $\mathcal{R}^I$ in the logical framework $\text{G-ARG}_{\mathcal{L}^*}$, whose sequents are generalised arguments of $\mathcal{L}^*$.[18] The rules for the sentential operators are, with two exceptions, exact counterparts of the rules in $\mathcal{R}^I$: they are obtained by adding schematic letters for fs-sets to all succedents. The only exceptions are the rules $(\neg_I^g)$ and $(\vee_E^g)$, which are simplifications of the 'canonical' rules for introducing $\neg$ and eliminating $\vee$:[19]

$$\frac{\Gamma_1 \succ \varphi, \Delta_1 \quad \Gamma_2 \succ \psi, \Delta_2}{\Gamma_1, \Gamma_2 \succ \varphi \wedge \psi, \Delta_1, \Delta_2} \ (\wedge_I^g) \qquad \frac{\Gamma \succ \varphi \wedge \psi, \Delta}{\Gamma \succ \varphi, \Delta} \ (1 \wedge_E^g) \quad \frac{\Gamma \succ \varphi \wedge \psi, \Delta}{\Gamma \succ \psi, \Delta} \ (2 \wedge_E^g)$$

$$\frac{\Gamma \succ \varphi, \Delta}{\Gamma \succ \varphi \vee \psi, \Delta} \ (1 \vee_I^g) \quad \frac{\Gamma \succ \psi, \Delta}{\Gamma \succ \varphi \vee \psi, \Delta} \ (2 \vee_I^g) \qquad \frac{\Gamma \succ \varphi \vee \psi, \Delta}{\Gamma \succ \varphi, \psi, \Delta} \ (\vee_E^g)$$

$$\frac{\Gamma, \varphi \succ \psi, \Delta}{\Gamma \succ \varphi \to \psi, \Delta} \ (\to_I^g) \qquad \frac{\Gamma_1 \succ \varphi \to \psi, \Delta_1 \quad \Gamma_2 \succ \varphi, \Delta_2}{\Gamma_1, \Gamma_2 \succ \psi, \Delta_1, \Delta_2} \ (\to_E^g)$$

$$\frac{\Gamma, \varphi \succ \Delta}{\Gamma \succ \neg\varphi, \Delta} \ (\neg_I^g) \qquad \frac{\Gamma_1 \succ \varphi, \Delta_1 \quad \Gamma_2 \succ \neg\varphi, \Delta_2}{\Gamma_1, \Gamma_2 \succ \psi, \Delta_1, \Delta_2} \ (\neg_E^g)$$

The calculus that comprises these rules and the rule (A) is denoted by $^g\mathcal{R}^C$. It is a standard calculus (see 5.1.3), which is easily seen to be $\text{ARG}_{\mathcal{L}^*}$-classical. Furthermore, it resembles $\mathcal{R}^I$ in every relevant respect except that it allows for sequents with more than one conclusion. The proof-theoretic arguments against classical logic thus depend on the assumption that a calculus with multiple conclusions cannot be adequate. In other words, the assumption that there is no permissible correlation between applications of its rules and informal argumentative moves in relation to which it yields an account of the meanings of the standard sentential operators. The purpose of this subsection is to discuss the nature of this assumption.

---

[18] Calculi of (a variant of) the logical framework G-ARG have been introduced by Gentzen (1934). (Gentzen adopted a framework with sequents whose antecedents and succedents are finite *sequences* of sentences instead of finite *sets* of sentences.) Natural deduction variants of Gentzen's calculus have been developed by Kneale (1956), von Kutschera (1962), Shoesmith & Smiley (1978), and Boričić (1985).

[19] The 'canonical' rules for introducing $\neg$ and eliminating $\vee$ are the following:

$$\frac{\Gamma_1, \varphi \succ \psi, \Delta_1 \quad \Gamma_2, \varphi \succ \neg\psi, \Delta_2}{\Gamma_1, \Gamma_2 \succ \neg\varphi, \Delta_1, \Delta_2} \qquad \frac{\Gamma_1 \succ \varphi \vee \psi, \Delta_1 \quad \Gamma_2, \varphi \succ \chi, \Delta_2 \quad \Gamma_3, \psi \succ \chi, \Delta_3}{\Gamma_1, \Gamma_2, \Gamma_3 \succ \chi, \Delta_1, \Delta_2, \Delta_3}$$

It is easily seen that the calculus which contains these rules instead of $(\neg_I^g)$ and $(\vee_E^g)$ produces the same sequents as $^g\mathcal{R}^C$.

The claim that calculi of natural deduction with multiple conclusions cannot yield an account of the relevant meanings has been questioned by Read (2000). Read argues that (a variant of) $^g\mathcal{R}^C$ can be used to rebut the proof-theoretic arguments against classical logic. (The differences between $^g\mathcal{R}^C$ and the calculus favoured by Read are irrelevant for present purposes.) He presents the following consideration as a justification for this claim:

> [Dummett and Prawitz] exclude multiple conclusions from consideration because they allow the assertion of disjunctions neither of whose disjuncts is assertible. But that is to beg the question. The question is whether intuitionistic logic is superior proof-theoretically to classical logic. To exclude forms of proof which are intuitionistically unacceptable is to introduce a circle in the reasoning (Read 2000: 145).

This suggests that Read believes that Dummett and Prawitz exclude calculi such as $^g\mathcal{R}^C$ because they produce sequents like $\succ p \lor \neg p$, which are not logically valid according to intuitionistic logic. (Someone who knows $p \lor \neg p$ to be logically true, but who neither knows p nor $\neg p$ to be true, would be entitled to assert a disjunction neither of whose disjuncts she would be entitled to assert.) Now, this would indeed introduce a circle in the reasoning of Dummett and Prawitz: according to this interpretation, they claim that intuitionistic logic is better than classical logic because intuitionistic calculi are better than classical calculi, and they claim that this is so because intuitionistic calculi correspond to intuitionistic logic, which is better than classical logic to which classical calculi correspond.

Uncontroversially, proponents of the proof-theoretic arguments against classical logic do not claim that no mathematical structure determines exactly those arguments of $\text{ARG}_{\mathcal{L}^*}$ which are logically valid according to classical sentential logic. They only claim that no such structure yields an account of the meanings of $\neg$, $\lor$, $\land$, $\rightarrow$. As I have emphasised in the previous subsection, this has to be understood as involving an implicit reference to a correlation between transitions of the relevant calculus (which here involve generalised arguments) and informal argumentative moves. Adherents of proof-theoretic arguments claim that there is no permissible correlation between transitions that involve generalised arguments and informal argumentative moves in relation to which an $\text{ARG}_{\mathcal{L}^*}$-classical calculus of the framework G-ARG yields an account of the meanings of the standard sentential operators. The dispute between Read and adherents of proof-theoretic arguments has to be understood against this background. It is a dispute about the existence of permissible correlations between transitions of the framework G-ARG and informal argumentative moves relative to which a calculus of this framework does or does not yield an account of the relevant meanings.

In his discussion of calculi of the logical framework G-ARG, Dummett considers only one possible interpretation of their sequents:

> In a succedent comprising more than one sentence, the sentences are connected disjunctively [...] (Dummett 1991: 187).

This is the standard interpretation of sequents of G-ARG given by the inventor of calculi of this framework (see Gentzen 1934: 180). Now, given this *disjunctive* interpretation of sequents of G-ARG, it is no more difficult to set up a correspondence between applications of rules of $^g\mathcal{R}^C$ and ordinary argumentative moves than to set up a correspondence between applications of rules of $\mathcal{R}^I$ and ordinary argumentative moves. If a multiple succedent is interpreted as the disjunction of its items, then applications of rules of $^g\mathcal{R}^C$ can be interpreted in perfect analogy to applications of rules of $\mathcal{R}^I$. In one kind of transitions, new assumptions are introduced; in a second kind of transitions, simple inferences in which the conclusion inherits the dependencies of the premises are drawn; in a third kind of transitions, simple inferences are combined with acts in which previously made assumptions are discharged. The question that remains is whether such an interpretation is permissible for an account of the meanings of $\neg, \vee, \wedge, \rightarrow$.

What could be the reason for claiming that this correlation between formal and informal rules is not permissible? According to Read, Dummett rejects this correlation because it licences the assertion of disjunctions neither of whose disjuncts is assertible. However, this is not correct. Dummett presents his reasons for rejecting this correlation by means of the following claims:

> Sequents with two or more sentences in the succedent [...] have no straightforwardly intelligible meaning, explicable without recourse to any logical constant. [...] In a succedent comprising more than one sentence, the sentences are connected disjunctively; and it is not possible to grasp the sense of such a connection otherwise than by learning the meaning of the constant "or" (Dummett 1991: 187).

The problem is not that systems of informal rules that correspond to certain calculi of the framework G-ARG permit intuitionistically unacceptable attitudes towards certain disjunctions. The problem is that the disjunctive interpretation of sequents of this framework presupposes what is to be explained: the meaning of one of the standard sentential operators.

As will be seen in 5.5, Dummett denies that rules such as the following can be part of an account of the meaning of $\rightarrow$:

$$\frac{\Gamma, \varphi \succ \psi \vee \chi}{\Gamma \succ (\varphi \rightarrow \psi) \vee \chi}$$

Simplifying slightly (see 5.5.1), he claims that rules which can account for the meaning of an operator have to be rules in which this operator only occurs in dominant positions, i.e. as the main operator of the sentences which contain it. Now, whether adequate calculi are only allowed to contain such rules is a matter of debate, but this is a debate that Read does not take part in. Possibly, he thinks that he does not have to because the rules he proposes only contain operators in dominant positions, e.g.:

$$\frac{\Gamma, \varphi \succ \psi, \Delta}{\Gamma \succ \varphi \rightarrow \psi, \Delta}$$

In this schema, $\to$ occurs in dominant position. However, if multiple conclusions are to be interpreted disjunctively, then the formal rules proposed by Read correspond to informal rules which do not have this property. For example, in the informal counterpart of the above rule, the introduced conditional is a disjunct of some disjunction. If rules like $\frac{\Gamma,\varphi\succ\psi\vee\chi}{\Gamma\succ(\varphi\to\psi)\vee\chi}$ cannot be part of adequate calculi because $\to$ is introduced into the scope of $\vee$, then rules like $\frac{\Gamma,\varphi\succ\psi,\Delta}{\Gamma\succ\varphi\to\psi,\Delta}$ also cannot be part of adequate calculi in whose sequents multiple succedents are interpreted disjunctively because they correspond to informal rules in which "if" is introduced into the scope of "or".[20]

I will not question Dummett's claim that calculi of the framework G-ARG cannot yield an account of the meanings of the standard sentential operators if its succedents are interpreted disjunctively. The only expression in a schematic transition of a rule of an adequate calculus to be interpreted as expressing the concept of disjunction is the operator $\vee$. In particular, representations of the rules of adequate calculi are not allowed to contain a new structural device which has to be interpreted in this way. Now, this might be taken to suggest the following desideratum on adequate calculi:

Desideratum$_1^a$ (D$_1^a$) If $\mathcal{R}$ is an adequate calculus, then $\mathcal{R}$ does not belong to a logical framework with sequents that have multiple succedents.

It should be noted, however, that D$_1^a$ does not follow from the assumption that calculi of the framework G-ARG cannot yield an account of the meanings of the sentential operators if its succedents are interpreted disjunctively. For all that has been said, there might be an alternative correlation between transitions of the framework G-ARG and informal argumentative moves, in relation to which a calculus like $^g\mathcal{R}^C$ yields an account of the meanings of the sentential operators.[21] It thus has to be kept in mind that the desideratum D$_1^a$ can at best be treated as an assumption with some initial plausibility.

---

[20]   Tennant seems to be a more appropriate target of Read's circularity objection than Dummett (or Prawitz):

> There is no acceptable interpretation of the 'validity' of a sequent $X : Q_1, \ldots, Q_n$ in terms of preservation of warrant to assert when X contains only sentences involving no disjunctions. If one is told that $X : Q_1, \ldots, Q_n$ is 'valid' in the extended sense for multiple-conclusion arguments just in case $X : Q_1 \vee \ldots \vee Q_n$ is valid in the usual sense for single-conclusion arguments, the intuitionist can demand to know precisely which disjunct Q_i, then, proves to be derivable from X. No answer to such a question can be provided in general with the multiple-conclusion sequent calculus of the classical logician (Tennant 1997: 320).

Tennant seems to reject the usual calculi of the framework G-ARG because they do not yield a notion of *preservation of warrant to assert* which intuitionists find acceptable. But then Read's criticism of Dummett and Prawitz can be levelled against Tennant: what is wanted is a proof-theoretic argument against classical logic which does not presuppose that intuitionistic standards on logical validity are correct.

[21]   Such an alternative proposal for interpreting transitions of the framework G-ARG has been made by Restall (2005). In my view, Restall's proposal has been successfully refuted by Rumfitt (2008).

To begin with, I will assume that proof-theoretic arguments against classical logic proceed from a strengthening of D_1^a:

Desideratum$_1$ (D$_1$) If $\mathcal{R}$ is an adequate calculus, then $\mathcal{R}$ is a standard calculus of the specialised logical framework ARG$_{\mathcal{L}^*}$.

According to this assumption, not only calculi of the framework G-ARG are inadequate. In addition, it is assumed that adequate calculi (i) must be schematic, (ii) must be calculi of natural deduction, (iii) must only contain the rule (A) and standard N-rules, (iv) are not allowed to belong to a framework with sequents without succedents, and (v) must belong to a specialised framework based on the language $\mathcal{L}^*$.

As was explained in 5.1.2, the assumption that adequate calculi are calculi of natural deduction is only a simplifying assumption on which nothing depends: there is a simple recipe for turning an arbitrary calculus into a calculus of natural deduction such that the new calculus inherits all properties that are relevant for the proof-theoretic arguments. The assumption that adequate calculi contain, apart from (A), only standard rules is also a simplifying assumption on which nothing depends. Although there are certain logical theories according to which it is important that not all rules of one's preferred calculus are standard (see 5.1.3), neither intuitionistic nor classical calculi require such rules. The assumption that the relevant sequents do not contain the empty set as their succedent and the assumption the calculus must belong to a specialised framework based on the language $\mathcal{L}^*$ will be discussed in 5.5 and 5.6.

5.3 Tennant's Proof-Theoretic Argument

In this section, I will discuss Tennant's (1997) proof-theoretic argument against classical logic. In the first subsection, I will introduce a desideratum for calculi, D_2, whose combination with the desideratum D_1 separates classical and intuitionistic logic: while there is an ARG$_{\mathcal{L}^*}$-*intuitionistic* calculus which satisfies D_1 and D_2, there is no such ARG$_{\mathcal{L}^*}$-*classical* calculus. According to Tennant, a calculus that yields an account of the meanings of the standard sentential operators has to satisfy the desiderata D_1 and D_2.[22] Since he also endorses the thesis EAC, according to which classical sentential logic is only sound if there is an adequate ARG$_{\mathcal{L}^*}$-classical, he concludes that the theory of classical sentential logic is not sound.

In 5.3.2, I will criticise Tennant's claim that an adequate calculus has to satisfy D_2. I will then present a weaker desideratum for calculi, the desideratum D_3, whose combination with D_1 still separates classical and intuitionistic

---

[22] More accurately, Tennant adopts a slight variant of the desideratum D_1 which allows for calculi of the logical framework ARG$_0$ of improper arguments:

Desideratum$_1^T$ (D$_1^T$)   If $\mathcal{R}$ is an adequate calculus, then $\mathcal{R}$ is a standard calculus of the specialised logical framework ARG$_{0\mathcal{L}^*}$.

But the differences between D_1 and D_1^T are irrelevant for the validity of the argument: there is also no ARG$_{\mathcal{L}^*}$-classical calculus which satisfies D_1^T and D_2 (cp. 5.5.3).

logic: there is no $\text{ARG}_{\mathcal{L}^*}$-classical calculus which satisfies D_1 and D_3. The proof-theoretic argument that is based on a combination of D_1 and D_3 is related to Dummett's proof-theoretic argument discussed in the remaining sections of this chapter.

5.3.1 Separability, Purity, Simplicity, and Directness

In this subsection, I will introduce the desideratum D_2, which Tennant claims to be necessary for adequate calculi and whose combination with D_1 suffices to separate classical and intuitionistic calculi (see Tennant 1997: 308-22, esp. 313-6). I will present D_2 as the final item of a list of desiderata of increasing strength. For every desideratum D of this list that is weaker than D_2, I will present an $\text{ARG}_{\mathcal{L}^*}$-classical calculus that satisfies D_1 and D. This will clarify the significance of the final desideratum D_2.

To begin with, I will introduce a property of calculi which concerns the existence of derivations that do not involve applications of certain rules. Suppose that $\mathcal{R}$ is some calculus and that s is an $\mathcal{R}$-derivable sequent. Then, an $\mathcal{R}$-derivation of s is called separable iff it does not involve applications of rules of $\mathcal{R}$ in whose schemata there are only operators not contained in s:

Separable $\mathcal{R}$-Derivations An $\mathcal{R}$-derivation of a sequent which contains precisely the operators $\sharp_1, \ldots, \sharp_n$ is called *separable* iff every transition is an application of a rule of $\mathcal{R}$ in whose schematic representation there is either no operator, or at least one of $\sharp_1, \ldots, \sharp_n$.

If, for example, the sequent s contains exactly one operator, $\sharp$, then a separable derivation of s is a derivation in which only such rules are applied that do not contain any operators or that contain $\sharp$. The property of having a separable derivation induces a fundamental property of calculi:

Separable Calculi A calculus $\mathcal{R}$ is called *separable* iff every $\mathcal{R}$-derivable sequent has a separable derivation in $\mathcal{R}$.

According to Tennant (1997: 315), a calculus which yields an account of the meanings of its operators has to be separable:

Desideratum$_2^a$ (D_2^a) If $\mathcal{R}$ is an adequate calculus, then $\mathcal{R}$ is separable.

I will here only hint at a justification of D_2^a (cp. Dummett 1991: 217-20). The basic idea is that in an adequate calculus, the meaning of a standard sentential operator is accounted for by precisely those rules which involve this operator. Now, if there is a derivable sequent in such a calculus which involves, say, only the operator $\sharp$, then its derivability (and, therefore, its logical validity) should only depend on the structure of this sequent and on the meaning of $\sharp$. Then, however, its derivation should only make use of 'structural rules' which do not contain any operators (see below) and the rules which account for the meaning of $\sharp$. Consequently, there should be a separable derivation of this sequent.

Unfortunately, the calculi $\mathcal{R}^I$ and $\mathcal{R}^C$ are both not separable. This is most easily illustrated by sequents such as $p, q \succ p$ which do not contain any operators at all. In a separable calculus, such sequents would have to be derivable by means of rules in whose schema *no* operator occurs, and, obviously, they are not derivable by means of (A) alone. Rules whose schemata contain no operators have a common name:

Structural Rules A rule is called *structural* iff no operator appears in its schema.

The rule (A) is the only structural rule considered thus far. If one wants to obtain a separable calculus for classical or intuitionistic logic, then one needs to have an additional structural rule. Here, I suggest the following rule:

$$\frac{\Gamma \succ \varphi \quad \Delta \succ \psi}{\Gamma, \Delta \succ \varphi} \ (\text{W})$$

The rule (W) is a standard N-rule, which allows one to make a derivable sequent $\Gamma \succ \varphi$ 'weaker' by introducing additional assumptions Δ (the acronym "W" stands for "weakening".) I will write $\mathcal{R}_W$ for the result of adding (W) to the calculus $\mathcal{R}$.[23]

There is then the following result: the calculus $\mathcal{R}_W^I$ is separable, but no $\text{ARG}_{\mathcal{L}^*}$-classical calculus of the logical framework ARG considered thus far is separable.[24] If separability is a virtue, then the intuitionistic calculus $\mathcal{R}_W^I$ is in some respect better than every classical calculus of the framework ARG considered thus far.

However, non-separability is not a general feature of $\text{ARG}_{\mathcal{L}^*}$-classical calculi of the framework ARG. There is a simple recipe for turning every calculus which has separable derivations of sequents without operators and which contains (A) into a separable calculus. One only has to alter its non-structural rules so that every operator of the language figures in their schematic representations. Every derivation of a sequent containing an operator is then obviously a separable derivation. Of course, one has to take care that the set of derivable sequents is not altered by this operation.

I will illustrate this recipe with respect to some calculus $\mathcal{R}$ based on the language $\mathcal{L}^*$. Suppose that $\mathcal{R}$ contains a non-structural rule with the following schema:

$$\frac{t_1 \dots t_n}{t}$$

---

[23] One might also consider the more usual rules $\frac{\Gamma \succ \varphi}{\Gamma, \Delta \succ \varphi}$, $\frac{\Gamma \succ \varphi}{\Gamma, \psi \succ \varphi}$, or $\frac{}{\Gamma, \varphi \succ \varphi}$ (see Humberstone 2011: 112, 124, and 153). I adhere to the rule (W) since it is a (standard) N-rule.

If one wants to obtain a separable standard calculus in the framework G-ARG that corresponds to $\mathcal{R}_W^I$, one has to use corresponding structural rules with multiple succedents:

$$\frac{\Gamma_1 \succ \Delta_1 \quad \Gamma_2 \succ \Delta_2}{\Gamma_1, \Gamma_2 \succ \Delta_1} \ (W_1) \qquad\qquad \frac{\Gamma \succ \Delta_1}{\Gamma \succ \Delta_1, \Delta_2} \ (W_2)$$

The rule (W_1) allows one to add further assumptions Γ_2 to a sequent $\Gamma_1 \succ \Delta_1$, and the rule (W_2) allows one to add further conclusions Δ_2 to a sequent $\Gamma_1 \succ \Delta_1$.

[24] The fact that $\mathcal{R}_W^I$ is separable follows from the fact that its derivations are *normalisable* (see Troelstra & Schwichtenberg 2000: 182). The non-separability of the $\text{ARG}_{\mathcal{L}^*}$-classical calculi of the framework ARG considered thus far is witnessed, for example, by Peirce's Law discussed below.

Now, alter the rule by adding a trivially derivable initial sequent for every logical operator to its schema:

$$\frac{t_1 \ldots t_n \quad \neg\varphi \succ \neg\varphi \quad \varphi \vee \psi \succ \varphi \vee \psi \quad \varphi \wedge \psi \succ \varphi \wedge \psi \quad \varphi \rightarrow \psi \succ \varphi \rightarrow \psi}{t}$$

Every derivation of the new calculus can be turned into a derivation of the old calculus: one only has to delete those parts of the derivation which lie above the added initial sequents. Furthermore, every derivation of the old calculus can be turned into a derivation of the new calculus: one only has to add (trivial) derivations of the added initial sequents at the relevant parts of the derivation. Consequently, one obtains an $\textsc{Arg}_{\mathcal{L}^*}$-classical separable calculus from every $\textsc{Arg}_{\mathcal{L}^*}$-classical calculus which contains (A) and (W).[25]

To obtain a result which sets intuitionistic and classical logic apart, one needs to exclude calculi separable in such an 'artificial' way. The simplest way to do this is by requiring that the operators are treated one by one:

Pure Rules A rule is called *pure* iff at most one operator appears in its schema.

Note that structural rules are counted as pure by this definition. (It is more usual to call only those rules pure in whose schema there is exactly one operator (see Dummett 1991: 257); however, the wider notion is more useful in what follows.) Now, Tennant (1997: 314) claims that calculi which yield an account of the relevant meanings are only allowed to contain pure rules. In combination with D_1^a, this would yield the following desideratum:

Desideratum$_2^b$ (D$_2^b$) If $\mathcal{R}$ is an adequate calculus, then $\mathcal{R}$ is separable and contains only pure rules.

I will discuss the plausibility of the claim that the rules of adequate calculi must be pure in the following subsection.

By combining the desiderata of separability and purity one excludes artificial separability: if there is a derivable sequent s in a separable calculus with only pure rules, then there is a derivation of s in which every transition conforms to a rule whose schematic representation includes only operators contained in s. In particular, for every derivable sequent s that contains only one operator, say $\sharp$, there is a derivation of s in which no rule is applied whose schema contains an operator distinct from $\sharp$. There is then the following improved result: no $\textsc{Arg}_{\mathcal{L}^*}$-classical calculus of the framework $\textsc{Arg}$ that has been considered thus far is separable and contains only pure rules. In particular, the above mentioned recipe for obtaining a separable calculus always yields a calculus whose rules are not pure. Furthermore, the $\textsc{Arg}_{\mathcal{L}^*}$-classical calculus $\mathcal{R}_W^C$ (which only contains pure rules) is not separable. This latter fact is witnessed by examples such as the following:

---

[25] Note that if it was not required that calculi are schematic (see 5.1.3), then there would be an even simpler recipe for obtaining separability: one could simply replace the original rules by a single new rule: the union of the original rules.

Peirce's Law $\succ ((p \to q) \to p) \to p$.

In the calculus $\mathcal{R}_W^C$, Peirce's Law can only be derived by means of the rule $\frac{\Gamma \succ \neg\neg\varphi}{\Gamma \succ \varphi}$ of double negation elimination. This follows from the fact that it is $\mathcal{R}_W^C$-derivable but not $\mathcal{R}_W^I$-derivable.[26]

However, the combination of D_1 and D_2^b does not yet preclude $\text{ARG}_{\mathcal{L}^*}$-classical calculi: there are standard calculi of the specialised logical framework $\text{ARG}_{\mathcal{L}^*}$ which are $\text{ARG}_{\mathcal{L}^*}$-classical, separable, and contain only pure rules. In particular, the union of $\mathcal{R}_W^C$ and each of the following rules will yield such a calculus (see Troelstra & Schwichtenberg 2000: 139-40):

$$\frac{\Gamma \succ (\varphi \to \psi) \to \varphi}{\Gamma \succ \varphi} \ (\text{PR}_1) \qquad \frac{}{\succ ((\varphi \to \psi) \to \varphi) \to \varphi} \ (\text{PR}_2)$$

I will call these rules *the first and the second version of Peirce's Rule*. (In a moment, I will mention the third version of Peirce's Rule.)

It thus follows that adherents of proof-theoretic arguments against classical logic have to claim that these two versions of Peirce's Rule are not allowed to figure in an adequate calculus. Note that the schema of the first version involves a premise of unusual specificity and that the schema of the second version involves a conclusion of unusual specificity. One idea is then to forbid such specificity. Following Dummett (1991: 257), I will call a rule *simple* iff the operators of its schema always occur in the most general position:

Simple Rules A rule is called *simple* iff an operator in a schematic sentence is always the main operator of this schematic sentence.

That is, in the schema of a simple rule, an operator is never in the scope of another operator. The two versions of Peirce's Rule are both not simple.

Now, Tennant (1997: 314) claims that rules of adequate calculi have to be simple. In combination with D_2^b, this yields the following desideratum:

Desideratum$_2^c$ (D$_2^c$) If $\mathcal{R}$ is an adequate calculus, then $\mathcal{R}$ is separable and contains only pure and simple rules.

While the intuitionistic calculus $\mathcal{R}_W^I$ satisfies this desideratum, no $\text{Arg}_{\mathcal{L}^*}$-classical calculus of the framework ARG considered thus far does. I will discuss the plausibility of the claim that the rules of adequate calculi must be simple in the following subsection.

However, the combination of D_1 and D_2^c does not preclude $\text{ARG}_{\mathcal{L}^*}$-classical calculi: there are standard calculi of the specialised logical framework $\text{ARG}_{\mathcal{L}^*}$ which are $\text{ARG}_{\mathcal{L}^*}$-classical, separable, and contain only pure and simple rules. In particular, the union of $\mathcal{R}_W^I$ and at least one of the following two rules yields such a calculus:

$$\frac{\Gamma, \varphi \to \psi \succ \varphi}{\Gamma \succ \varphi} \ (\text{PR}_3) \qquad \frac{\Gamma, \varphi \succ \psi \quad \Delta, \neg\varphi \succ \psi}{\Gamma, \Delta \succ \psi} \ (\text{Dil})$$

---

[26]   The fact that Peirce's Law is not $\mathcal{R}_W^I$-derivable again follows from the fact that the derivations of this calculus are normalisable (see n. 24).

The rule (PR$_3$) is called the *third version of Peirce's rule* (see Humberstone 2011: 521 and 1057-67), and (Dil) is called the rule of *dilemma*. Instead of these rules one could also add the rule of *classical reductio*:

$$\frac{\Gamma, \neg\varphi \succ \psi \quad \Delta, \neg\varphi \succ \neg\psi}{\Gamma, \Delta \succ \varphi} \ (\text{CR})$$

However, the rules (PR$_3$) and (Dil) are not only pure and simple. They also contain only one token of the relevant operator, and Tennant (1997: 314) claims that operational rules of adequate calculi should only contain one token of one operator.

Just as adherents of proof-theoretic arguments have to claim that the first two versions of Peirce's Rule and the rule of double negation elimination must not figure in adequate calculi, so they have to claim that the third version of Peirce's Rule and the rules (Dil) and (CR) must not figure in adequate calculi. Note that the schemata of these latter rules contain a discharged assumption of unusual specificity. One idea is then to forbid such specificity. Following Dummett (1991: 257), I will call *N*-rules without such assumptions direct:

Direct Rules An *N*-rule is called *direct* iff every discharged assumption is represented by a schematic letter for sentences in its schema.

Rules which are not direct are called *oblique*. Given the restriction to standard rules of the framework ARG, the requirement of directness amounts to the requirement that rules can be represented by a schema of the following form:

$$\frac{[\chi_1^1] \cdots [\chi_{n_1}^1] \quad [\chi_1^m] \cdots [\chi_{n_m}^m]}{\underset{A}{\overset{\vdots \qquad\qquad \vdots}{B_1 \quad \cdots \quad B_m}}}$$

where every χ_j^i is a schematic letter for sentences.[27]

Why should it be claimed that an adequate calculus must not contain oblique rules? Note that applications of oblique rules correspond to argumentative moves which are quite common. In particular, the rule of *classical reductio* corresponds to a standard method of proof in classical mathematics. Dummett gives the following reason:

> Should oblique rules be admitted as self-justifying, either as elimi-
> nation rules (as here) or as introduction rules? We did not formally
> ban them; but reflection on the foregoing example [*classical reductio*]
> will surely lead us to do so. For no way is apparent to formulate a
> corresponding introduction rule [...] (Dummett 1991: 297).

Dummett's line of thought can be spelled out as follows: "The rule of *classical reductio* could only be part of an account of the meaning of ¬ if it were a

---

[27]   Note that it is not required that the schematic letters χ_j^i are distinct. This would exclude rules like (¬$_I$).

self-justifying elimination rule. However, it could be a self-justifying elimination rule only if there were a corresponding introduction rule, and there is no
introduction rule which corresponds to *classical reductio*. Therefore, *classical
reductio* cannot be part of an account of the meaning of ¬." Dummett's reasoning involves several notions which have not been introduced yet. I will introduce
these notions in 5.4 and discuss the plausibility of Dummett's reasoning in 5.6.

Now, instead of invoking the desideratum that rules must be direct, I will
here employ two 'weaker' desiderata. Consider the following concept:

♯-Direct Rules An N-rule is called *♯-direct* iff no discharged assumption is
   represented by a schematic sentence that contains ♯ in its schema.

Rules which are not ♯-direct are called *♯-oblique*. Now, instead of requiring that
rules of adequate calculi have to be direct it suffices to require that such rules
be ¬-direct and →-direct. There is a simple method for reducing conjunctive
and disjunctive assumptions. Conjunctive assumptions $C_1 \wedge C_2$ can simply be
replaced by their conjuncts without altering the set of derivable sequents:

$$
\begin{array}{ccc}
[Z_1, C_1 \wedge C_2] \quad [Z_m] & & [Z_1, C_1, C_2] \quad [Z_m] \\
\vdots \qquad\qquad \vdots & & \vdots \qquad\qquad \vdots \\
\dfrac{B_1 \quad \dots \quad B_m}{A} & \Rightarrow & \dfrac{B_1 \quad \dots \quad B_m}{A}
\end{array}
$$

Given the rules $(\wedge_I)$ and $(\wedge_E)$, the displayed rules are interderivable. Similarly,
an initial sequent with a disjunctive assumption $C_1 \vee C_2$ can be replaced by two
copies of this sequent in which the disjunctive assumption is replaced by their
disjuncts:

$$
\begin{array}{ccc}
[Z_1, C_1 \vee C_2] \quad [Z_m] & & [Z_1, C_1] \quad [Z_1, C_2] \quad [Z_m] \\
\vdots \qquad\qquad \vdots & & \vdots \qquad\qquad \vdots \qquad\qquad \vdots \\
\dfrac{B_1 \quad \dots \quad B_m}{A} & \Rightarrow & \dfrac{B_1 \qquad B_1 \quad \dots \quad B_m}{A}
\end{array}
$$

Given the rules $(\vee_I)$ and $(\vee_E)$, the displayed rules are interderivable.

In combination with D_2^c, the desiderata of ¬-directness and →-directness
then yield the following desideratum:

Desideratum₂ (D₂) If $\mathcal{R}$ is an adequate calculus, then $\mathcal{R}$ is separable and
   contains only rules that are pure, simple, ¬-direct, and →-direct.

The combination of D_1 and D_2 sets intuitionistic and classical logic apart:

Anti-Cla₁ No ARG$_{\mathcal{L}^*}$-classical calculus satisfies D_1 and D_2.

Pro-Int₁ At least on ARG$_{\mathcal{L}^*}$-intuitionistic calculus satisfies D_1 and D_2.

If the property of being a separable standard calculus in the specialised logical
framework ARG$_{\mathcal{L}^*}$ with pure, simple, ¬-direct, and →-direct rules is a virtue,
then some ARG$_{\mathcal{L}^*}$-intuitionistic calculi have a virtue which no ARG$_{\mathcal{L}^*}$-classical
calculus has.[28]

---

[28]   The thesis Pro-Int₁ simply follows from the fact that $\mathcal{R}_W^I$ satisfies D_1 and D_2. The thesis
   Anti-Cla₁ is implied by the fact that every simple and direct standard N-rule of the
   specialised framework ARG$_{\mathcal{L}^*}$ in whose schema only ¬ figures is either admissible in $\mathcal{R}^I$
   or not admissible in $\mathcal{R}^C$.

5.3.2   Failure and Improvement

In this subsection, I will criticise Tennant's proof-theoretic argument. In particular, I will question his claim that rules of adequate calculi have to be pure and simple. I will then present a strengthening of the result Anti-Cla$_1$; that is, I will introduce a desideratum (D$_3$) which is weaker than D$_2$ but which still suffices, in combination with D$_1$, to exclude classical calculi (there is no ARG$_{\mathcal{L}^*}$-classical calculus which satisfies D$_1$ and D$_3$).

Consider Tennant's claim that adequate calculi are only allowed to contain pure and simple rules, and recall that Tennant's proof-theoretic argument breaks down if the requirement of simplicity is simply given up. The ARG$_{\mathcal{L}^*}$-classical calculus that results from $\mathcal{R}_W^I$ by adjoining the first version of Peirce's Rule and the rule of double negation elimination:

$$\frac{\Gamma \succ (\varphi \to \psi) \to \varphi}{\Gamma \succ \varphi} \ (\text{PR}_1) \qquad\qquad \frac{\Gamma \succ \neg\neg\varphi}{\Gamma \succ \varphi} \ (\neg_E)$$

satisfies all the other mentioned virtues of $\mathcal{R}_W^I$. In particular, this calculus is separable, and (PR$_1$) and ($\neg_E$) are pure and direct rules containing only two tokens of the relevant operator.

What speaks against non-simple and impure rules? Note that applications of informal rules of inference that correspond to non-simple rules like the rule of double negation elimination and impure rules like the rule of disjunctive syllogism are quite common. Tennant presents the following consideration to substantiate his claim that non-simple rules like double negation elimination cannot belong to an adequate calculus:

> [The rule of double negation elimination] is contentious as a basic rule because it involves two occurrences of the logical operator in question. Why should we have to deal with two occurrences simultaneously? (When an intuitionist searches for systematicity, the search is in earnest.) Surely, the intuitionist maintains, whatever disagreement there may be about the very meaning of negation should be able to be brought into the open in the context of differing (schematizable) logical practices with regard to single occurrences of the logical operator concerned. After all, if meaning is compositional, and an operator makes the same systematic contribution to sentence meaning at each of its occurrences therein, surely any difference in the meanings imputed by, say, the classicist and the intuitionist, should make themselves manifest in single occurrences of the operator in question (Tennant 1997: 310)?

I have presented Tennant's complete justification of the claim that rules which contain more than one token of a logical operator cannot be part of adequate calculi to make it clear that it does not contain a convincing argument. He poses the question of why it is not allowed that a rule of an adequate calculus contains two tokens of an operator. Referring to the compositionality of meaning, he then answers this question by asserting that any difference in the meanings imputed by the classicist and the intuitionist should make themselves manifest in single occurrences of the operator in question.[29] I cannot see how his assertion can be used to provide an answer to the question he posed before (cp. Milne 2002: 523-4).

Dummett rejects the claim that an adequate calculus must not contain impure and non-simple rules:

> An impure **c**-introduction rule will make the understanding of **c** depend on the prior understanding of the other logical constants figuring in the rule. Certainly we do not want such a relation of dependence to be cyclic; but there would be nothing in principle objectionable if we could so order the logical constants that the understanding of each depended only on the understanding of those preceding it in the ordering. Given such an ordering, we could not demand that each rule be simple, either (Dummett 1991: 257).

According to Dummett, it cannot be required that the meaning of every operator is accounted for by pure and simple rules. The meaning of some operator $\sharp_2$ may depend on the meaning of some other operator $\sharp_1$; in this case, nothing speaks against an account of the meaning of $\sharp_2$ which involves rules that mention $\sharp_1$ (a violation of purity). Furthermore, nothing even speaks against an account of the meaning of $\sharp_2$ which involves rules in which $\sharp_2$ takes scope over $\sharp_1$ (a violation of simplicity). (Like Tennant, Dummett claims that the calculus $\mathcal{R}_W^C$ which differs from $\mathcal{R}_W^I$ only in containing the rule of double negation elimination cannot be adequate. However, unlike Tennant, he does not base this claim on the thesis that adequate calculi must only contain pure and simple rules.) As far as I can see, Dummett's reason for rejecting Tennant's requirements of purity and simplicity are compelling. Tennant's proof-theoretic argument against classical logic must therefore be rejected.

In the rest of this subsection, I will introduce a weakened version of Tennant's desideratum D_2, which leads to an improved proof-theoretic argument against classical logic. As far as I know, this argument has not been singled out before.

The improved argument is obtained by presenting suitable weakened versions of the requirements of purity and simplicity. For ease of exposition, however, I will not develop a *general* account of how these requirements should be weakened. Since I only deal with calculi of the specialised logical framework $\mathrm{ARG}_{\mathcal{L}^*}$, I can here exploit the fact that proof-theoretic disputes between intuitionists and classicists do not concern disjunction and conjunction; i.e., for present purposes, it is unproblematic to assume that there are pure and simple rules for $\vee$ and $\wedge$ which account for their meaning.

To state this more precisely, let $\mathcal{L}_{\sharp_1,\ldots,\sharp_n}$ be the sub-language of $\mathcal{L}$ which contains only the operators $\sharp_1,\ldots,\sharp_n$, and let $\mathcal{R}_{\sharp_1,\ldots,\sharp_n}$ be the sub-calculus of $\mathcal{R}$ which comprises the pure rules that govern the operators $\sharp_1,\ldots\sharp_n$. (Recall that structural rules are pure.) Then, for an evaluation of the proof-theoretic

---

[29]   As an aside, one may note that Tennant here claims that classicists and intuitionists attach different meanings to the logical operators. See Chapter 2, for my reasons for rejecting this claim.

disputes between intuitionists and classicists, it is unproblematic to assume that the following desideratum for adequate calculi holds:

Desideratum$_3^a$ (D$_3^a$) If $\mathcal{R}$ is an adequate calculus, then its sub-calculus $\mathcal{R}_{\vee,\wedge}$ comprises only direct and simple rules, and every sequent of one of the languages $\mathcal{L}_\vee$, $\mathcal{L}_\wedge$, $\mathcal{L}_{\vee,\wedge}$ is derivable in $\mathcal{R}$ iff it is derivable in the corresponding sub-calculus $\mathcal{R}_\vee$, $\mathcal{R}_\wedge$, $\mathcal{R}_{\vee,\wedge}$.

In short, there are pure, direct, and simple rules for $\vee$ and $\wedge$, which suffice to yield an account of their meaning. (Note that D$_3^a$ incorporates a requirement related to the requirement of separability. The reason for using this more complicated alternative is that the requirement of separability loses all its force if the requirement of purity is given up; see 5.3.1.)

Now, I follow Dummett in assuming that adequate calculi may contain rules for $\neg$ and $\rightarrow$ which are neither pure nor simple. What kinds of rules for $\neg$ and $\rightarrow$ may figure in adequate calculi? Recall that a proponent of a proof-theoretic argument against classical logic has to discredit rules like (PR$_1$) and ($\neg_E$). In each of these (pure and direct) rules, one of the operators $\neg$ and $\rightarrow$ takes scope over itself. Similarly, an intuitionistic proponent of a proof-theoretic argument against classical logic has to discredit rules such as $\dfrac{\Gamma \succ \neg(\varphi \rightarrow \psi)}{\Gamma \succ \varphi}$ and $\dfrac{\Gamma \succ \neg\neg\varphi \rightarrow \varphi}{\Gamma \succ \varphi}$, in which one of these operators takes scope over the other. Are there plausible desiderata which disqualify these rules? Consider the following proposal:

(?) If $\mathcal{R}$ is an adequate calculus, then every rule of $\mathcal{R} - \mathcal{R}_{\vee,\wedge}$ has to be a direct rule in whose schema at most one token of $\neg$ or $\rightarrow$ is allowed to figure.

The idea would be that since the meanings of $\vee$ and $\wedge$ can be taken for granted, these operators can be freely used in every further schematic representation of a rule. Furthermore, by requiring that at most one token of $\neg$ and $\rightarrow$ is allowed to figure in the representation of a rule of $\mathcal{R} - \mathcal{R}_{\vee,\wedge}$, one ensures that they cannot take scope over each other, thereby precluding the possibility that these rules only yield a circular account of their meanings.

However, as Milne has pointed out, a proponent of a proof-theoretic argument against classical logic cannot be satisfied with (?). Milne (2002: 514) presents the following two rules which satisfy this requirement:

$$\dfrac{\Gamma, \varphi \succ \psi \vee \chi}{\Gamma \succ (\varphi \rightarrow \psi) \vee \chi} \ (\rightarrow_I^M) \qquad\qquad \dfrac{\Gamma, \varphi \succ \psi}{\Gamma \succ \neg\varphi \vee \psi} \ (\neg_I^M)$$

According to Milne, this calculus can be referred to in order to show that proof-theoretic arguments against classical logic are not convincing. It is an $\text{ARG}_{\mathcal{L}^*}$-classical calculus which satisfies not only the desiderata D$_1$ and D$_3^a$, but also the desideratum (?). (Note that the pairs of rules $(\rightarrow_I),(\rightarrow_I^M)$ and $(\neg_I)$, $(\neg_I^M)$ are precisely those that correspond to the rules $(\rightarrow_I^g)$ and $(\neg_I^g)$ of the calculus $^g\mathcal{R}^C$ of the framework G-ARG; see 5.2.2.)

It thus follows that the desiderata D$_1$, D$_3^a$, (?) do not suffice to preclude $\text{ARG}_{\mathcal{L}^*}$-classical calculi. The desideratum (?) has to be replaced by one that does not allow for rules like the ones proposed by Milne. To this end, consider the following notion:

♯-Simple Rules A rule is ♯-*simple* iff ♯ occurs only as a main operator in its
    schema.

Note that $(\rightarrow_I^M)$ is not $\rightarrow$-simple and that $(\neg_I^M)$ is not ¬-simple. The notion
of ♯-simplicity can be used to present a replacement of (?) whose combination
with D_1 and D_3^a precludes $\text{ARG}_{\mathcal{L}^*}$-classical calculi:

(!) If $\mathcal{R}$ is an adequate calculus, then every rule of $\mathcal{R} - \mathcal{R}_{\vee,\wedge}$ has to be $\rightarrow$-
    direct, ¬-direct, $\rightarrow$-simple, and ¬-simple.

The combination of D_3^a and (!) yields the following desideratum for calculi:

Desideratum₃ (D₃) If $\mathcal{R}$ is an adequate calculus, then the following holds:

$(\vee, \wedge)$ the sub-calculus $\mathcal{R}_{\vee,\wedge}$ of $\mathcal{R}$ comprises only direct and simple rules,
        and every sequent of one of the languages $\mathcal{L}_\vee$, $\mathcal{L}_\wedge$, $\mathcal{L}_{\vee,\wedge}$ is derivable
        in $\mathcal{R}$ iff it is derivable in the corresponding sub-calculus $\mathcal{R}_\vee$, $\mathcal{R}_\wedge$,
        $\mathcal{R}_{\vee,\wedge}$;

$(\rightarrow, \neg)$ every rule of $\mathcal{R} - \mathcal{R}_{\vee,\wedge}$ is $\rightarrow$-direct, ¬-direct, $\rightarrow$-simple, and ¬-simple.

The desideratum D_3 is weaker than the desideratum D_2 in allowing for impure
rules like $\frac{\Gamma \succ \varphi \vee \psi \quad \Delta \succ \neg \psi}{\Gamma, \Delta \succ \varphi}$ and non-simple rules like $\frac{\Gamma \succ \neg(\varphi \vee \psi)}{\Gamma \succ \neg \varphi}$, but it suffices to
yield an intuitionistically acceptable proof-theoretic argument against classical
logic:

Anti-Cla₂ No $\text{ARG}_{\mathcal{L}^*}$-classical calculus satisfies D_1 and D_3.

Pro-Int₂ At least one $\text{ARG}_{\mathcal{L}^*}$-intuitionistic calculus satisfies D_1 and D_3.

The combination of Anti-Cla₂, EAC (according to which the soundness of classical sentential logic depends on the existence of an adequate calculus), and
the thesis that an adequate calculus must satisfy D_1 and D_3 implies that classical sentential logic is not sound. This then completes the presentation of the
improved version of Tennant's proof-theoretic argument.[30]
    There remains the question of whether the requirements stated in $(\rightarrow, \neg)$ are
reasonable. However, I will not pursue this question here. In 5.6, I will criticise
Dummett's proof-theoretic argument, which will be introduced in the following
two parts. The criticism presented in 5.6.2 will also pertain to the improved
version of Tennant's argument.

---

[30] The thesis Pro-Int₂ is a trivial consequence of Pro-Int₁. The thesis Anti-Cla₂ follows from
the fact that every rule of a calculus $\mathcal{R}$ satisfying D_1 and D_3 that is not admissible in $\mathcal{R}^I$
is also not admissible in $\mathcal{R}^C$. This is trivial for every rule of the sub-calculus $\mathcal{R}_{\vee,\wedge}$. The
fact that every rule of $\mathcal{R} - \mathcal{R}_{\vee,\wedge}$ has this property can be seen by a successive reduction
process. Here is a short sketch. Suppose that ρ is a rule of $\mathcal{R} - \mathcal{R}_{\vee,\wedge}$ that is $\rightarrow$-direct,
¬-direct, $\rightarrow$-simple, and ¬-simple, and that is not admissible in $\mathcal{R}^I$. One then constructs
a sequence of rules $\rho = \rho_0, \rho_1, \ldots, \rho_n$ such that (i) ρ_n is not admissible in $\mathcal{R}^C$, and such
that (ii) ρ_{i-1} is not admissible in $\mathcal{R}^C$ if ρ_i is not admissible in $\mathcal{R}^C$ $(1 < i \leq n)$. To this
end, one moves from ρ_i to ρ_{i+1} by reducing the complexity of the schematic conclusion or
a schematic premise of ρ_i. For example, if the schema of ρ_i contains a schematic premise
of the form $\neg(\varphi \vee \psi)$ (depending on some discharged schematic assumptions), then the
schema of ρ_{i+1} contains two corresponding schematic premises, one of the form $\neg \varphi$ and
the other of the form $\neg \psi$ (depending on the same discharged schematic assumptions).

5.4   Harmony

Tennant's proof-theoretic argument essentially involves the premise that the
rules of an adequate calculus have to be pure and simple, and this premise was
seen to be unwarranted. I have presented a strengthening of Tennant's argument
which does not involve the requirements of purity and simplicity. However,
the crucial premise of this strengthening is in need of further justification. In
the remaining sections, I will discuss Dummett's proof-theoretic argument. In
contrast to Tennant, Dummett presents an initially plausible justification of the
crucial premise of his argument.

Dummett's proof-theoretic argument is based on the idea that the rules of
adequate calculi have to exhibit a certain kind of *harmony*. It is claimed that
(certain) intuitionistic calculi are adequate because they are constituted by pairs
of introduction and elimination rules which match each other perfectly, and it
is held that no classical calculus can be adequate because no classical calculus is
constituted by such harmonious pairs of rules. This idea has been developed in
different ways. I will introduce what I take to be the most plausible development
and discuss whether it yields a convincing proof-theoretic argument against
classical logic.

In the first subsection, I will introduce the basic idea of Dummett's argument
in general terms. In the following two subsections, this idea will be made precise.
In 5.4.2, I will present the relevant notions of introduction and elimination rules,
and in 5.4.3, I will introduce the relevant notion of harmony between collections
of introduction and elimination rules.

5.4.1   The Idea of Dummett's Proof-Theoretic Argument

For Dummett's views about a sound meaning theory, it is crucial that there
are linguistic acts of different types. Dummett believes that different classes of
linguistic behaviour are governed by different norms, and he thinks that these
norms may be in tension with one another. If a linguistic practice is governed
by such incompatible norms, then Dummett describes the practice as being
incoherent and in need of reform (cp. Dummett 1991: 210):

> An existing practice in the use of a certain fragment of language is
> capable of being subjected to criticism if it is impossible to system-
> atise it [...]. What makes it possible that such a practice may prove
> to be incoherent and therefore in need of revision is that there are
> different aspects of the use of a sentence; if the whole practice is to
> be capable of systematisation in the present sense, there must be a
> certain harmony between these different aspects (Dummett 1973b:
> 220).

Dummett's proof-theoretic argument is based on the claim that our current
linguistic practice is incoherent, i.e. governed by incompatible norms, and that
a coherent alternative practice agrees with the principles of intuitionistic logic.

To spell out this claim, one needs to take a closer look at the different kinds
of linguistic behaviour. According to Dummett, there are two central classes of

norms which concern the assertoric use of language, i.e. norms which concern assertions performed by utterances of (declarative) sentences. First, there is a collection of norms that state under which conditions an assertive utterance of a sentence S is correct. Second, there is a collection of norms that state under which conditions certain reactions to an assertive utterance of S are correct (see Dummett 1973b: 221 and 1991: 210-2). The fundamental idea of his proof-theoretic argument is that these two classes of norms have to be in a kind of equilibrium: the conditions for being entitled to make an assertion by uttering S must match the conditions for being entitled to react in certain ways to an assertion made by uttering S.

With respect to a mathematical sentence S, the relevant conditions pertain to argumentative situations involving S. Restricting attention to simple argumentative moves, the idea is the following.[31] Norms of the first kind, which are called *I-norms*, state from which sentences the sentence S can be inferred. Norms of the second kind, which are called *E-norms*, state which sentences can be inferred from S together with further sentences (see Dummett 1973b: 221). According to Dummett, these norms have to be in harmony lest they induce incoherence into linguistic practice: the conditions for being entitled to infer S from some premises must match the conditions for being entitled to infer a sentence from S together with further sentences.

Following Dummett, I will assume that I-norms have a certain priority in accounts of meaning. This does not mean that a proof-theoretic argument based on the idea of harmonious rules *depends* on the assumption that I-norms have priority in accounts of meaning (see Dummett 1983: 142). However, since it simplifies the following discussion, I will assume that Dummett's proof-theoretic argument involves this assumption. My criticism of the argument (see 5.6) will not rest on this simplification. The essential ingredients of his views are then the following:

I-Norms for Sentences For every mathematical sentence S, there is a collection of I-norms which determines the meaning of S.

E-Norms for Sentences There is a relation between collections of I-norms and E-norms for mathematical sentences which relates every collection of I-norms to at least one collection of E-norms and which relates a collection of I-norms that determines the meaning of a mathematical sentence S only to collections of E-norms that precisely capture the meaning of S.

According to these claims, sentential meanings are determined by collections of I-norms, and these collections of I-norms correspond to collections of E-norms which are in perfect conformity to the determined sentential meanings. Now, Dummett's proof-theoretic argument concerns sentential meanings only in a derivative way; primarily, it concerns the meanings of the logical operators. Consequently, the notions of I-norms and E-norms for sentences are replaced by

---

[31] Below it will be taken into account that argumentative moves can be complex acts in which an act of inference is combined with an act in which a previously made assumption is discharged.

notions of I-norms and E-norms for logical operators for which corresponding
principles are assumed to hold (see Dummett 1991: 215-7):

I-Norms for Operators For every logical operator $\sharp$, there is a collection of
I-norms which determines the meaning of $\sharp$.

E-Norms for Operators There is a relation between collections of I-norms
and E-norms for logical operators which relates every collection of I-
norms to at least one collection of E-norms and which relates a collection
of I-norms that determines the meaning of a logical operator $\sharp$ only to
collections of E-norms that precisely capture the meaning of $\sharp$.

According to these claims, meanings of logical operators are determined by
collections of I-norms, and these collections of I-norms correspond to collections of E-norms which are in perfect conformity to the determined meanings.
Dummett's proof-theoretic argument is then based on the claim that our current linguistic practice is incoherent because it is governed by incompatible
norms for the logical operators (norms that correspond to the basic principles
of classical logic), and that a coherent alternative practice is governed by norms
for the logical operators that correspond to the basic principles of intuitionistic
logic.

If one wants to develop a proof-theoretic argument based on Dummett's
claims about I-norms and E-norms for logical operators, then one has to carry
out two tasks: the relevant notions of I-norms and E-norms for logical operators
have to be made precise, and a relation H between collections of I-norms and
collections of E-norms has to be specified. This has to be done in such a way
that (i) every possible meaning of a logical operator is determined by a collection
I of I-norms and (ii) exactly those collections of E-norms are H-related to I
which precisely capture the meaning determined by I.

The idea for how to carry out these tasks stems from Gentzen. In the famous
paper in which he introduces calculi of natural deduction, Gentzen suggests that
the *introduction rules* for a logical operator determine its meaning:[32]

> The introductions represent, as it were, the 'definitions' of the sym-
> bols concerned [...] (Gentzen 1934: 189).

Of course, introduction rules are neither explicit definitions nor ordinary recursive definitions. However, Gentzen nevertheless takes them to endow the
relevant operators with their meanings. Furthermore, he claims that the *elimination rules* for logical operators are uniquely determined by their introduction
rules and that they should be in perfect conformity to the meanings given by
the introduction rules:

> [T]he eliminations are no more, in the final analysis, than the con-
> sequences of these definitions. [...] It should be possible to dis-
> play the E-inferences as unique functions of their corresponding
> I-inferences (ibid.).

---

[32]   The following translations are taken from Szabo 1969: 80-1.

Of course, elimination rules are not 'consequences' of introduction rules in the sense of being derivable from them. But Gentzen nevertheless takes them to be justified through their relation to the meaning conferring introduction rules. Taking up Gentzen's idea, the fundamental ingredient of Dummett's proof-theoretic argument then amounts to the claim that our current linguistic practice is incoherent because our use of the logical operators is governed by introduction and elimination rules which do not cohere with one another. Furthermore, it is claimed that meaning conferring introduction rules, in combination with harmonious elimination rules, yield intuitionistic but not classical calculi. In the remainder of this section, I will make the central notions of this claim more precise.

5.4.2 Introduction and Elimination Rules

The notions of an introduction rule and of an elimination rule are of fundamental significance for Dummett's proof-theoretic argument. In this subsection, I will introduce these notions and two important refinements of them.

I will use $\overline{A}$, $\overline{B}$, $\overline{C}$ for schematic lists without schematic letters for fs-sets, i.e. for finite (possibly empty) and alternating sequences of schematic sentences and commas which begin and end with a schematic sentence. Furthermore, I will use $\overline{\varphi}$, $\overline{\psi}$, $\overline{\chi}$ for schematic lists which contain only schematic letters for sentences. Finally, I will use $\overrightarrow{\varphi}$, $\overrightarrow{\psi}$, $\overrightarrow{\chi}$ for schematic lists which contain only *distinct* schematic letters for sentences. Now, if $\sharp$ is an n-ary sentential operator, then the notions of a $\sharp$-introduction rule and the notion of a $\sharp$-elimination rule are defined as follows:

$\sharp$-**Introduction Rules** A $\sharp$-*introduction rule* is a standard rule in whose schema the conclusion has $\sharp$ as its main operator:

$$\frac{\begin{array}{ccc}[\overline{C}_1] & & [\overline{C}_m] \\ \vdots & & \vdots \\ B_1 & \cdots & B_m\end{array}}{\sharp(\overline{A})}$$

$\sharp$-**Elimination Rules** A $\sharp$-*elimination rule* is a standard rule in whose schema some premise has $\sharp$ as its main operator:

$$\frac{\begin{array}{cccc}[\overline{C}_0] & [\overline{C}_1] & & [\overline{C}_m] \\ \vdots & \vdots & & \vdots \\ \sharp(\overline{A}) & B_1 & \cdots & B_m\end{array}}{D}$$

In an application of a $\sharp$-introduction rule, one infers a sentence in which $\sharp$ is dominant from other sentences. In an application of a $\sharp$-elimination rule, one infers a sentence from a sentence in which $\sharp$ is dominant and further sentences. The premises of a $\sharp$-elimination rule in whose representation $\sharp$ is dominant are called *major premises*; the other premises are called *minor premises*.

A calculus may contain several $\sharp$-introduction rules. I will then employ schematic representations of these rules in which distinct schematic conclusions are not *orthographic variants* of one another; i.e., it has to be impossible to turn one schematic conclusion into another schematic conclusion by a systematic replacement of schematic letters for sentences. (Such a replacement is called *systematic* iff identity and distinctness of schematic letters are preserved.) Thus, for example, it is not allowed that there are $\sharp$-introduction rules with schematic conclusions $\sharp\varphi$ and $\sharp\psi$ for some unary operator $\sharp$ since these are distinct orthographic variants of one another.

Now, according to the fundamental idea of Dummett's proof-theoretic argument, for every adequate calculus $\mathcal{R}$ and for every operator $\sharp$ of the language $\mathcal{L}(\mathcal{R})$ of $\mathcal{R}$, there is a set $I_{\mathcal{R}}(\sharp)$ of $\sharp$-introduction rules and a set $E_{\mathcal{R}}(\sharp)$ of $\sharp$-elimination rules such that the following three conditions are satisfied:

(i) $\mathcal{R}$ is the union of structural rules and the collections $I_{\mathcal{R}}(\sharp)$ and $E_{\mathcal{R}}(\sharp)$;

(ii) for every operator $\sharp$ of $\mathcal{L}(\mathcal{R})$, $I_{\mathcal{R}}(\sharp)$ determines the meaning of $\sharp$;

(iii) for every operator $\sharp$ of $\mathcal{L}(\mathcal{R})$, $E_{\mathcal{R}}(\sharp)$ precisely captures the meaning of $\sharp$.

Dummett's proof-theoretic argument will be based on a precise version of this fundamental idea. To this end, the concepts of an *independent family* of $\sharp$-introduction rules and the concept of *respecting a complexity ordering* have to be introduced.

The notion of an independent family of introduction rules is a generalisation of the notion of a family of general introduction rules:

General $\sharp$-Introduction Rules A *general* $\sharp$-introduction rule is a $\sharp$-introduction rule with a schema of the following form:

$$\frac{\overset{[\overline{C_1}]}{\vdots}\atop B_1 \quad \ldots \quad \overset{[\overline{C_m}]}{\vdots}\atop B_m}{\sharp(\vec{\varphi})}$$

In a general $\sharp$-introduction rule, the conclusion is represented by a schematic sentence of the most general kind: a schematic sentence in which the operator $\sharp$ is applied to distinct schematic letters for sentences. (A notion of a *general $\sharp$-elimination rule* can be defined analogously.[33]) Families of general $\sharp$-introduction rules are special cases of independent families of $\sharp$-introduction rules:

Independent Families of Rules A family x of rules is called *independent* iff distinct schematic conclusions of schematic representations of members of x are *incompatible*; i.e., no sentence is an instance of distinct schematic conclusions of schematic representations of members of x.

---

[33]   Humberstone (2011: 521) calls the property of being a general introduction or elimination rule the property of *generality in respect of constituent-formulas*.

For example, if $\sharp$ is a unary operator, then an independent family of $\sharp$-introduction rules does not contain members whose schematic conclusions are $\sharp\varphi$ and $\sharp\sharp\varphi$, because a sentence such as $\sharp\sharp p$ is an instance of both of these schemata. As far as I know, Dummett has not explicitly singled out the notion of an independent family of introduction rules. However, it will be seen to be of crucial importance for his proof-theoretic argument (see 5.5.1).

The notion of an introduction rule that respects a certain complexity ordering concerns the relative complexities of the various schematic sentences that figure in the schema of the rule. Dummett introduces this notion as follows:

> Hence the minimal demand we should make on an introduction rule
> intended to be self-justifying is that its form be such as to guarantee
> that, in any application of it, the conclusion will be of higher logical
> complexity than any of the premises and than any discharged hypo-
> thesis. We may call this the 'complexity condition' (Dummett 1991:
> 258).

(Introduction rules of adequate calculi are called *self-justifying* because they determine the meanings of the relevant operators.)

However, Dummett's definition has to be refined. The problem is that he does not state what *logical complexity* amounts to. For example, he does not state which of the following two schematic sentences is logically the more complex one: $((\varphi \vee \psi) \vee \chi) \vee \varphi$ or $\neg(\varphi \rightarrow \psi)$. (The former contains more tokens of logical operators; the latter contains more types of logical operators.) To obtain a precise definition, one has to specify an ordering of the schematic sentences which is meant to be respected by an introduction rule. To this end, I will introduce a class of relations for comparing the complexities of schematic sentences:

Complexity Orderings A well-quasi-ordering $\leq$ of the schematic sentences is called a *complexity ordering* iff it extends the containment relation: if a schematic sentence A is a constituent of a schematic sentence B, then $A \leq B$.[34]

A complexity ordering induces a corresponding class of introduction rules:

$\leq$-Respecting $\sharp$-Introduction Rules A $\sharp$-introduction rule which *respects* $\leq$ is an introduction rule with a schema of the following form:

$$
\begin{array}{ccc}
[\overline{C_1}] & & [\overline{C_m}] \\
\vdots & & \vdots \\
B_1 & \cdots & B_m \\
\hline
& \sharp(\overline{A}) &
\end{array}
$$

such that for every item D of the list $\overline{C}_1, \ldots, \overline{C}_m, B_1, \ldots, B_m$, it holds that $D < \sharp(\overline{A})$.

---

[34] I will write $A < B$ for $A \leq B \wedge \neg(B \leq A)$. Recall that a well-quasi-ordering $\leq$ on a set X is a reflexive and transitive (binary) relation such that the following holds for every $Y \subseteq X$: if for every $x \in X$ with the property that every $y < x$ is a member of Y, x is a member of Y, then $X = Y$.

In an introduction rule which respects a complexity ordering $\leq$, the discharged assumptions and the premises are all less complex, *in the sense of* $\leq$, than the conclusion.

Now, Dummett's proof-theoretic argument is based on the assumption that for every adequate calculus $\mathcal{R}$ there is a 'suitable' complexity ordering $\leq$ respected by the rules of the sets $I_{\mathcal{R}}(\sharp)$. I will here adopt the minimal assumption that a complexity ordering is suitable iff it treats orthographic variants of schematic letters alike:[35]

Standard Complexity Orderings A complexity ordering $\leq$ is called *standard* iff for every pair A, B of orthographic variants and for every schematic sentence C the following holds: $A \leq C$ iff $B \leq C$, and $C \leq A$ iff $C \leq B$.

In relation to a standard complexity ordering, no schematic sentence which contains an operator $\sharp$ is simpler than a schematic sentence in which $\sharp$ is applied to a schematic letter for sentences. However, there are standard complexity orderings in which any schematic sentence which does not contain $\sharp$ is simpler than any schematic sentence containing $\sharp$.

I will conclude by introducing two special classes of introduction rules. As to the first one, note that the fewer schematic sentences are related by a complexity ordering, the more unlikely it is that an introduction rule respects it. Introduction rules which respect even the coarsest complexity orderings are of special importance:

Strict $\sharp$-Introduction Rules A *strict* $\sharp$-introduction rule is an introduction rule which respects the containment relation: the discharged assumptions and the premises are all proper constituents of the conclusion.

In the schema of a strict introduction rule, the conclusion is literally built up from the discharged assumptions, the premises, and, possibly, further schematic sentences. The containment relation itself is not a standard complexity ordering, but strict introduction rules respect every standard complexity ordering.

Second, there are introduction rules of a certain ideal kind:

Perfect $\sharp$-Introduction Rules A *perfect* $\sharp$-introduction rule is a strict and general $\sharp$-introduction rule.

A perfect $\sharp$-introduction rule has a schema of the following form:

$$\begin{array}{ccc} [\overline{\chi}_1] & & [\overline{\chi}_m] \\ \vdots & & \vdots \\ \psi_1 & \cdots & \psi_m \\ \hline & \sharp(\overrightarrow{\varphi}) & \end{array}$$

where every schematic letter of the list $\overline{\chi}_1, \ldots, \overline{\chi}_m, \psi_1, \ldots, \psi_m$ is a schematic letter of $\overrightarrow{\varphi}$. Perfect introduction rules are pure, direct, and simple, and their

---

[35] Two schematic sentences A and B are orthographic variants of one another iff A can be turned into B by a systematic replacement of schematic letters.

schemata contain only one token of an operator. Inspection shows that the rules $(^1\vee_I)$, $(^2\vee_I)$, $(\wedge_I)$, $(\rightarrow_I)$ are perfect introduction rules, while, for example, $(\neg_I)$ is not a perfect introduction rule since it does not respect the containment relation, or, indeed, any standard complexity ordering.

5.4.3 The Relation of Harmony

Following Read (2010), I will now introduce the relation of harmony between collections of introduction and elimination rules.[36] In what follows, I will use the expression "asserting a sentence S" to refer to acts of assertions performed by uttering S and the term "$\sharp$-sentence" to refer to sentences in which the operator $\sharp$ is dominant.

Suppose that $\sharp$ is some sentential operator and that $I(\sharp)$ is a set of $\sharp$-introduction rules which determines the meaning of $\sharp$. Under what conditions does a set $E(\sharp)$ of $\sharp$-eliminations rules precisely capture the meaning determined by $I(\sharp)$. Read presents the following characterisations of the pertinent notion of harmony:

> I propose, therefore, to say that the introduction- and elimination-rules are in harmony when the E-rules do no more and no less than spell out what may be inferred from the assertion of the conclusion of the I-rules, given the grounds for its assertion (Read 2010: 562).

> The idea of ge-harmony is that we may infer from an assertion all and only what follows from the various grounds for that assertion (Read 2010: 563).

I will make the simplifying assumption that the grounds for asserting the conclusion of a $\sharp$-introduction rule ρ can be identified with the sets of pairs of premises and discharged assumptions of the corresponding application of ρ.[37] Read's claim that the collection $I(\sharp)$ of introduction rules gives $\sharp$ its meaning by displaying the grounds for asserting a $\sharp$-sentence then amounts to the claim that the collection $I(\sharp)$ gives $\sharp$ its meaning by displaying the sets of pairs of premises and discharged assumptions of the applications of the rules in $I(\sharp)$. Read now calls the collection $E(\sharp)$ of elimination rules *harmonious* with $I(\sharp)$ iff the rules in $E(\sharp)$ express that a sentence S^* can be inferred from a $\sharp$-sentence S if S^* holds on the assumption that a component of each of the grounds for asserting S holds (where these grounds are given by the sets of pairs of premises and discharged assumptions displayed by the rules in $I(\sharp)$).[38] Here, a pair of

---

[36] Read (2010) uses the term "general-elimination harmony" for this relation (cp. Francez & Dyckhoff 2012).

[37] According to Read (2010: 563), these grounds are rather given by the sets of *derivations* of the premises from the discharged assumptions. However, this additional level of complexity can be set aside here.

[38] One should not confuse Read's and Prawitz's uses of the term "ground" (see 3.4.3 for the latter). For example, Read claims that (derivations of) the conjuncts represent the ground for asserting a conjunction (see below), while Prawitz claims that a ground for asserting a conjunction is the result of applying a certain operation to the grounds for asserting its conjuncts.

premises X and assumptions Y *holds* iff each of X holds on the assumption that each of Y holds (see below).

Suppose, for example, that the perfect $\wedge$-introduction rule ($\wedge_I$) gives $\wedge$ its meaning; i.e., that it displays the (complex) ground for asserting a conjunction. According to Read (2010: 565), the following two elimination rules are in harmony with ($\wedge_I$):

$$\frac{\varphi \wedge \psi \quad \overset{[\varphi]}{\underset{\chi}{\vdots}}}{\chi} \; (^1\wedge_E^*) \qquad\qquad \frac{\varphi \wedge \psi \quad \overset{[\psi]}{\underset{\chi}{\vdots}}}{\chi} \; (^2\wedge_E^*)$$

The meaning of $\wedge$ is taken to be such that the components of the ground for asserting a conjunction $S_1 \wedge S_2$ are constituted by its conjuncts.[39] This is taken to mean that harmonious $\wedge$-elimination rules must express that a sentence S^* can be inferred from a conjunction $S_1 \wedge S_2$ if it holds on the assumption that a conjunct S_i ($i \in \{1, 2\}$) holds. (At the end of this subsection, I will clarify the relation between the ordinary elimination rules of the standard sentential operators and those that are harmonious with the ordinary introduction rules.)

Following Read (2010), I will now define a function h between collections of introduction rules and collections of elimination rules, which is meant to capture a strong version of the informal relation of harmony. (The function h will map the singleton $\{(\wedge_I)\}$ to the set $\{(^1\wedge_E^*), (^2\wedge_E^*)\}$.) Two sets of rules will then be called *strictly harmonious* iff they are related by the function h. Furthermore, this strong notion of harmony can also be used to define a weaker notion, the notion of *harmony*, which allows for more pairs of sets of rules to be related (see the end of this subsection).

Let $\sharp$ be some sentential operator, and let $I(\sharp)$ be a set of $\sharp$-introduction rules which gives the meaning of $\sharp$ by displaying the grounds for asserting a $\sharp$-sentence. The image of $I(\sharp)$ under h is obtained in accordance with the types of $\sharp$-introduction rules that are members of $I(\sharp)$. There are three cases of increasing generality and a fourth case of a somewhat special nature.

First, suppose that $I(\sharp)$ contains only one member and that in applications of this rule no assumption is discharged:

$$\frac{B_1 \quad \cdots \quad B_m}{\sharp(\overline{A})} \; (\sharp_I^a)$$

In applications of $(\sharp_I^a)$, one infers a $\sharp$-sentence of the form $\sharp(\overline{A})$ from sentences of the forms $B_1, \ldots, B_m$. According to Read, the elements of a harmonious collection of elimination rules express that a sentence S^* can be inferred from a $\sharp$-sentence S of the form $\sharp(\overline{A})$ if S^* holds on the assumption that a component of the ground for asserting S holds, where these components are identified with the corresponding premises of the forms $B_1, \ldots, B_m$. That is, the set $E(\sharp) := h(I(\sharp))$

---

[39]   If there are no discharged assumptions, then the ground for asserting the conclusion of a $\sharp$-introduction rule ρ can simply be identified with the set of premises of the corresponding application of ρ.

of harmonious $\sharp$-elimination rules is taken to contain precisely the following m members (where φ is a schematic letter that is not contained in the list $\sharp(\overline{A}), B_1, \ldots, B_m$):

$$
\begin{array}{c}
\quad\quad [B_1] \\
\quad\quad \vdots \\
\dfrac{\sharp(\overline{A}) \quad \varphi}{\varphi} \ (^1\sharp_E^a)
\end{array}
\qquad \cdots \qquad
\begin{array}{c}
\quad\quad [B_m] \\
\quad\quad \vdots \\
\dfrac{\sharp(\overline{A}) \quad \varphi}{\varphi} \ (^m\sharp_E^a)
\end{array}
$$

These rules express that a sentence S^* can be inferred from a $\sharp$-sentence S if S^* holds on the assumption that a premise of a corresponding application of $(\sharp_I^a)$ holds.

I would like to draw attention to the limiting case of a single $\sharp$-introduction rule in whose applications no assumptions are discharged: the case in which there are no premises. A set that contains precisely one $\sharp$-introduction rule without schematic initial sequents is mapped to the empty set (of $\sharp$-elimination rules). As an example, recall the second version of Peirce's Rule $\dfrac{}{((\varphi \to \psi) \to \varphi) \to \varphi}$ or the $\vee$-introduction rule $\dfrac{}{\varphi \vee \neg\varphi}$ that corresponds to the principle of the excluded third.

Second, suppose that $I(\sharp)$ contains several members in none of whose applications assumptions are discharged. I will deal with the example of a set $I(\sharp)$ with two $\sharp$-introduction rules:

$$
\dfrac{B_1}{\sharp(\overline{A})} \ (^1\sharp_I^b)
\qquad\qquad
\dfrac{B_2^1 \quad B_2^2}{\sharp(\overline{A})} \ (^2\sharp_I^b)
$$

In an application of the rule $(^1\sharp_I^b)$, one infers a sentence of the form $\sharp(\overline{A})$ from a sentence of the form B_1, and in applications of the rule $(^2\sharp_I^b)$, one infers a sentence of the form $\sharp(\overline{A})$ from sentences of the forms B_2^1 and B_2^2. According to Read, the elements of a harmonious collection of elimination rules express that a sentence S^* can be inferred from a $\sharp$-sentence of the form $\sharp(\overline{A})$ if S^* holds on the assumption that the grounds displayed by the corresponding premises hold. (Since there are two $\sharp$-introduction rules, there are two grounds for asserting a $\sharp$-sentence of the form $\sharp(\overline{A})$.) That is, the set $E(\sharp) := h(I(\sharp))$ of harmonious $\sharp$-elimination rules is taken to contain precisely the following two rules (where φ is a schematic letter that is not contained in the list $\sharp(\overline{A}), B_1, B_2^1, B_2^2$):

$$
\begin{array}{c}
\quad\quad [B_1] \ [B_2^1] \\
\quad\quad\quad \vdots \ \ \vdots \\
\dfrac{\sharp(\overline{A}) \quad \varphi \quad \varphi}{\varphi} \ (^1\sharp_E^b)
\end{array}
\qquad\qquad
\begin{array}{c}
\quad\quad [B_1] \ [B_2^2] \\
\quad\quad\quad \vdots \ \ \vdots \\
\dfrac{\sharp(\overline{A}) \quad \varphi \quad \varphi}{\varphi} \ (^2\sharp_E^b)
\end{array}
$$

These rules express that a sentence S^* can be inferred from a $\sharp$-sentence if S^* holds on the assumption that a premise of a corresponding application of $(^1\sharp_I^b)$ and a premise of a corresponding application of $(^2\sharp_I^b)$ hold.[40] As an example, note that the ordinary $\vee$-elimination rule $(\vee_E)$ is precisely the rule whose

---

[40] It is straightforward to extend the indicated mapping to sets of introduction rules with more than two members. Note that if the set has m members and the ith member, $i \in \{1, \ldots, m\}$, has n_i premises, then there will be $\prod_{i=1}^{m} n_i$ elimination rules each of which has $m + 1$ premises (cp. Read 2010: 563).

singleton is obtained from the set that contains the two perfect ∨-introduction rules $(^1\vee_I)$ and $(^2\vee_I)$; i.e., $h(\{(^1\vee_I),(^2\vee_I)\})$ equals $\{(\vee_E)\}$.

Third, suppose that $I(\sharp)$ contains a member in whose applications assumptions are discharged. I will deal with the example of a set $I(\sharp)$ in which there is only one member with one initial sequent:

$$\frac{\begin{array}{c}[\overline{C}]\\ \vdots\\ B\end{array}}{\sharp(\overline{A})}\ (\sharp_I^c)$$

In applications of the rule $(\sharp_I^c)$, one infers a sentence of the form $\sharp(\overline{A})$ from a sentence of the form B, while, at the same time, one discharges assumptions of the forms of the items of the list $\overline{C}$. According to Read, the elements of a harmonious collection of elimination rules express that a sentence S^* can be inferred from a $\sharp$-sentence S of the form $\sharp(\overline{A})$ if S^* holds on the assumption that the ground for asserting S holds. Now, if one were to write $\overline{C}\Rightarrow B$ for a sentence that expresses that B holds on the assumption that the items of $\overline{C}$ hold, then one would obtain the following harmonious elimination rule (where φ is a schematic letter that is not contained in the list $\sharp(\overline{A}),B,\overline{C}$):

$$\frac{\sharp(\overline{A})\quad \begin{array}{c}[\overline{C}\Rightarrow B]\\ \vdots\\ \varphi\end{array}}{\varphi}$$

This rule is meant to express that a sentence S^* can be inferred from a $\sharp$-sentence if S^* holds on the (conditional) assumption that B holds on the assumption that the items of $\overline{C}$ hold. Now, this is obviously not a rule of a kind allowed here.[41] Fortunately, however, it can be replaced by an ordinary rule which captures its significance:[42]

$$\frac{\sharp(\overline{A})\quad \overline{C}\quad \begin{array}{c}[B]\\ \vdots\\ \varphi\end{array}}{\varphi}\ (\sharp_E^c)$$

This rule expresses that a sentence S^* can be inferred from a $\sharp$-sentence and the sentences discharged in a corresponding application of $(\sharp_I^c)$ if S^* holds on the assumption that the premise of this application of $(\sharp_I^c)$ holds.

As a first example, consider the elimination rule that corresponds to the perfect →-introduction rule $(\rightarrow_I)$:

$$\frac{\varphi\rightarrow\psi\quad \varphi\quad \begin{array}{c}[\psi]\\ \vdots\\ \chi\end{array}}{\chi}\ (\rightarrow_E^*)$$

[41] For an approach which permits such 'higher-order' rules, see Schroeder-Heister 1984.
[42] In the following schema, the notation $\overline{C}$ is slightly inaccurate. What is meant is that the items of $\overline{C}$ represent premises.

This rule expresses that a sentence S^* can be inferred from a conditional and its antecedent if S^* holds on the assumption that the consequent of the conditional holds. As a second example, consider the elimination rules that correspond to the ¬-introduction rule $(\neg_I)$:

$$\cfrac{\neg\varphi \quad \varphi \quad \begin{array}{c}[\psi]\\ \vdots \\ \chi\end{array}}{\chi} \; (^1\neg^*_E) \qquad\qquad \cfrac{\neg\varphi \quad \varphi \quad \begin{array}{c}[\neg\psi]\\ \vdots \\ \chi\end{array}}{\chi} \; (^2\neg^*_E)$$

The rule $(^1\neg^*_E)$ expresses that a sentence S^* can be inferred from a sentence and its negation if S^* holds on the assumption that S holds, where S is allowed to be any sentence. (The rule $(^2\neg^*_E)$ is a sub-rule of $(^1\neg^*_E)$; in its applications, it is required that the sentence S is a ¬-sentence.)

Finally, suppose that $I(\sharp)$ is the empty set (of $\sharp$-introduction rules). In this case, the collection of harmonious $\sharp$-elimination rules contains precisely the following element:

$$\frac{\sharp(\overrightarrow{\varphi})}{\chi} \; (\sharp^d_E)$$

The fact that there are no $\sharp$-introduction rules implies that there is no ground for asserting a $\sharp$-sentence. Therefore, since one may infer from a $\sharp$-sentence S every sentence which holds on the assumption that its ground holds, one may infer every sentence from S.

   This completes the presentation of the mapping h between sets of rules. It may be noted that the function h is applicable to any set of rules whose members are represented by schematic transitions, all of which have the same schematic conclusion. In particular, it is applicable even if this schematic conclusion contains no operator at all and, therefore, is not an introduction rule.

I will now define two relations between sets of rules, a narrow one and a broader one. To begin with, I will deal with the case in which the rules of the first set are represented by schematic transitions, all of which have the same schematic conclusion. Thus, suppose that x and y are sets of rules such that there are representations of the rules in x, all of which have the same schematic conclusion:

Strict Harmony* (SH*) The collections x and y are called *strictly harmonious* iff $y = h(x)$.

That is, to be strictly harmonious is to be linked by the mapping h.
   Now, Dummett's proof-theoretic argument can be presented in terms of the notion of strict harmony, but such a presentation has the disadvantage that calculi which contain the ordinary introduction and elimination rules for $\wedge$ and $\rightarrow$ could not be adequate. It is thus more convenient to introduce a broader notion of harmony. To this end, recall the following structural rules:

$$\frac{}{\varphi \succ \varphi} \; (A) \qquad \frac{\Gamma \succ \varphi \quad \Delta \succ \psi}{\Gamma, \Delta \succ \varphi} \; (W) \qquad \frac{\Gamma, \varphi \succ \psi \quad \Delta \succ \varphi}{\Gamma, \Delta \succ \psi} \; (T)$$

which are classically and intuitionistically acceptable (cp. 2.2.3). Two collections of rules are called *equivalent* iff they are interderivable in a calculus that contains (A), (W), (T). For example, the set $\{(^1\wedge_E), (^2\wedge_E)\}$ of ordinary $\wedge$-elimination rules is equivalent to the set $\{(^1\wedge_E^*), (^2\wedge_E^*)\}$ of $\wedge$-elimination rules that is strictly harmonious to $\{(\wedge_I)\}$.[43] Similarly, the singleton $\{(\rightarrow_E)\}$ is equivalent to the set $\{(\rightarrow_E^*)\}$, and the singleton $\{(\neg_I)\}$ is equivalent to the set $\{(^1\neg_E^*), (^2\neg_E^*)\}$. Now, suppose that x and y are sets of rules such that there are representations of the rules in x, all of which have the same schematic conclusion:

Harmony* (H*) The collections x and y are called *harmonious* iff y is equivalent to the set $h(x)$.

It follows that for each of the standard sentential constants $\sharp$, the collections of $\sharp$-introduction and $\sharp$-elimination rules in the calculus $\mathcal{R}^I$ are harmonious.

Now, in general, it cannot be assumed that the $\sharp$-introduction rules of some calculus have representations with the same schematic conclusion. To deal with this complication, one has to partition the set of introduction rules into subsets whose elements have representations with the same schematic conclusion.[44] To this end, I will call a partition $x_i, i \in I$ of a set x of rules *acceptable* iff the following holds: two rules of x belong to the same partition class x_i, for some $i \in I$, iff they have representations with the same schematic conclusion. Then, one obtains the following general definitions of the notion of harmony and the notion of strict harmony:

(Strict) Harmony Two collections of rules x and y are called *harmonious* (*/strictly harmonious*) iff there is an acceptable partition of $x_i, i \in I$ of x and a partition $y_i, i \in I$ of y such that for every $i \in I$, the sets x_i and y_i are harmonious (/strictly harmonious) in the sense of H* (/SH*).

That is, the collections x and y are (strictly) harmonious iff there is an acceptable partition of x and a partition of y such that the partition classes are (strictly) harmonious in the previously specified sense.

5.5   Dummett's Proof-Theoretic Argument

Recall the general form of a proof-theoretic argument against classical logic. Its first premise is a mathematical result of the form Anti-Cla$_F$, according to which no ARG$_{\mathcal{L}^*}$-classical calculus is F. Its second premise is the thesis EAC, which states that the soundness of classical sentential logic depends on the existence of an adequate ARG$_{\mathcal{L}^*}$-classical calculus. Its third premise is a claim of the

---

[43]   The following two derivations display the equivalence of $(^1\wedge_E)$ and $(^1\wedge_E^*)$:

$$\frac{\Gamma \succ \varphi \wedge \psi \quad \varphi \succ \varphi}{\Gamma \succ \varphi} \qquad \frac{\dfrac{\Gamma \succ \varphi \wedge \psi}{\Gamma \succ \varphi} \quad \Delta, \varphi \succ \chi}{\Gamma, \Delta \succ \chi}$$

The equivalence of $(^2\wedge_E)$ and $(^2\wedge_E^*)$ can be displayed by analogous derivations.

[44]   Read (2010) does not deal with this complication, but it is crucial for an evaluation of Dummett's proof-theoretic argument.

form AIF, according to which an adequate calculus has to be F. Specific proof-theoretic arguments are obtained by presenting a predicate that instantiates the predicate symbol "F", which figures in the first and the third premise.

According to Tennant's proof-theoretic argument, the predicate that replaces "F" applies to precisely those calculi that satisfy the desiderata D_1 and D_2:

Desideratum₁ (D₁) If $\mathcal{R}$ is an adequate calculus, then $\mathcal{R}$ is a standard calculus of the specialised logical framework $\textsc{Arg}_{\mathcal{L}^*}$.

Desideratum₂ (D₂) If $\mathcal{R}$ is an adequate calculus, then $\mathcal{R}$ is separable and contains only rules that are pure, simple, $\neg$-direct, and $\rightarrow$-direct.

However, it was seen that the desideratum D_2 must be rejected: the rules of an adequate calculus do not have to be pure and simple. It was then shown how Tennant's argument could be improved by weakening D_2:

Desideratum₃ (D₃) If $\mathcal{R}$ is an adequate calculus, then the following holds:

$(\vee, \wedge)$ the sub-calculus $\mathcal{R}_{\vee,\wedge}$ of $\mathcal{R}$ comprises only direct and simple rules, and every sequent of one of the languages $\mathcal{L}_\vee$, $\mathcal{L}_\wedge$, $\mathcal{L}_{\vee,\wedge}$ is derivable in $\mathcal{R}$ iff it is derivable in the corresponding sub-calculus $\mathcal{R}_\vee$, $\mathcal{R}_\wedge$, $\mathcal{R}_{\vee,\wedge}$;

$(\rightarrow, \neg)$ every rule of $\mathcal{R} - \mathcal{R}_{\vee,\wedge}$ is $\rightarrow$-direct, $\neg$-direct, $\rightarrow$-simple, and $\neg$-simple.

The requirement $(\vee, \wedge)$ seems to be unproblematic: it seems as if classicists and intuitionists could agree, e.g., that the sub-calculus $\mathcal{R}_{\vee,\wedge}$ of an adequate calculus $\mathcal{R}$ can comprise the usual introduction and elimination rules for $\vee$ and $\wedge$ plus the structural rules (A) and (W). Rather, the crucial question about the improved version of Tennant's argument is whether the requirement $(\rightarrow, \neg)$ is plausible.

In this and the following section, I will deal with two versions of Dummett's proof-theoretic argument against classical logic. As far as I know, these arguments have not been presented in detail elsewhere, and the involved mathematical premises have not been proven. The major goal of this section is to develop precise versions of the two arguments. In 5.6, I will try to show that they are not convincing.

Dummett's proof-theoretic argument is based on the idea that in an adequate calculus, the introduction rules determine the meanings of the relevant operators and the elimination rules precisely capture these meanings. In the first subsection, I will introduce a desideratum, D_4, which is meant to be a formal counterpart of this fundamental idea of Dummett's. Assuming that an adequate calculus has to satisfy this desideratum, it then remains to be seen which further desideratum D can accompany D_4 so that a plausible proof-theoretic argument can be based on the premise that adequacy implies the satisfaction of D and D_4. A central observation will be that there is no $\textsc{Arg}_{\mathcal{L}^*}$-intuitionistic calculus that satisfies D_1 and D_4. Consequently, an intuitionistically acceptable proof-theoretic argument cannot be based on these two desiderata.

In the remaining two subsections, I will introduce two weakened versions of D_1 which fare better in this respect. In 5.5.2, I will deal with the weakening D_1^D, which allows for languages that contain additional logical operators, and in 5.5.3, I will discuss the weakening D_1^T, which allows for sequents with an empty succedent. It will be easy to see that there is an $\text{ARG}_{\mathcal{L}^*}$-intuitionistic calculus that satisfies D_1^D and D_4 and a closely related $\text{ARG}_{\mathcal{L}^*}$-intuitionistic calculus that satisfies D_1^T and D_4. I will call the argument based on the assumption that an adequate calculus must satisfy D_1^D and D_4 the *first version of Dummett's proof-theoretic argument*, and I will call the argument based on the assumption that an adequate calculus must satisfy D_1^T and D_4 the *second version of Dummett's proof-theoretic argument*.

5.5.1 The Fundamental Idea of Dummett's Argument

In what follows, I will use $\mathcal{L}(\mathcal{R})$ to refer to the language that underlies some given calculus $\mathcal{R}$ and will use $\mathcal{R}_{\text{str}}$ to refer to the structural rules of $\mathcal{R}$. According to the fundamental idea of Dummett's proof-theoretic argument, for every adequate calculus $\mathcal{R}$ and for every operator $\sharp$ of $\mathcal{L}(\mathcal{R})$, there is a set $I_{\mathcal{R}}(\sharp)$ of $\sharp$-introduction rules and a set $E_{\mathcal{R}}(\sharp)$ of $\sharp$-elimination rules such that the following three conditions are satisfied:

(i) $\mathcal{R}$ is the union of $\mathcal{R}_{\text{str}}$ and $\bigcup_{\sharp \in \mathcal{L}(\mathcal{R})} I_{\mathcal{R}}(\sharp) \cup E_{\mathcal{R}}(\sharp)$,

(ii) for every operator $\sharp$ of $\mathcal{L}(\mathcal{R})$, $I_{\mathcal{R}}(\sharp)$ determines the meaning of $\sharp$,

(iii) for every operator $\sharp$ of $\mathcal{L}(\mathcal{R})$, $E_{\mathcal{R}}(\sharp)$ precisely captures the meaning of $\sharp$.

I will now present a precise desideratum that corresponds to this idea.

The third condition will be replaced by the desideratum that the collections $I_{\mathcal{R}}(\sharp)$ and $E_{\mathcal{R}}(\sharp)$ are harmonious. It is less clear, however, what a suitable replacement of the second condition might be. Recall that Dummett (1991: 258) claims that an introduction rule which helps to determine the meaning of its operator must satisfy a *complexity condition*: there has to be a standard complexity ordering which every introduction rule that helps to determine the meaning of its operator respects (see 5.4.2). It might now be proposed that the combination of this requirement and the requirement of harmony suffice to yield a strong enough desideratum:

Desideratum$_4^a$ (D_4^a) If $\mathcal{R}$ is an adequate calculus, then, for every operator $\sharp$ of $\mathcal{L}(\mathcal{R})$, there is a set $I_{\mathcal{R}}(\sharp)$ of $\sharp$-introduction rules and a set $E_{\mathcal{R}}(\sharp)$ of $\sharp$-elimination rules such that the following holds:

(C$_1$) $\mathcal{R}$ is the union of $\mathcal{R}_{\text{str}}$ and $\bigcup_{\sharp \in \mathcal{L}(\mathcal{R})} I_{\mathcal{R}}(\sharp) \cup E_{\mathcal{R}}(\sharp)$,

(C$_2$) there is a standard complexity ordering $\leq$ such that for every operator $\sharp$ of $\mathcal{L}(\mathcal{R})$ and for every ρ of $I_{\mathcal{R}}(\sharp)$, ρ respects $\leq$,

(C$_3$) for every operator $\sharp$ of $\mathcal{L}(\mathcal{R})$, $I_{\mathcal{R}}(\sharp)$ and $E_{\mathcal{R}}(\sharp)$ are harmonious.

Note that it is not only required that for every rule ρ in $\bigcup_{\sharp \in \mathcal{L}(\mathcal{R})} I_{\mathcal{R}}(\sharp)$, there is a standard complexity ordering respected by ρ. Rather, it is required that there is a standard complexity ordering respected by every rule in $\bigcup_{\sharp \in \mathcal{L}(\mathcal{R})} I_{\mathcal{R}}(\sharp)$. The idea that underlies Dummett's complexity condition is the following: the rules of $I_{\mathcal{R}}(\sharp)$ determine the meaning of a $\sharp$-sentence S in terms of the meanings of the premises and discharged assumptions of applications whose conclusion is S. The complexity condition is then meant to preclude circular accounts of meaning, i.e. accounts in which the meanings of two sentences S_1 and S_2 are explained in such a way that the first explanation takes the meaning of S_2 for granted, while the second explanation takes the meaning of S_1 for granted. Consequently, if one wants to use the complexity condition to preclude circular accounts of meaning, then the same measure of complexity must be assumed for all rules.

However, the combination of D_4^a and desiderata such as D_1, D_1^P, or D_1^T does not separate the classes of intuitionistic and classical calculi. To see this, suppose that $\mathcal{R}$ is some $\text{ARG}_{\mathcal{L}^*}$-intuitionistic calculus that satisfies D_4^a and (a weakened version of) D_1. Then, it is not difficult to find collections of introduction and elimination rules for $\rightarrow$ and $\neg$ of the specialised framework $\text{ARG}_{\mathcal{L}^*}$ whose combination with the rules of $\mathcal{R}$ constitutes an $\text{ARG}_{\mathcal{L}^*}$-*classical* calculus that also satisfies D_4^a. As regards $\rightarrow$, one adds the first version of Peirce's Rule and a harmonious $\rightarrow$-introduction rule:

$$\frac{\Gamma \succ \varphi}{\Gamma \succ (\varphi \to \psi) \to \varphi} \ (\text{PR}_1^*) \qquad\qquad \frac{\Gamma \succ (\varphi \to \psi) \to \varphi}{\Gamma \succ \varphi} \ (\text{PR}_1)$$

As regards $\neg$, one adds the rule of double negation elimination and a harmonious $\neg$-introduction rule:

$$\frac{\Gamma \succ \varphi}{\Gamma \succ \neg\neg\varphi} \ (\neg_I) \qquad\qquad \frac{\Gamma \succ \neg\neg\varphi}{\Gamma \succ \varphi} \ (\neg_E)$$

It thus follows that if there is an $\text{ARG}_{\mathcal{L}^*}$-intuitionistic calculus that satisfies D_4^a and (a weakened version of) D_1, then there is also an $\text{ARG}_{\mathcal{L}^*}$-classical calculus that satisfies D_4^a and (the same weakened version of) D_1.

As far as I can see, the only plausible suggestion for strengthening the desideratum D_4^a is based on the idea that a collection x of $\sharp$-introduction rules which determines the meaning of $\sharp$ has to be independent; i.e., there have to be schematic representations of the rules of x such that no sentence is an instance of distinct schematic conclusions (see 5.4.2). In combination with D_4^a, this requirement yields the following desideratum:

Desideratum$_4$ (D$_4$) If $\mathcal{R}$ is an adequate calculus, then, for every operator $\sharp$ of $\mathcal{L}(\mathcal{R})$, there is a set $I_{\mathcal{R}}(\sharp)$ of $\sharp$-introduction rules and a set $E_{\mathcal{R}}(\sharp)$ of $\sharp$-elimination rules such that the following holds:

(C$_1$) $\mathcal{R}$ is the union of $\mathcal{R}_{\text{str}}$ and $\bigcup_{\sharp \in \mathcal{L}(\mathcal{R})} I_{\mathcal{R}}(\sharp) \cup E_{\mathcal{R}}(\sharp)$,

(C$_2^a$) for every operator $\sharp$ of $\mathcal{L}(\mathcal{R})$, $I_{\mathcal{R}}(\sharp)$ is an independent family,

(C$_2^b$) there is a standard complexity ordering $\leq$ such that for every operator $\sharp$ of $\mathcal{L}(\mathcal{R})$ and for every ρ of $I_{\mathcal{R}}(\sharp)$, ρ respects $\leq$,

(C$_3$) for every operator $\sharp$ of $\mathcal{L}(\mathcal{R})$, $I_{\mathcal{R}}(\sharp)$ and $E_{\mathcal{R}}(\sharp)$ are harmonious.

I will assume that D_4 is an acceptable formal counterpart of the fundamental assumption of Dummett's proof-theoretic argument.

Now, the combination of D_1 and D_4 yields a proof-theoretic argument against classical logic:

Anti-Cla$_3$ No $\text{ARG}_{\mathcal{L}^*}$-classical calculus satisfies D_1 and D_4.

In combination with EAC and the relevant instance of AIF, the claim Anti-Cla$_3$ implies that the theory of classical sentential logic is not sound. Unfortunately, this proof-theoretic argument is not intuitionistically acceptable, because there is a corresponding result about $\text{ARG}_{\mathcal{L}^*}$-intuitionistic calculi:

Anti-Int$_3$ No $\text{ARG}_{\mathcal{L}^*}$-intuitionistic calculus satisfies D_1 and D_4.

Consequently, if the proof-theoretic argument based on Anti-Cla$_3$ shows that classical sentential logic is not sound, then a parallel argument based on Anti-Int$_3$ shows that intuitionistic sentential logic is not sound.

As a matter of fact, these results can even be strengthened:

Anti-Cla$_3^+$ No $\text{ARG}_{\mathcal{L}^*}$-classical calculus satisfies D_1 and D_4^a.

Anti-Int$_3^+$ No $\text{ARG}_{\mathcal{L}^*}$-intuitionistic calculus satisfies D_1 and D_4^a.

That is, the assumption (C_2^a) of desideratum D_4, which will be of crucial significance once the desideratum D_1 is weakened, can be dispensed with here.

I do not know whether Anti-Cla$_3^+$ and Anti-Int$_3^+$ have already been proven. In the remainder of this subsection, I have therefore included a proof of these results. This proof is based on the following theorem:

Normalisability Suppose that for every operator $\sharp$ of $\mathcal{L}(\mathcal{R})$, $I_{\mathcal{R}}(\sharp)$ is a collection of $\sharp$-introduction rules, $E_{\mathcal{R}}(\sharp)$ is a collection of $\sharp$-elimination rules that is strictly harmonious with $I_{\mathcal{R}}(\sharp)$, $\mathcal{R} = \mathcal{R}_{\text{str}} \cup \bigcup_{\sharp \in \mathcal{L}(\mathcal{R})} I_{\mathcal{R}}(\sharp) \cup E_{\mathcal{R}}(\sharp)$, and there is a standard complexity ordering $\leq$ respected by every rule of $\bigcup_{\sharp \in \mathcal{L}(\mathcal{R})} I_{\mathcal{R}}(\sharp)$. Then, every $\mathcal{R}$-derivable sequent has an $\mathcal{R}$-derivation in which the major premises of applications of rules of $\bigcup_{\sharp \in \mathcal{L}(\mathcal{R})} E_{\mathcal{R}}(\sharp)$ are undischarged assumptions.

Note that the antecedent of this conditional involves a strengthening of the condition (C_3) of D_4^a: it is required that the sets $I_{\mathcal{R}}(\sharp)$ and $E_{\mathcal{R}}(\sharp)$ are *strictly harmonious*. This theorem follows from a simple variant of Prawitz's (1965) main result (see Troelstra & Schwichtenberg 2000: 182 and 226).[45]

Proof. Suppose that $\mathcal{R}$ is a standard calculus of the specialised framework $\text{ARG}_{\mathcal{L}^*}$ which equals the following union:

$$\mathcal{R}_{\text{str}} \cup I_{\mathcal{R}}(\vee) \cup E_{\mathcal{R}}(\vee) \cup I_{\mathcal{R}}(\wedge) \cup E_{\mathcal{R}}(\wedge) \cup I_{\mathcal{R}}(\to) \cup E_{\mathcal{R}}(\to) \cup I_{\mathcal{R}}(\neg) \cup E_{\mathcal{R}}(\neg),$$

---

[45]   The combination of strict harmony and satisfaction of Dummett's complexity condition implies that the 'conversion operations' (see Troelstra & Schwichtenberg ibid.) terminate. Therefore, they can be applied to any $\mathcal{R}$-derivation in such a way that an $\mathcal{R}$-derivation of the same sequent is produced in which the major premises of applications of rules of $\bigcup_{\sharp \in \mathcal{L}(\mathcal{R})} E_{\mathcal{R}}(\sharp)$ are undischarged assumptions.

where there is a standard complexity ordering respected by the rules of the union $\bigcup_{\sharp \in \mathcal{L}(\mathcal{R})} I_{\mathcal{R}}(\sharp)$, and where each of the pairs $I_{\mathcal{R}}(\sharp)$ and $E_{\mathcal{R}}(\sharp)$ are *strictly harmonious*. I will now show that the assumption that the calculus $\mathcal{R}$ is either $\text{ARG}_{\mathcal{L}^*}$-classical or $\text{ARG}_{\mathcal{L}^*}$-intuitionistic leads to a contradiction.[46]

On the one hand, it is shown that if $\mathcal{R}$ is either $\text{ARG}_{\mathcal{L}^*}$-classical or $\text{ARG}_{\mathcal{L}^*}$-intuitionistic, then $I_{\mathcal{R}}(\neg)$ contains a general $\neg$-introduction rule, i.e. a $\neg$-introduction rule whose schematic conclusion has the form $\neg\varphi$ (see 5.4.2). From the assumption that $\mathcal{R}$ is either $\text{ARG}_{\mathcal{L}^*}$-classical or $\text{ARG}_{\mathcal{L}^*}$-intuitionistic, it follows that the sequent $p, \sharp p \succ q$ is $\mathcal{R}$-derivable. The normalisability theorem therefore implies that there is a derivation d of this sequent in which the major premises of applications of rules of $\bigcup E_{\mathcal{R}}(\sharp)$ are undischarged assumptions. Since p and $\neg p$ are the only undischarged assumptions of d, this has the consequence that every rule of $\bigcup E_{\mathcal{R}}(\sharp)$ that was applied is a rule whose major premise is represented by a schematic sentence of the form $\neg\varphi$. Furthermore, since d cannot consist solely of applications of structural rules and rules of $\bigcup I_{\mathcal{R}}(\sharp)$, it follows that there is a rule of $E_{\mathcal{R}}(\neg)$ that has been applied and whose major premise is represented by a schematic sentence of the form $\neg\varphi$. However, a collection of $\neg$-elimination rules which contains such a rule is only harmonious to a collection of $\neg$-introduction rules which contains a general $\neg$-introduction rule (in which the conclusion is represented by $\neg\varphi$).

On the other hand, it is shown that if $\mathcal{R}$ is either $\text{ARG}_{\mathcal{L}^*}$-classical or $\text{ARG}_{\mathcal{L}^*}$-intuitionistic, then $\mathcal{R}$ cannot contain a general $\neg$-introduction rule. This directly follows from the fact that general $\neg$-introduction rules of the specialised framework $\text{ARG}_{\mathcal{L}^*}$ which satisfy a standard complexity ordering are not admissible in $\text{ARG}_{\mathcal{L}^*}$-classical or $\text{ARG}_{\mathcal{L}^*}$-intuitionistic calculi. To see this, suppose that $\mathcal{R}$ contains such a rule. This rule has the following form:

$$\frac{\overset{[\overline{C}_1]}{\vdots} \qquad \overset{[\overline{C}_m]}{\vdots}}{\frac{B_1 \quad \ldots \quad B_m}{\neg\varphi}}$$

where no schematic sentence of the list $B_1, \ldots, B_m, \overline{C}_1, \ldots, \overline{C}_m$ contains $\neg$. (The fact that these schematic sentences are not allowed to contain $\neg$ directly follows from the assumption that this rule respects some standard complexity ordering.) Now, consider the application of this rule obtained from instantiating every schematic sentence with $p \rightarrow p$. It then follows that the sequents $\overline{C}_1 \succ B_1, \ldots, \overline{C}_m \succ B_m$ are $\mathcal{R}$-derivable, and, therefore, with an application of this rule, it follows that the sequent $\succ \neg(p \rightarrow p)$ is $\mathcal{R}$-derivable. However, this contradicts the assumption that $\mathcal{R}$ is either $\text{ARG}_{\mathcal{L}^*}$-classical or $\text{ARG}_{\mathcal{L}^*}$-intuitionistic. $\square$

---

[46]   It is not difficult to see that if there are no $\text{ARG}_{\mathcal{L}^*}$-classical or $\text{ARG}_{\mathcal{L}^*}$-intuitionistic calculi with strictly harmonious collections of rules that satisfy D_1 and D_4^a, then there are no $\text{ARG}_{\mathcal{L}^*}$-classical or $\text{ARG}_{\mathcal{L}^*}$-intuitionistic calculi that satisfy D_1 and D_4^a.

5.5.2   The First Version of Dummett's Argument

In the previous subsection, it was seen that the desiderata D_1 and D_4 are too strong for Dummett's purposes: there is no $\text{ARG}_{\mathcal{L}^*}$-intuitionistic calculus which satisfies them. Since the desideratum D_4 corresponds to the fundamental idea of Dummett's reasoning, it remains to be seen whether there is a suitable weakened version of the desideratum D_1 which can be used for an intuitionistically acceptable argument against classical logic. I will consider two such proposals for replacing D_1 with a weaker desideratum.

In his discussion of the problem that the 'intuitionistic' $\neg$-introduction rule $(\neg_I)$ does not satisfy the complexity condition, Dummett makes the following proposal:

> This strongly suggests the well-known device of treating $\ulcorner\neg\mathbf{A}\urcorner$ as a definitional abbreviation of $\ulcorner\mathbf{A} \to \bot\urcorner$, where $\bot$ is a constant sentence (Dummett 1991: 295).

In this subsection, I will introduce the first version of Dummett's proof-theoretic argument, which is based on this proposal for treating negation.

For a finite number of sentential operators $\sharp_1, \ldots, \sharp_n$, let $\mathcal{L}^{\sharp_1, \ldots, \sharp_n}$ be the language that is based on the sentential constants of $\mathcal{L}^*$ (i.e. the standard sentential operators) and the operators $\sharp_1, \ldots, \sharp_n$. Then, consider the following replacement of the desideratum D_1:

Desideratum$_1^D$ (D_1^D) If $\mathcal{R}$ is an adequate calculus, then there are sentential operators $\sharp_1, \ldots, \sharp_n$ such that $\mathcal{R}$ is a standard calculus of the specialised logical framework $\text{ARG}_{\mathcal{L}^{\sharp_1, \ldots, \sharp_n}}$.

According to D_1^D, it is not precluded that an adequate calculus $\mathcal{R}$ belongs to a specialised framework based on an 'ordinary' extension of $\mathcal{L}^*$, i.e. an extension obtained by the addition of further sentential operators.

The combination of D_1^D and D_4 constitutes the third premise of the first version of Dummett's proof-theoretic argument. The fact that this argument is an intuitionistically acceptable argument against classical logic follows from the following two mathematical theorems:

Anti-Cla$_4$ No $\text{ARG}_{\mathcal{L}^*}$-classical calculus satisfies D_1^D and D_4.

Pro-Int$_4$ At least one $\text{ARG}_{\mathcal{L}^*}$-intuitionistic calculus satisfies D_1^D and D_4.

As fas as I know, the claim Anti-Cla$_4$ has not been proven. However, since its proof is more involved than the one of Anti-Cla$_3$, I will not present it here. (In the following subsection, I will introduce the second version of Dummett's proof-theoretic argument based on a mathematical result less difficult to prove. I will confine myself to presenting a sketch of this simpler proof.)

In the remainder of this subsection, I will show that there is an $\text{ARG}_{\mathcal{L}^*}$-intuitionistic calculus that satisfies D_1^D and D_4 and which belongs to a specialised framework that is based on the language $\mathcal{L}^\bot$ obtained from $\mathcal{L}^*$ by adding the nullary sentential operator $\bot$, the so-called *falsum constant*. This calculus will be referred to as $^{\perp}\mathcal{R}_W^I$. The central idea is that the meaning of $\perp$ is determined by the empty set (of $\perp$-introduction rules); i.e., $I_{\perp\mathcal{R}_W^I}(\perp) = \emptyset$. Then, the singleton of the following rule is a harmonious set of $\perp$-elimination rules (see 5.4.3):

$$\frac{\perp}{\varphi} \ (\perp_E)$$

That is, $E_{\perp\mathcal{R}_W^I}(\perp) = \{(\perp_E)\}$. For trivial reasons, the set of $\perp$-introduction rules contains only perfect $\perp$-introduction rules. Moreover, the only $\perp$-elimination rule is pure, simple, and direct, and it satisfies a condition that is the obvious counterpart of Dummett's complexity condition for introduction rules: φ is of lower logical complexity than $\perp$ (see Dummett 1991: 283). Given the falsum constant, it is then easily possible to introduce a negation operator by means of the following introduction rule and a harmonious elimination rule:[47]

$$\frac{\begin{array}{c}[\varphi]\\ \vdots\\ \perp\end{array}}{\neg\varphi} \ (\neg_I^{\perp}) \qquad \blacktriangleright \qquad \frac{\neg\varphi \quad \varphi \quad \begin{array}{c}[\perp]\\ \vdots\\ \chi\end{array}}{\chi} \qquad \approx \qquad \frac{\neg\varphi \quad \varphi}{\perp} \ (\neg_E^{\perp})$$

These rules are simple and direct but not pure. Furthermore, the rule $(\neg_I^{\perp})$ respects every complexity ordering in which $\perp$ is less complex that $\neg\varphi$.[48] Now, $^{\perp}\mathcal{R}_W^I$ is the calculus of the specialised framework $\mathrm{ARG}_{\mathcal{L}^{\perp}}$ obtained from $\mathcal{R}_W^I$ by deleting the rules $(\neg_I)$ and $(\neg_E)$ and adding the rules $(\perp_E)$, $(\neg_I^{\perp})$, and $(\neg_E^{\perp})$. It is easily seen to be $\mathrm{ARG}_{\mathcal{L}^*}$-intuitionistic. Note that there are only two rules of $^{\perp}\mathcal{R}_W^I$ which can be used to introduce a sentence that contains $\perp$ into a derivation: the structural rule (A) and the rule $(\neg_E^{\perp})$.

The rule $(\neg_E^{\perp})$ is not only a $\neg$-elimination rule; it is also a $\perp$-introduction rule. To stress this property of $(\neg_E^{\perp})$, I will also refer to it as $(\perp_I^{\neg})$. One might think that the calculus $^{\perp}\mathcal{R}_W^I$ does not satisfy the desideratum D_4 because there is no complexity ordering respected by the following two introduction rules:

$$\frac{\Gamma, \varphi \succ \perp}{\Gamma \succ \neg\varphi} \ (\neg_I^{\perp}) \qquad\qquad \frac{\Gamma \succ \varphi \quad \Delta \succ \neg\varphi}{\Gamma, \Delta \succ \perp} \ (\perp_I^{\neg})$$

The rule $(\neg_I^{\perp})$ respects only those complexity orderings in which $\perp$ is simpler than $\neg\varphi$, while the rule $(\perp_I^{\neg})$ respects only those complexity orderings in which $\neg\varphi$ is simpler than $\perp$. Now, it is important to note that this does *not* conflict with desideratum D_4 because the $\perp$-introduction rule $(\perp_I^{\neg})$ does not belong to $\bigcup I_{\mathcal{R}}(\sharp)$, and, according to D_4, only those introduction rules which do belong to $\bigcup I_{\mathcal{R}}(\sharp)$ have to respect some complexity ordering.

---

[47] The symbol $\blacktriangleright$ expresses that the singleton of the rule represented by the term to its left is strictly harmonious with the singleton of the rule represented by the term to its right. The symbol $\approx$ expresses that the adjacent terms represent equivalent rules. (Recall that two rules are equivalent iff they are interderivable in a calculus that contains (A), (W), and (T).)

[48] As Gentzen (1934: 189) notes, the rules $(\neg_I^{\perp})$ and $(\neg_E^{\perp})$ are satisfied by a defined operator of the language $\mathcal{L}^{\perp}$: $\neg S := S \to \perp$. Accordingly, many intuitionists treat $\neg$ as a defined symbol.

By way of conclusion, I would like to make a remark about the question of whether there is an intuitive explanation of the meaning of $\bot$ (which does not presuppose knowledge of the meanings of the standard sentential operators). In the context of arithmetical theories, it has been observed that $\bot$ can be replaced by the sentence $0 = S0$ (see Troelstra & van Dalen 1988: 121 and 126). As Tennant (1999) has stressed, however, such a definition does not seem to be acceptable in non-arithmetical contexts. More plausible seems to be the suggestion to define $\bot$ in terms of sentential quantification. Following Humberstone (2011: 1282-3), one might propose to interpret $\bot$ as the sentence $\forall \nu.\nu$ (in which ν functions as a sentential variable). The $\bot$-elimination rule ($\bot_E$) could then be treated as a special case of the standard rule for eliminating a universal (sentential) quantifier, namely as the rule $\frac{\forall \nu.\nu}{\varphi}$.

It thus seems as if it were possible to provide an intuitive explanation of the meaning of $\bot$. However, according to Dummett's idea of meaning conferring collections of introduction rules, it is not *necessary* to provide such an explanation. According to this idea, it suffices to explain the meaning of $\bot$ by saying that it is determined by the empty set (of $\bot$-introduction rules).

5.5.3   The Second Version of Dummett's Argument

In this subsection, I will introduce the second version of Dummett's proof-theoretic argument. It is based on a weakened version of D_1 according to which it is not required that an adequate calculus must belong to the framework ARG:

Desideratum$_1^T$ (D_1^T) If $\mathcal{R}$ is an adequate calculus, then $\mathcal{R}$ is a standard calculus of the specialised logical framework $\text{ARG}_{0\mathcal{L}^*}$.

Note that every calculus of the framework ARG is also a calculus of the framework ARG_0. (A calculus $\mathcal{R}$ belongs to a framework $\mathcal{F}$ iff every sequent of an application of a rule of $\mathcal{R}$ belongs to the framework $\mathcal{F}$.) Thus, the desideratum D_1^T does not preclude adequate calculi of the framework $\text{ARG}_{\mathcal{L}^*}$, but it allows for calculi with rules whose applications involve transitions without succedents.

By way of motivation, consider the following sketch of an indirect proof of the claim that there is no greatest prime number:

> Suppose that there were a greatest prime number M. Then the number $N = (2 \times 3 \times \ldots \times M) + 1$ would either be prime itself or else would be divisible by a prime number between M and N. Either way, there would be a prime number greater than M. Contradiction. So: there is no greatest prime number (Rumfitt 2000: 793).

Rumfitt characterises this argumentation as follows:

> [I]t would be perverse to try to assign a propositional content to the expression 'contradiction'. [...] It marks the point where the supposition that there is a greatest prime number has been shown to lead to a logical dead end, and is thus discharged, prior to an assertion of its negation (Rumfitt 2000: 793-4).

According to this view, informal argumentations may contain expressions which indicate that a logical dead end has been reached, where these expressions are not to be understood as having a propositional content. (I will return to this claim in 5.6.1.) The obvious choice of a logical framework whose derivations correspond to informal argumentations with logical dead ends is the framework ARG_0 in which sequents with empty succedents are allowed for. Dummett claims that calculi of this framework can be adequate:

> A sequent with an empty succedent expresses that the sentences comprised by the antecedent are contradictory [...]. We may therefore allow the notion of a sequent to be extended to cover those with empty succedents, since, although it is greatly at variance with our practice in natural language, it is readily intelligible [...] (Dummett 1991: 187).

While Dummett denies that sequents with multiple succedents may figure in adequate calculi, he allows for sequents with no succedents.

In representations of derivations of standard calculi of the framework ARG_0, I will use the symbol $\overset{*}{\underset{\,}{\bullet}}$ to indicate a logical dead end. (Alternatively, one might use an empty space.) It is important to note that $\overset{*}{\underset{\,}{\bullet}}$, unlike $\bot$, is not an embeddable symbol of the object language. Using this device to indicate logical dead ends, the indirect proof that was sketched above can be captured by a derivation of the following form:

$$
\begin{array}{c}
A^{(1)} \\
\vdots \\
\overset{*}{\bullet} \\
\hline
\neg A^{(1)}
\end{array}
$$

In the penultimate step of this argumentation, the assumption that there is a greatest prime number, which is expressed by the sentence A, was seen to lead to a logical dead end. In the final step, this assumption is then discharged, and its negation is asserted.

The combination of D_1^T and D_4 constitutes the third premise of the second version of Dummett's proof-theoretic argument. The fact that this variant is an intuitionistically acceptable argument against classical logic is based on the following two mathematical theorems:

Anti-Cla$_5$ No $\text{ARG}_{\mathcal{L}^*}$-classical calculus satisfies D_1^T and D_4.

Pro-Int$_5$ At least one $\text{ARG}_{\mathcal{L}^*}$-intuitionistic calculus satisfies D_1^T and D_4.

As fas as I know, the claim Anti-Cla$_5$ has not been proven. First, I will present a calculus which shows that Pro-Int$_5$ is true. Then, I will indicate how the claim Anti-Cla$_5$ can be seen to hold.

The calculus $^\bullet\mathcal{R}_W^I$, to be introduced now, is a calculus of the specialised framework $\text{ARG}_{0\mathcal{L}^*}$ which closely resembles the calculus $^\bot\mathcal{R}_W^I$. It contains three structural rules: (A), (W), and the following rule resembling $(\bot_E)$:

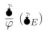

$$
\frac{\overset{*}{\bullet}}{\varphi} \; (\overset{*}{\bullet}_E)
$$

This rule expresses that if a logical dead end has been reached, then one can assert any sentence, conditional on the assumptions that have not been discharged yet.

Second, the calculus $^{\bullet}\mathcal{R}_W^I$ contains the introduction and elimination rules for the operators $\vee$, $\wedge$, $\to$ that belong to $\mathcal{R}^I$. In addition, there is the following supplementation to the rule $(\vee_E)$:

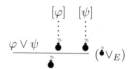

Note that there cannot be similar supplementations to the other introduction and elimination rules for the operators $\vee$, $\wedge$, $\to$ of $\mathcal{R}^I$.

Finally, there are rules for $\neg$ which correspond to $(\neg_I^{\perp})$ and $(\neg_E^{\perp})$:

$$
\frac{\begin{array}{c}[\varphi]\\\vdots\\\bullet\end{array}}{\neg\varphi}\ (\neg_I^{\,\bullet})
\qquad\qquad
\frac{\varphi\quad\neg\varphi}{\bullet}\ (\neg_E^{\,\bullet})
$$

The rule $(\neg_I^{\,\bullet})$ expresses that if a logical dead end has been reached on the basis of an assumption S, then one can assert $\neg S$, while, at the same time, one can discharge the assumption S. The rule $(\neg_E^{\,\bullet})$ expresses that a sentence together with its negation constitute a logical dead end. Note that the rules $(\neg_I^{\,\bullet})$ and $(\neg_E^{\,\bullet})$ are not only direct and simple but also pure. Note also that applications of the rule $(\neg_E^{\,\bullet})$ constitute the only possibilities of reaching a logical dead end in a derivation. This completes the presentation of $^{\bullet}\mathcal{R}_W^I$. Inspection shows that the calculus $^{\bullet}\mathcal{R}_W^I$ satisfies the desiderata D_1^T and D_4. In particular, each of its introduction rules is perfect and thus respects even the containment relation (see 5.4.2).

I will conclude by presenting a proof sketch of Anti-Cla5.

Sketch. Suppose that $\mathcal{R}$ is a standard calculus of the specialised framework $\text{ARG}_{0\mathcal{L}^*}$ which equals the following union:

$$
\mathcal{R}_{\text{str}}\cup I_{\mathcal{R}}(\vee)\cup E_{\mathcal{R}}(\vee)\cup I_{\mathcal{R}}(\wedge)\cup E_{\mathcal{R}}(\wedge)\cup I_{\mathcal{R}}(\to)\cup E_{\mathcal{R}}(\to)\cup I_{\mathcal{R}}(\neg)\cup E_{\mathcal{R}}(\neg),
$$

where each of the collections $I_{\mathcal{R}}(\sharp)$ is independent, where there is a standard complexity ordering $\le$ that is respected by the rules of the union $\bigcup_{\sharp\in\mathcal{L}(\mathcal{R})} I_{\mathcal{R}}(\sharp)$, and where each of the pairs $I_{\mathcal{R}}(\sharp)$ and $E_{\mathcal{R}}(\sharp)$ is strictly harmonious. I will now indicate that the assumption that the calculus $\mathcal{R}$ is $\text{ARG}_{\mathcal{L}^*}$-classical leads to a contradiction.

As in the proof of Anti-Cla3, it is shown that $I_{\mathcal{R}}(\neg)$ contains a general $\neg$-introduction rule (see 5.5.1). Then, since the family $I_{\mathcal{R}}(\neg)$ is independent, it follows that each of its members is a general $\neg$-introduction rule. Let $\mathcal{C}^*$ be the collection of general $\neg$-introduction rules of the specialised framework $\text{ARG}_{0\mathcal{L}^*}.$

which respect the standard complexity ordering $\leq$. It follows that every rule of $\mathcal{C}^*$ has the following form:

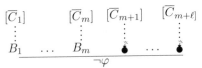

where m and ℓ are greater than or equal to 0, and where the schematic sentences of the list $B_1, \ldots, B_m, \overline{C}_1, \ldots, \overline{C}_{m+\ell}$ only contain the operators $\vee$, $\wedge$, $\rightarrow$. Let $(I_0, E_0) := (I_\mathcal{R}(\neg), E_\mathcal{R}(\neg))$.

Now, a sequence of pairs (I_0, E_0), (I_1, E_1), $\ldots$, (I_n, E_n) with the following properties is constructed: (i) for every $j \leq n$, I_j is a collection of rules of $\mathcal{C}^*$ and E_j is strictly harmonious with I_j; (ii) for every $j \leq n$ and for ever sequent s of $\mathrm{ARG}_{0\mathcal{L}^*}$, s is derivable in $\mathcal{R}$ iff s derivable in $(\mathcal{R} - (I_0 \cup E_0)) \cup (I_j \cup E_j)$; (iii) $(\mathcal{R} - (I_0 \cup E_0)) \cup (I_n \cup E_n)$ is not $\mathrm{ARG}_{\mathcal{L}^*}$-classical, which is a contradiction. Here, I will only present a sketch of the construction of this sequence.

First, one replaces the rules of I_0 with rules that do not contain schematic letters distinct from φ; this yields I_1. Second, one replaces the rules of I_1 with rules that do not contain $\vee$, $\wedge$, or $\rightarrow$; this yields I_2. Third, one replaces the rules of I_2 with rules that do not contain initial schematic sequents of the form $\Gamma, \varphi \succ \varphi$; this yields I_3. It follows that the rules of I_3 have initial schematic sequents of the following forms: $\Gamma \succ \varphi$; $\Gamma \succ \overset{*}{\bullet}$; $\Gamma, \varphi \succ \overset{*}{\bullet}$. In the fourth step, one then replaces the rules of I_3 with rules that do not have an initial schematic sequent of the form $\Gamma \succ \varphi$; this yields I_4. Then, it is not difficult to see that $(\mathcal{R} - (I_0 \cup E_0)) \cup (I_4 \cup E_4)$ is not $\mathrm{ARG}_{\mathcal{L}^*}$-classical. $\qquad\square$

5.6 Against Dummett's Proof-Theoretic Argument

The two versions of Dummett's proof-theoretic argument are based on the claim that the theory of classical sentential logic is not sound because there is no suitable $\mathrm{ARG}_{\mathcal{L}^*}$-classical calculus that satisfies D_4, i.e., because there is no suitable $\mathrm{ARG}_{\mathcal{L}^*}$-classical calculus constituted by independent families of introduction rules that respect some complexity ordering and families of elimination rules that are in harmony with them. In this final section, I will argue that both arguments are not convincing. In the first subsection, I will present extensions of the specialised logical framework $\mathrm{ARG}_{\mathcal{L}^*}$ which allow for standard $\mathrm{ARG}_{\mathcal{L}^*}$-classical calculi that satisfy the desideratum D_4. It will be argued that Dummett has not shown that such calculi are inadequate. In the second subsection, I will question the assumption EAC, according to which the soundness of classical sentential logic depends on the existence of an adequate $\mathrm{ARG}_{\mathcal{L}^*}$-classical calculus, and the assumption that an adequate calculus must satisfy the desideratum D_4.

5.6.1   Reflection and Conjugation

I will introduce two non-standard extensions of the language $\mathcal{L}^*$: the languages $^\frown\mathcal{L}$ and $^{+/}\mathcal{L}$. I will argue that proponents of proof-theoretic arguments have not shown that adequate calculi must not belong to specialised logical frameworks based on such languages. Consequently, for all they have shown, the desideratum D_1^D has to be replaced by a weaker one:

Desideratum$_1^R$ (D$_1^R$) If $\mathcal{R}$ is an adequate calculus, then $\mathcal{R}$ is a standard calculus of the framework ARG.

In contrast to D_1^D, the desideratum D_1^R does not preclude calculi that are not based on $\mathcal{L}^*$ or an ordinary extension of $\mathcal{L}^*$.

I will present two ARG$_{\mathcal{L}^*}$-classical calculi that satisfy the desiderata D_1^R and D_4: the calculus $\sim\!\mathcal{R}_W^C$, which is based on the language $\sim\!\mathcal{L}$, and the calculus $^{+/-}\mathcal{R}_W^C$, which is based on the language $^{+/-}\mathcal{L}$. If one assumes that the ARG$_{\mathcal{L}^*}$-intuitionistic calculus $^\perp\mathcal{R}_W^I$ is adequate because it satisfies the desiderata D_1^D and D_4, then, for all that has been shown, the ARG$_{\mathcal{L}^*}$-classical calculi $\sim\!\mathcal{R}_W^C$ and $^{+/-}\mathcal{R}_W^C$ might be adequate because they satisfy the desiderata D_1^R and D_4. It thus follows that Dummett's proof-theoretic argument has not been shown to be convincing.

By way of motivation, recall the calculus $^g\mathcal{R}_W^C$ of the specialised framework G-ARG$_{\mathcal{L}^*}$, whose sequents are generalised arguments of the language $\mathcal{L}^*$. This calculus not only meets Tennant's desideratum D_2, but it also satisfies Dummett's desideratum D_4. More precisely, its introduction rules are *perfect* introduction rules which are in harmony with its elimination rules. The only reason why the calculus $^g\mathcal{R}_W^C$ could not be used to undermine proof-theoretic arguments is that it has not been shown that there is a permissible correlation between its rules and informal rules of inference. (In particular, the standard interpretation according to which multiple succedents are interpreted disjunctively was rejected as not permissible.) However, the existence of calculi such as $^g\mathcal{R}_W^C$ might motivate the search for further possibilities for extending the specialised framework ARG$_{\mathcal{L}^*}$ so that there are adequate ARG$_{\mathcal{L}^*}$-classical calculi of these extensions.

The oldest proposal for such an extension that I am aware of stems from Frank Ramsey. Ramsey introduces the idea of a language in which sentences are negated by reflecting them along a horizontal axis:

> We might, for instance, express negation not by inserting a word "not," but by writing what we negate upside down (Ramsey 1927: 161-2).

Slightly extending Ramsey's proposal, I will now introduce a language in which reflections of sentences are again sentences, and which also contains a negation operator. It will be seen that a specialised framework that is based on such a language allows for ARG$_{\mathcal{L}^*}$-classical calculi that satisfy D_1^R and D_4.

The pertinent language will be denoted by $\sim\!\mathcal{L}$. For ease of exposition, I will assume that it contains only six sentential constants: $p, \text{ɓ}, q, \text{ɖ}, r, \text{ʟ}$. In addition, it contains the usual brackets and the symbols $\vee, \wedge, \nearrow, \searrow, \circ$ for the sentential operators. (I will employ schematic letters for sentences that also come in pairs of reflected symbols: $\varphi, \text{ɔ}, \ldots$) Note that $\vee, \wedge$ and $\nearrow, \searrow$ are reflections of one another and that $\circ$ is identical to its reflection. Note also that the reflection of any sentence is again a sentence and that the reflection of the reflection of a sentence S is identical to S.

There is then the following calculus $\mathcal{R}_W^C$ of the specialised logical framework $\mathrm{ARG}_{\frown\mathcal{L}}$. It is a standard calculus with three structural rules: (A), (W), and the following rule of *Ramseian Reductio*:

$$
\frac{
\begin{array}{cc}
[\varphi] & [\varphi] \\
\vdots & \vdots \\
\psi & \frown\!\psi
\end{array}
}{\frown\!b}\ \text{(RR)}
$$

The rule (RR) expresses that one can infer the reflection of an assumption S from a sentence and its reflection, while, at the same time, one is allowed to discharge the assumption S.

The rules for the binary sentential operators $\wedge$ and $\nearrow$ are represented by the usual schematic transitions (for $\wedge$ and $\to$):

$$
\frac{\Gamma \succ \varphi \quad \Delta \succ \psi}{\Gamma, \Delta \succ \varphi \wedge \psi} \quad
\frac{\Gamma \succ \varphi \wedge \psi}{\Gamma \succ \varphi} \quad
\frac{\Gamma \succ \varphi \wedge \psi}{\Gamma \succ \psi} \quad
\frac{\Gamma, \varphi \succ \psi}{\Gamma \succ \varphi \nearrow \psi} \quad
\frac{\Gamma \succ \varphi \nearrow \psi \quad \Delta \succ \varphi}{\Gamma, \Delta \succ \psi}
$$

The only difference is, of course, that the schematic letters here represent sentences of $\frown\mathcal{L}$. (Unlike before, the rules for $\vee$ are derivable in $\mathcal{R}_W^C$.) Finally, there are two rules for the unary sentential operator $\circ$:

$$
\frac{\varphi}{\circ\!\frown\!b} \qquad \frac{\circ\varphi}{\frown\!b}
$$

These rules express that a sentence is equivalent to the result of prefixing $\circ$ to its reflection. This completes the presentation of $\mathcal{R}_W^C$.[49]

Now, relative to a suitable translation function, the calculus $\mathcal{R}_W^C$ can be said to be $\mathrm{ARG}_{\mathcal{L}^*}$-classical. To spell this out, let $\mathcal{L}^*(p,q,r)$ be the fragment of $\mathcal{L}^*$ that is based on the sentential constants: p, q, r,[50] and let π_a be the function which maps a sequent s of $\mathcal{L}^*(p,q,r)$ to the sequent $\pi_a(s)$ of $\frown\mathcal{L}$ which one obtains from s by replacing $\to$ and $\neg$ with $\nearrow$ and $\circ$. The following result is then easily seen to hold:

Translation$_a$ For every sequent s of $\mathcal{L}^*(p,q,r)$, s is $\mathcal{R}^C$-derivable iff $\pi_a(s)$ is $\mathcal{R}_W^C$-derivable.[51]

---

[49] It is straightforward to obtain an extension of $\mathcal{R}_W^C$ that pertains to a language with quantifiers. In such a language, there would be a pair of symbols, say $\bigwedge$ and $\bigvee$, for the two quantifiers and a countable supply of variables and parameters, each of which is identical to its own reflection. Furthermore, individual constants and function symbols would also be identical to their own reflections, while relation symbols would come in pairs of a symbol and the distinct reflection of this symbol. (In particular, if one wanted to use a symbol for identity, then it would have to be a symbol which is distinct from its reflection.)

[50] For the general case, one would need to employ an extension of $\frown\mathcal{L}$ with countably many pairs of sentential constants.

[51] It poses no problem to show that if s is $\mathcal{R}^C$-derivable, then $\pi_a(s)$ is $\mathcal{R}_W^C$-derivable. The converse is most easily seen by a model-theoretic argument: one proves that the rules of $\mathcal{R}_W^C$ preserve truth in models of an appropriate class, and one exploits the fact that $\mathcal{R}^C$ is complete with respect to a corresponding notion of truth-preservation.

The image of the translation function π_a is the set of those sequents of $^\frown\!\mathcal{L}$ whose sentences do not contain the symbols $\searrow$, $\flat$, $\natural$, λ. One may now say that the calculus $^\frown\!\mathcal{R}_W^C$ is ARG$_{\mathcal{L}^*}$-*classical relative to the translation function* π_a. (Note that such a translation function is needed only because the language $^\frown\!\mathcal{L}$ has to contain a symbol corresponding to $\to$ which is *not* identical to its reflection and a symbol corresponding to $\neg$ which *is* identical to its reflection.) Finally, inspection shows that $^\frown\!\mathcal{R}_W^C$ satisfies the desiderata D_1^R and D_4.

The second proposal for an extension of the specialised framework ARG$_{\mathcal{L}^*}$ which allows for an ARG$_{\mathcal{L}^*}$-classical calculus that satisfies D_1^R and D_4 resembles the previous one. It is based on a language that consists of pairs of *signed* sentences. Unlike Ramsey's proposal, it has played some role in recent discussions of proof-theoretic arguments against classical logic (see Bendall 1978, Smiley 1996, Humberstone 2000, and Rumfitt 2000).

The pertinent language will be denoted by $^{+/-}\!\mathcal{L}$. One obtains this language from $\mathcal{L}^*$ by replacing every sentence S by the pair of *signed sentences* $+S$ and $-S$.[52] Following Rumfitt (2000: 804), I will call two sentences $+S$ and $-S$ *conjugates* of one another.

There is then the following calculus $^{+/-}\!\mathcal{R}_W^C$ of the specialised logical framework ARG$_{+/-\mathcal{L}}$. It is a standard calculus with four structural rules: (A), (W), and the following two rules whose union Rumfitt (2000: 804) has called the rule of *Smileian Reductio*:

$$
\cfrac{\begin{array}{cc} [+\varphi] & [+\varphi] \\ \vdots & \vdots \\ +\psi & -\psi \end{array}}{-\varphi} \; (\text{SR}_1)
\qquad\qquad
\cfrac{\begin{array}{cc} [-\varphi] & [-\varphi] \\ \vdots & \vdots \\ +\psi & -\psi \end{array}}{+\varphi} \; (\text{SR}_2)
$$

These rules express that one may infer the conjugate of an assumption S from a sentence and its conjugate, while, at the same time, it allows one to discharge the assumption S.

There are many different possibilities for choosing rules for the operators, all of which yield calculi with the same set of derivable sequents. I will assume here that $^{+/-}\!\mathcal{R}_W^C$ contains the following introduction and elimination rules:

$$
\frac{+\varphi \quad +\psi}{+\,\varphi\wedge\psi}
\qquad
\frac{-\varphi \quad -\psi}{-\,\varphi\vee\psi}
\qquad
\frac{+\varphi \quad -\psi}{-\,\varphi\to\psi}
\qquad
\frac{-\varphi}{+\,\neg\varphi}
$$

$$
\frac{+\,\varphi\wedge\psi}{+\varphi}\quad\frac{+\,\varphi\wedge\psi}{+\psi}
\qquad
\frac{-\,\varphi\vee\psi}{-\varphi}\quad\frac{-\,\varphi\vee\psi}{-\psi}
\qquad
\frac{-\,\varphi\to\psi}{+\varphi}\quad\frac{-\,\varphi\to\psi}{-\psi}
\qquad
\frac{+\,\neg\varphi}{-\varphi}
$$

This completes the presentation of the calculus $^{+/-}\!\mathcal{R}_W^C$.[53]

---

[52] In what follows, I will continue to use schematic letters such as φ, ψ, χ for sentences of $\mathcal{L}^*$. For some purposes, it would be simpler if one were to use another type of schematic letters: schematic letters for sentences of $^{+/-}\!\mathcal{L}$. For present purposes, however, it seems more convenient to dispense with them.

[53] The calculus $^{+/-}\!\mathcal{R}_W^C$ differs from the one proposed by Smiley (1996) only in containing the displayed $\to$-introduction rule instead of another $\to$-elimination rule: $\frac{+\varphi\to\psi \quad +\varphi}{+\psi}$. Since $^{+/-}\!\mathcal{R}_W^C$ is meant to show that Dummett's proof-theoretic argument fails, it is necessary to use a pair of (harmonious) collections of introduction and elimination rules for each of the standard sentential operators.

Similar to the case of $^\frown\mathcal{R}_W^C$, there is a translation function relative to which $^{+/-}\mathcal{R}_W^C$ is $\mathrm{ARG}_{\mathcal{L}^*}$-classical. Let π_b be the function which maps a sequent s from $\mathcal{L}^*$ to the sequent $\pi_b(s)$ of $^{+/-}\mathcal{L}$ which one obtains from s by prefixing $+$ to its components. The following result is then easily seen to hold:

Translation$_b$ For every sequent s of $\mathcal{L}^*$, s is $\mathcal{R}^C$-derivable iff $\pi_b(s)$ is $^{+/-}\mathcal{R}_W^C$-derivable.

The image of the translation function π_b is the set of those sequents of $^{+/-}\mathcal{L}$ which do not contain the sign $-$. One may now say that the calculus $^{+/-}\mathcal{R}_W^C$ is $\mathrm{ARG}_{\mathcal{L}^*}$-*classical relative to the translation function* π_b. (Note that such a translation function would not have been needed if the sign $+$ had not been used: instead of using pairs like $(+p, -p)$, one could equally use pairs like $(p, -p)$. In this way, a language could have been used whose set of sentences comprises the sentences of $\mathcal{L}^*$.) Again, inspection shows that $^{+/-}\mathcal{R}_W^C$ satisfies the desiderata D_1^R and D_4.

Now, the differences between the calculi $^\frown\mathcal{R}_W^C$ and $^{+/-}\mathcal{R}_W^C$ are rather marginal. Like the $\mathrm{ARG}_{\mathcal{L}^*}$-intuitionistic calculus $^\perp\mathcal{R}_W^I$, they are both separable standard calculi of the framework ARG with perfect introduction rules and harmonious elimination rules. It remains to be seen whether there is a reason for adhering to D_1^P (or D_1^T); i.e., it remains to be seen whether there is a reason for precluding calculi that are based on non-standard extensions of $\mathcal{L}^*$ such as $^\frown\mathcal{L}$ or $^{+/-}\mathcal{L}$.

According to Ramsey, although we do not actually use a language like $^\frown\mathcal{L}$, we could easily do so:

> Such a symbolism is only inconvenient because we are not trained to perceive complicated symmetry about a horizontal axis [...] (Ramsey 1927: 161-2).

On the face of it, the language $^\frown\mathcal{L}$ and the calculus $^\frown\mathcal{R}_W^C$ do not seem to be more problematic than $\mathcal{L}^\perp$ and $^\perp\mathcal{R}_W^I$. As a matter of fact, we do not use a (syntactically and semantically) simple sentence involved in the determination of the meaning of "not", and we do not use a reflection relation among sentences that is involved in the determination of the meaning of "not". However, just as it seems *possible* to employ the language $\mathcal{L}^\perp$ in accordance with the rules of $^\perp\mathcal{R}_W^I$, so it seems *possible* to employ the language $^\frown\mathcal{L}$ in accordance with the rules of $^\frown\mathcal{R}_W^C$.

In reaction to this, it might be replied that an adequate calculus is not allowed to be based on an extension of $\mathcal{L}^*$ that involves resources which have no counterparts in ordinary linguistic practice. Rather, it might be claimed, the applications of the rules of an adequate calculus must constitute a systematisation of basic argumentative moves which are really employed (cp. 5.2.1). Then, it seems questionable that calculi such as $^\perp\mathcal{R}_W^I$ and $^\frown\mathcal{R}_W^C$ can be adequate because in natural languages there are no (syntactically and semantically) simple sentences that are logically false, and there are no contradictory pairs of sentences which are reflections of one another.

This might suggest that the second version of Dummett's proof-theoretic argument is better than the first one, and that one should only claim that the calculus $\overset{\bullet}{\mathcal{R}}_W^I$ is adequate, but not that the calculus $^\perp\mathcal{R}_W^I$ is adequate. Recall that proponents of the second version of Dummett's proof-theoretic argument defend the claim that the calculus $\overset{\bullet}{\mathcal{R}}_W^I$ is adequate by pointing out that in informal mathematical proofs, expressions without propositional content are used to signal that a logical dead end has been reached. As an example, they mention freestanding uses of the word "contradiction" (cp. 5.5.3). Can this observation be used to defend the desideratum D_1^T and, therefore, the second version of Dummett's proof-theoretic argument?

I am skeptical about the thesis that freestanding uses of the word "contradiction" in informal mathematical proofs are non-propositional indicators of logical dead ends. This word is not a shorthand for a logical falsehood (to be formalised by $\perp$), but this does not imply that it does not have propositional content. Rather, it seems to be a shorthand for a sentence which characterises the present argumentative situation, a sentence such as "a contradiction has been reached" or "*this* sentence contradicts *that* one", where the phrase "this sentence" is used to refer to the sentence that was just uttered and where the phrase "that sentence" is used to refer to a salient sentence uttered before.

The thesis that informal mathematical argumentations contain *non-propositional* expressions which indicate that a logical dead end has been reached seems to be intuitively implausible and is, therefore, in need of defence. The thesis that informal mathematical argumentations contain shorthands for sentences which characterise the relevant argumentative situation, shorthands for sentences that are part of many mathematical proofs anyway, seems to be intuitively plausible and should, therefore, only be given up in view of convincing criticism. As far as I know, the first thesis has not been defended and the second thesis has not been criticised. Consequently, it has not been shown that derivations of calculi such as $\overset{\bullet}{\mathcal{R}}_W^I$ correspond to (idealised) informal argumentations.

However, even if intuitionists were to succeed in showing that the calculus $\overset{\bullet}{\mathcal{R}}_W^I$ not only satisfies the desideratum D_4 but also constitutes a systematisation of informal rules of inference, it would not follow that there are no $\text{ARG}_{\mathcal{R}^*}$-classical calculi that satisfy D_4 and constitute a systematisation of informal rules of inference. By way of conclusion, I will indicate that the $\text{ARG}_{\mathcal{L}^*}$-classical calculus $^{+/-}\mathcal{R}_W^C$ is at least as plausible a candidate for a systematisation of informal rules of inference as the calculus $\overset{\bullet}{\mathcal{R}}_W^I$.

As the proponents of calculi such as $^{+/-}\mathcal{R}_W^C$ have pointed out, we actually *do* use pairs of sentences which can plausibly be taken to correspond to conjugated pairs of signed sentences of $^{+/-}\mathcal{L}$, namely pairs of positive and negative answers to yes-or-no questions. (For example, a pair of answers to a question such as "Is there a largest prime number?" is taken to correspond to a pair $(+S, -S)$ of conjugates of $^{+/-}\mathcal{L}$.) Now, the idea is that both kinds of speech acts, i.e. acts in which a yes-or-no question is answered positively and acts in which a yes-or-no question is answered negatively, have to be referred to in an account of the meanings of atomic sentences and in an account of the meanings of the standard sentential operators:

> [M]astering the sense of an atomic sentence A will involve learning
> methods whose deployment might entitle one *either* to affirm it

or to reject it. Successful application of these methods, one might say, will put one in a position to answer the question whether *A* either by 'Yes' or 'No' as appropriate. [...] The acts of answering a propositional question affirmatively and of answering it negatively - the acts of accepting its content and of rejecting it - are conceived to be on all fours (Rumfitt 2000: 797).

[W]hat ultimately makes [it possible to give a direct specification of the classically intended senses for the connectives] is the existence of a readily comprehensible variety of actual deductive practice in which the components of arguments express the assignation of affirmative or negative force to propositional contents. For [...] we can readily comprehend - and evaluate the validity of - arguments whose premises and conclusions are yes-or-no questions with one of other of the expected answers attached (Rumfitt 2000: 799).

For present purposes, Rumfitt's claim that there is a variety of *actual* deductive practice whose arguments are made up of answers to yes-or-no questions is most important. Here, Rumfitt thinks of arguments such as the following (see Rumfitt 2000: 799):

P_1 Is there a smallest prime number? Yes

P_2 Is there a largest prime number? No

C   Thus: Is there a smallest and a largest prime number? No

The premises and the conclusion of this argument are positively or negatively answered yes-or-no questions.

Now, it might be debatable to which extent our actual deductive practice involves such arguments, but, as far as I can see, the claim that it does is no more implausible than the claim that our actual deductive practice involves argumentations in which certain words are non-propositional indicators of logical dead ends. Furthermore, even if it turned out that 'yes-or-no arguments' do not play any significant role in our actual deductive practice, one surely has to agree with Rumfitt that arguments such as the one presented above are readily comprehensible and evaluable. It seems to require more effort to get used to argumentative moves that involve logical dead ends such as the inference ($\overset{*}{\spadesuit}E$), in which an arbitrary sentence is 'inferred from' a logical dead end.

It thus seems that there is an $\text{ARG}_{\mathcal{L}^*}$-classical calculus, the calculus $^{+/-}\mathcal{R}_W^C$ to wit, that is at least as close to a collection of basic informal rules of inference as the $\text{ARG}_{\mathcal{L}^*}$-intuitionistic calculi $^{\perp}\mathcal{R}_W^I$ and $^{\spadesuit}\mathcal{R}_W^I$. Consequently, a plausible replacement of the desideratum D_1 that does not preclude the calculi $^{\perp}\mathcal{R}_W^I$ and $^{\spadesuit}\mathcal{R}_W^I$ should not preclude the calculus $^{+/-}\mathcal{R}_W^C$ either. Therefore, since $^{+/-}\mathcal{R}_W^C$ also satisfies the desideratum D_4 which captures Dummett's fundamental idea of the requirement of meaning determining introduction rules and harmonious elimination rules, it follows that it has not been shown that either version of Dummett's proof-theoretic argument is convincing.

5.6.2   Classical Logic and Non-Circular Classical Rules

In the previous subsection, I argued that there are $\textsc{Arg}_{\mathcal{L}^*}$-classical calculi which
equal those put forward by proponents of proof-theoretic arguments in every re-
spect that can plausibly be taken to be relevant for the question of whether they
are adequate. Thus, even if one accepts Dummett's crucial claim that the theory
of classical sentential logic is only sound if there is a suitable $\textsc{Arg}_{\mathcal{L}^*}$-classical
calculus which satisfies D_4, it does not follow that classical sentential logic is
not sound. In this subsection, I will conclude by questioning Dummett's crucial
claim itself. First, I will argue that Dummett has not established the thesis
EAC, according to which the soundness of classical sentential logic depends on
the existence of an adequate $\textsc{Arg}_{\mathcal{L}^*}$-classical calculus. Second, I will argue that
he has not shown that an adequate calculus must satisfy the desideratum D_4.
It follows that Dummett's proof-theoretic argument must be rejected.

A theory that classifies the arguments of $\mathcal{L}^*$ into those that are logically
valid and those that are not will be called a *logical theory for* $\textsc{Arg}_{\mathcal{L}^*}$.[54] A
calculus $\mathcal{R}$ is said to *correspond* to a logical theory for $\textsc{Arg}_{\mathcal{L}^*}$ iff the following
holds: a sequent of $\textsc{Arg}_{\mathcal{L}^*}$ is $\mathcal{R}$-derivable iff it is logically valid according to $\mathcal{T}$.
Now, the thesis EAC follows from two simpler claims which might be taken to
justify it:

Adequacy$_1$ (A$_1$) If $\mathcal{T}$ is a sound logical theory for $\textsc{Arg}_{\mathcal{L}^*}$ and $\mathcal{R}$ is an adequate calculus, then $\mathcal{R}$ corresponds to $\mathcal{T}$.

Adequacy$_2$ (A$_2$) There is an adequate calculus.

The combination of A$_1$ and A$_2$ implies EAC. To see this, suppose that the theory
of classical sentential logic is sound. According to A$_2$, there is an adequate
calculus $\mathcal{R}$. Therefore, A$_1$ implies that $\mathcal{R}$ corresponds to the theory of classical
sentential logic; i.e., A$_1$ implies that $\mathcal{R}$ is $\textsc{Arg}_{\mathcal{L}^*}$-classical.

Here, I will take a critical look at the thesis A$_1$. It seems to be motivated
by the following consideration. If $\mathcal{T}$ is a sound logical theory for $\textsc{Arg}_{\mathcal{L}^*}$, then
it declares precisely those arguments of $\mathcal{L}^*$ to be logically valid that are truth-
preserving in virtue of the meanings of the standard sentential operators. In
addition, if $\mathcal{R}$ is adequate, then it yields an account of the meanings of the
standard sentential operators, i.e. an account of those meanings in virtue of
which the logically valid arguments of $\mathcal{L}^*$ are truth-preserving. This suggests
that $\mathcal{R}$ must produce precisely those arguments of $\mathcal{L}^*$ which are truth-preserving
in virtue of the meanings of the standard sentential operators. Consequently,
the theory $\mathcal{T}$ and the calculus $\mathcal{R}$ must deliver the same verdicts about sequents
of $\textsc{Arg}_{\mathcal{L}^*}$; i.e., a sequent of $\textsc{Arg}_{\mathcal{L}^*}$ must be $\mathcal{R}$-derivable iff it is logically valid
according to $\mathcal{T}$.

The crucial step in this reasoning is the inference from the assumption that
$\mathcal{R}$ yields an account of the meanings in virtue of which the logically valid arguments of $\mathcal{L}^*$ are truth-preserving to the conclusion that $\mathcal{R}$ produces precisely

---

[54]   In the terminology of section 1.1, it can be conceived of as the union of the formal basic
part and the logical part of a partial logical theory that deals with arguments that are
logically valid with respect to the standard sentential operators.

those arguments of $\mathcal{L}^*$ which are truth-preserving in virtue of the meanings it yields an account of. Now, with respect to what I take to be a plausible conception of accounts of meanings, one would say that a calculus $\mathcal{R}$ yields an account of the meanings of logical operators if it comprises rules of certain types that are strong enough to uniquely characterise the logical properties of these operators. On such a conception, however, an adherent of classical sentential logic will claim that $\text{ARG}_{\mathcal{L}^*}$-*intuitionistic* calculi can be adequate. As was seen in section 2.2.3, the rules of $\text{ARG}_{\mathcal{L}^*}$-intuitionistic calculi uniquely characterise the standard sentential operators. Consequently, an adherent of classical sentential logic might say, for example, that the $\text{ARG}_{\mathcal{L}^*}$-intuitionistic calculus $^{\perp}\mathcal{R}_W^I$ yields an account of the meaning of $\neg$ according to which an argument such as $\neg\neg p \succ p$ is truth-preserving, although it is not derivable in $^{\perp}\mathcal{R}_W^I$.

The pertinent question is whether a calculus which yields an account of the meanings of its operators must produce *every* argument that is truth-preserving in virtue of the meanings it accounts for. Consider 'arithmetical theories' which classify the arguments of an arithmetical language $\mathcal{L}^{Ar}$ into those that are truth-preserving in virtue of meaning and those that are not, and consider 'arithmetical calculi' in which some of the arguments of $\mathcal{L}^{Ar}$ are derivable. Now, it might be plausible to assume that an arithmetical calculus $\mathcal{R}$ which yields an account of the meanings of the terms of $\mathcal{L}^{Ar}$ must produce *only* arguments that are truth-preserving in virtue of the meanings it accounts for. It is less clear, however, that such a calculus must produce *every* argument that is truth-preserving in virtue of the meanings it accounts for. For example, consider the intuitionistic arithmetical calculus $\mathcal{R}^{Ar}$ which corresponds to the theory of Heyting arithmetic (see Appx. A.1). As far as I can see, an intuitionist might be justified in claiming that such a calculus yields an account of the meanings of its terms. Furthermore, if S is a sentence of $\mathcal{L}^{Ar}$ which is equivalent (in $\mathcal{R}^{Ar}$) to a sentence that expresses that $\mathcal{R}^{Ar}$ is consistent, then, I think, she might also be justified in claiming that a sound arithmetical theory declares the argument $\succ S$ (without premises and with conclusion S) to be truth-preserving in virtue of the meanings of its terms. But then, as is shown by Gödel's second incompleteness theorem, the calculus $\mathcal{R}^{Ar}$ does *not* produce every argument of $\mathcal{L}^{Ar}$ that is truth-preserving in virtue of the meanings of its terms.

Now, I do not want to suggest that this view about truth-preservation in virtue of meaning is mandatory. Alternatively, it might be said that a sentence of $\mathcal{L}^{Ar}$ which is equivalent (in $\mathcal{R}^{Ar}$) to a sentence expressing that $\mathcal{R}^{Ar}$ is consistent, is true but not in virtue of the meanings of its terms. Similarly, it might be claimed that if an $\text{ARG}_{\mathcal{L}^*}$-intuitionistic calculus such as $^{\perp}\mathcal{R}_W^I$ yields an account of the meaning of $\neg$, then an argument such as $\neg\neg p \succ p$ might be truth-preserving but not in virtue of the meaning of $\neg$. Here, I will not try to decide which of these views on truth-preservation in virtue of meaning is more plausible. I only wanted to point out that a thesis such as A_1 is not wholly unproblematic, and, as the arithmetical example shows, that this is so independently of the dispute between classicists and intuitionists.

I will now turn to Dummett's claim that an adequate calculus must satisfy the desideratum D_4. This claim combines two theses of different character (see

5.4.1). According to the first thesis, a coherent linguistic practice has to be governed by collections of norms that are in harmony with one another. In particular, it is claimed that the norms that state under which conditions a mathematical sentence S can be inferred from other sentences (in combination with acts in which assumptions are discharged) must match the norms that state under which conditions a sentence can be inferred from S and other sentences (in combination with acts in which assumptions are discharged). This thesis leads to the desideratum that adequate calculi must consist of collections of harmonious rules. According to the second thesis, the collections of harmonious rules have to be collections of introduction and elimination rules such that the collections of introduction rules are independent families of rules whose members respect some standard complexity ordering.

The first thesis has controversial implications about the relation between linguistic practice and linguistic meaning,[55] but it certainly has intuitive appeal. The second thesis, by contrast, presupposes strong theoretical assumptions about the possible basic kinds of (harmonious) norms that may govern coherent linguistic practice. In particular, it presupposes that the basic norms that govern a coherent practice of using some logical operator $\sharp$ have to consist of two harmonious sub-collections for inferring to and for inferring from $\sharp$-sentences such that *the premises and the discharged assumptions of an application of a rule of the former sub-collection are less complex than the conclusion of this application.*[56] Such a thesis is clearly not supported by common sense but rather, if at all, by theoretical argument.

To bring out more clearly the contrast between these two theses, I would like to mention a calculus that coheres with the former but not with the latter. Recall that two collections of rules may be *harmonious* although neither of them is a collection of introduction rules (see 5.4.3), and that a collection of rules may be *independent* although it is not a collection of introduction rules (see 5.4.2). This observation suggests a weakened version of the desideratum D$_4$ which captures only the first ingredient of Dummett's fundamental idea. To this end, a rule whose schema contains no operator besides $\sharp$ will be called a $\sharp$-rule. (Since the calculus to be introduced contains only pure rules, I will incorporate the desideratum of purity, although this is *not* part of Dummett's fundamental idea). Now, consider the following desideratum:

---

[55] See 2.4 for the view that the current linguistic practices of classicists *and* intuitionists are coherent, in the sense that all participants understand the expressions they use (and do so in the same way), although classicists and intuitionists follow different rules in their respective practices.

[56] In addition, it presupposes that the collections of introduction rules must be independent. This implies, for example, that one does not obtain an adequate calculus by adding the following rules to $^{\perp}\mathcal{R}_W^I$:

$$\frac{\Gamma \succ \varphi}{\Gamma \succ (\varphi \to \psi) \to \varphi} \qquad \frac{\Gamma \succ (\varphi \to \psi) \to \varphi}{\Gamma \succ \varphi} \qquad \frac{\Gamma \succ \varphi}{\Gamma \succ \neg\neg\varphi} \qquad \frac{\Gamma \succ \neg\neg\varphi}{\Gamma \succ \varphi}$$

because one would obtain families of $\to$-introduction rules and $\neg$-introduction rules that are not independent. In what follows, I will not discuss this additional presupposition.

Desideratum$_5$ (D$_5$) If $\mathcal{R}$ is an adequate calculus, then, for every operator $\sharp$ of $\mathcal{L}(\mathcal{R})$, there are sets $I_\mathcal{R}(\sharp)$ and $E_\mathcal{R}(\sharp)$ of $\sharp$-rules such that the following holds:

(C$_1$) $\mathcal{R}$ is the union of $\mathcal{R}_{\text{str}}$ and $\bigcup_{\sharp \in \mathcal{L}(\mathcal{R})} I_\mathcal{R}(\sharp) \cup E_\mathcal{R}(\sharp)$,

(C$_2^a$) for every operator $\sharp$ of $\mathcal{L}(\mathcal{R})$, $I_\mathcal{R}(\sharp)$ is an independent family,

(C$_3$) for every operator $\sharp$ of $\mathcal{L}(\mathcal{R})$, $I_\mathcal{R}(\sharp)$ and $E_\mathcal{R}(\sharp)$ are harmonious.

In contrast to the desideratum D$_4$, it is not required that the rules of the collections $I_\mathcal{R}(\sharp)$ are $\sharp$-introduction rules and that the rules of the collections $E_\mathcal{R}(\sharp)$ are $\sharp$-elimination rules, and it is not required that the rules of the union $\bigcup_{\sharp \in \mathcal{L}(\mathcal{R})} I_\mathcal{R}(\sharp)$ respect some complexity ordering.

A calculus that satisfies D$_5$ must be a calculus that coheres with Dummett's first thesis, but it does not have to be a calculus that coheres with Dummett's second thesis. Remarkably, there is an ARG$_{\mathcal{L}^*}$-classical calculus that not only satisfies D$_5$ but also D$_1$. (Recall that there is no ARG$_{\mathcal{L}^*}$-classical or ARG$_{\mathcal{L}^*}$-intuitionistic calculus that satisfies D$_4$ and D$_1$.) To see this, note that the singleton of the rule of *classical reductio* is harmonious with the singleton of the usual $\neg$-elimination rule ($\neg_E$) (cp. Milne 1994 and Slater 2008):

$$\frac{\Gamma, \neg\varphi \succ \psi \quad \Delta, \neg\varphi \succ \neg\psi}{\Gamma, \Delta \succ \varphi} \ (\text{CR})$$

Now, let $'\mathcal{R}_W^C$ be the calculus that results from $\mathcal{R}_W^I$ by replacing ($\neg_I$) with (CR). The calculus $'\mathcal{R}_W^C$ is easily seen to be an ARG$_{\mathcal{L}^*}$-classical calculus that satisfies D$_1$ and D$_5$. (It violates D$_4$ because the rule (CR) is not a $\neg$-introduction rule.)

It follows that the idea that our use of the standard sentential operators must be governed by coherent collections of norms does not show classical sentential logic to be not sound. Dummett's proof-theoretic argument is based on the intuitively plausible assumption that coherent linguistic practice has to be governed by harmonious collections of norms. However, it also presupposes a special form of an inferentialist account of meaning. It thus presupposes the falsity of non-inferentialist meaning theories such as classical truth conditional semantics. More strongly, it even depends on an argument in favour of a peculiar form of an inferentialist meaning theory. To defend a proof-theoretic argument with this radical consequence, one needs to justify a strong version of an 'anti-holistic' inferential account of meaning. And such a justification has not been provided by adherents of proof-theoretic arguments. Therefore, Dummett's proof-theoretic argument is also not convincing.

By way of conclusion, Dummett's proof-theoretic argument has not been shown to undermine the plausibility of classical sentential logic. On the one hand, it presupposes a radical inferential thesis which has not been made plausible. On the other hand, it can be countered by presenting classical calculi such as $^{+/-}\mathcal{R}_W^C$, which satisfy the formal counterpart of this inferential thesis, desideratum D$_4$ to wit, and which are at least as close to a collection of basic informal rules for the relevant vocabulary as intuitionistic calculi are. Philosophical proof-theory does not seem to be an appropriate tool for refuting classical logic.

Chapter 6

Conclusion

Four kinds of arguments against classical logic have been examined: (i) quasi-mathematical justifications of set-theoretic and function-theoretic principles that contradict classical logical truths; (ii) epistemological arguments about informal provability that were meant to show that there are simple classical logical truths that are presently not known to be true; (iii) the manifestation argument against classical truth conditional semantics (a theory which, according to Dummett, would have to be true if classical logic were sound); (iv) proof-theoretic arguments to the conclusion that classical logic is not sound because there are no classical calculi that yield an account of the meanings of the logical operators. I argued that each of these arguments incorporates an assumption that is neither intuitively compelling nor made plausible by a convincing argument.

(i) The quasi-mathematical arguments presuppose the existence of certain finite quasi-mathematical objects. McCarty's justification of the Uniformity Principle is based on the assumption that if sets of numbers are φ-related to numbers, then there is a rule which can be applied to set of numbers and which can be used to determine, if it is applied to such a set, a number to which the set is φ-related. Similarly, de Swart's justification of the Continuity Principle presupposes that if sequences of numbers are φ-related to numbers, then there is an agent who can present, for every sequence of numbers, a number and a proof that the sequence is φ-related to the number. However, neither McCarty nor de Swart presents any reason for accepting these crucial existential claims.

(ii) The epistemological arguments are based on the claim that a plausible account of informal provability and correct deductive inferences incorporates the assumption that an arithmetical disjunction is provable only if at least one of its disjuncts is provable. It was shown that the proposed intuitionistic theory of informal provability and correct inferences, namely Prawitz's theory of grounds, could be modified in such a way that disjunctions that instantiate the principle of the excluded third can trivially be known to be true.

(iii) The manifestation argument turned out to depend on an unwarrantedly strong version of the manifestation thesis, according to which someone's understanding of a sentence must be manifestable in her use of this very sentence. It was argued that a plausible replacement of the manifestation thesis has to take into account that meaning and understanding are compositional, and it was shown that compositional manifestation of understanding is compatible with classical truth conditional semantics. In short, if someone's knowledge of the truth condition of a sentence S derives from her knowledge of what the parts of S contribute to truth conditions of sentences, then she can manifest her knowledge of the truth condition of S by manifesting her knowledge of what the parts of S contribute to truth conditions of sentences. Furthermore, someone can manifest her knowledge of what a number of expressions contribute to truth conditions of sentences even if there is a true sentence that can be formed by means of these expressions for which she could not be in a position to know that it is true.

(iv) The proof-theoretic arguments are based on unjustified assumptions about which collections of rules of inference can yield an account of the meanings of the logical operators. In particular, it was seen that there are classical calculi with harmonious rules of inference that satisfy all reasonable requirements that are satisfied by intuitionistic calculi. To preclude classical calculi, the proponents of proof-theoretic arguments set up additional requirements, but these were seen to amount to radical and unwarranted inferentialist theses about the meanings of the logical operators.

I would like to conclude by indicating what I take to be the most plausible view on the dispute between classicists and intuitionists. The philosophical intuitionists that I have discussed believe that there are convincing and intuitionistically acceptable arguments against classical logic and classical mathematics. Complementarily, some philosophical classicists believe that there are convincing and classically acceptable arguments against intuitionistic logic and intuitionistic mathematics. In contrast to these *optimistic* views, I take a *pessimistic* view to be more plausible, according to which there are no convincing intuitionistic arguments against classicism and no convincing classical arguments against intuitionism.

Now, this pessimistic claim has to be qualified. Of course, every intuitionist and every classicist can present mathematical 'refutations' of fundamental principles of her opponent. However, this is only an alternative way of expressing that intuitionism and classicism are incompatible. The interesting question is whether there are arguments against either view that do not involve a logical or mathematical assumption that contradicts a basic principle of this view. I refer to such arguments as *non-questing begging* arguments. My conjecture is that there no non-question arguments which either side can use to refute the other side. (Of course, I have not discussed *every* intuitionistically acceptable argument against classicism, and I did not present my reasons for rejecting classically acceptable arguments against intuitionism. Consequently, this claim is put forward as a *conjecture* here.)

In several works, and especially in his 1991 book, Dummett proposes to resolve the dispute between classicists and intuitionists by carrying out a meaning-theoretic project. He claims that philosophers have to find out which form a meaning theory for an entire natural language ought to have, that they can do so without presupposing which logical and mathematical theories are correct, and that the resulting meaning theory will show which logical and mathematical theories are correct. The presented results undermine Dummett's proposal: they suggest that meaning-theoretic investigations cannot be used to resolve the logical and mathematical disputes between classicists and intuitionists. First, the evaluation of the manifestation argument suggests that the idea of a truth conditional semantics is neutral between classical and intuitionistic logic. In particular, truth-theory and refined truth-theory incorporate only axioms that are classically and intuitionistically acceptable (see 4.2.3 and 4.2.5),[1] and classicists and intuitionists can draw on the same linguistic behaviour in their accounts of the understanding of sentences such as Goldbach's Conjecture (see 4.5). Second, the idea of Prawitz's theory of grounds can be accepted by intuitionists and also by classicists. The claim that there must be plausible theories of correct inferences and informal provability and Prawitz's suggestion that such theories could be based on the idea of applying operations to grounds for accepting sentences does not rule out either logical theory (see 3.4.3). Third, the claim that there must be a system of rules of inference that yields an account of the meanings of the logical operators does not preclude either logical theory. In particular, there are collections of harmonious rules that correspond to intuitionistic logic and collections of harmonious rules that correspond to classical logic (see 5.6).

This is not to say that a fully developed truth conditional semantics, a Prawitzian theory of grounds, or an inferential account of the meanings of the logical operators *must* be neutral between classical and intuitionistic logic. In each case, one can develop the relevant ideas into theories that are either classical or intuitionistic. However, this does not imply that there is a logically and mathematically neutral path for developing a semantic theory such that the result can be used to decide which logical and mathematical theories are correct. Rather, there seem to be neutral skeletons of the relevant theories, and the only way to decide how these skeletons are to be completed into theories that decide the logical and mathematical questions depends on a prior choice as to whether one wants to end up with a classical theory or with an intuitionistic theory.

Dummett (1973b) argues that there are essentially only two possible strategies for solving the logical and mathematical conflict between intuitionists and classicists. According to the first strategy, one tries to reduce it to an ontological conflict: a conflict about the ontological status of mathematical objects. I agree with Dummett that this strategy is not likely to be successful: an intuitionist can believe that mathematical objects are abstract entities that are independent of human activity and a classicist can believe that mathematical objects are mental entities which owe their nature and existence to mental processes. (For example, McCarty (1987) and Prawitz (1998) defend platonist versions of intuitionism.) According to the second strategy, one tries to reduce the logical and mathematical conflict to a semantic one: a conflict about the correct meaning theory for the language that contains the relevant logical and mathematical terms. Now, *pace* Dummett, the present findings suggest that

---

[1]   Compare McCarty (1987: 556), who stresses that the same can be said, for example, about Montague semantics.

the prospects of a meaning-theoretic solution to the logical and mathematical conflict might be as bad as the prospects of an ontological solution to it. In addition, I also discussed the strategy of solving the logical and mathematical conflict by presenting a quasi-mathematical justification of a fundamental intuitionistic principle and the strategy of solving it by presenting an epistemological theory about correct inferences. These strategies were also not convincing.

What would follow if there were no non-question begging arguments that intuitionists can put forward against classicists and also no non-question begging arguments that classicists can put forward against intuitionists? According to dualists such as Carnap, it is not to be expected that there are such arguments, because they think that intuitionists and classicists attach different senses to logical and mathematical terms. But dualism was seen to be false (see Chapter 2). The fact that there is an argumentative stalemate between intuitionists and classicists does not mean that there is no real conflict between them. They put forward contradictory claims, and since two contradictory claims cannot both be true, it follows that intuitionists and classicists cannot both be correct.

For all that, however, one should not become a logical and mathematical agnosticist. For example, nothing of what has been said suggests that one should withhold judgement as to the question of whether every real number is or is not rational. Rather, I plead for *logical/mathematical fundamentalism*. The dispute between classicists and intuitionists cannot be solved by reducing logical and mathematical questions to more fundamental questions of other subjects. It can thus only be solved logically/mathematically. However, if one solves it logically/mathematically, then one must beg the question against one's opponents. For example, I think that everyone can and should accept the following disjunction: either it is known that every real number is or is not rational or it is known that not every real number is or is not rational. As a consequence, logical and mathematical agnosticism must be rejected: either it is correct to judge that every real number is or is not rational, or it is correct to judge that not every real number is or is not rational. Furthermore, whichever of these two judgements is correct, it is based on logical/mathematical insight that conflicts with basic assumptions of intuitionists or classicists. Either classicists or intuitionists are justified in their respective logical and mathematical claims, but they have to beg the question when they present refutations of the fundamental logical and mathematical views of their opponents.

Appendix A

Appendix to Chapter 2

A.1   Arithmetical Uniqueness

In this appendix, I will first present the arithmetical calculus $\mathcal{R}^{Ar}$. Then, I will mention the rules that govern the new binary predicate symbol π which expresses the correspondence between the objects that satisfy N_1 and the objects that satisfy N_2. Finally, I will indicate how the theorem T^{Ar} can be proven.

Apart from the rules of the intuitionistic calculus $\mathcal{R}^I$, the calculus $\mathcal{R}^{Ar}$ contains the following arithmetical rules.[1] First, there are rules which express that the predicates ID, SC, SM, PR apply only to natural numbers:

$$\frac{ID(t_1,t_2)}{Nt_1} \qquad \frac{ID(t_1,t_2)}{Nt_2} \qquad \frac{SC(t_1,t_2)}{Nt_1} \qquad \frac{SC(t_1,t_2)}{Nt_2} \qquad \frac{SM(t_1,t_2,t_3)}{Nt_1}$$

$$\frac{SM(t_1,t_2,t_3)}{Nt_2} \qquad \frac{SM(t_1,t_2,t_3)}{Nt_3} \qquad \frac{PR(t_1,t_2,t_3)}{Nt_1} \qquad \frac{PR(t_1,t_2,t_3)}{Nt_2} \qquad \frac{PR(t_1,t_2,t_3)}{Nt_3}$$

These rules are not essential for the following proof, but they allow for certain simplifications.

Second, there are rules to the effect that the predicates SC, SM, PR stand for the graphs of functions on the natural numbers:

$$\frac{SC(t_1,t_2) \quad SC(t_1,t_3)}{ID(t_2,t_3)} \qquad \frac{Nt}{\exists a SC(t,a)} \qquad \frac{SM(t_1,t_2,t_3) \quad SM(t_1,t_2,t_4)}{ID(t_3,t_4)}$$

$$\frac{Nt_1 \quad Nt_2}{\exists a SM(t_1,t_2,a)} \qquad \frac{PR(t_1,t_2,t_3) \quad PR(t_1,t_2,t_4)}{ID(t_3,t_4)} \qquad \frac{Nt_1 \quad Nt_2}{\exists a PR(t_1,t_2,a)}$$

The first two rules express that every number has exactly one successor. The following rules express that every two numbers have exactly one sum and exactly one product.

---

[1]   In the schematic representations of the rules of $\mathcal{R}^{Ar}$, I will make use of a usual simplification. In *simple* rules of inference, in whose applications no assumptions are discharged, I will only mention the sentences to the right of $\succ$. In *complex* rules of inference, in whose applications assumptions are discharged, I will mention the sentences to the right of $\succ$ and, in square brackets, the discharged assumptions. See 5.1 for details.

Third, there are rules which state that the predicate ID expresses an equivalence relation on the set of natural numbers which is respected by the predicates SC, SM, PR:

$$\frac{Nt}{ID(t,t)} \qquad \frac{ID(t_1,t_2)}{ID(t_2,t_1)} \qquad \frac{ID(t_1,t_2) \quad ID(t_2,t_3)}{ID(t_1,t_3)}$$

$$\frac{SC(t_1,t_2) \quad ID(t_1,t_3)}{SC(t_3,t_2)} \qquad \frac{SC(t_1,t_2) \quad ID(t_2,t_3)}{SC(t_1,t_3)}$$

$$\frac{SM(t_1,t_2,t_3) \; ID(t_1,t_4)}{SM(t_4,t_2,t_3)} \quad \frac{SM(t_1,t_2,t_3) \; ID(t_2,t_4)}{SM(t_1,t_4,t_3)} \quad \frac{SM(t_1,t_2,t_3) \; ID(t_3,t_4)}{SM(t_1,t_2,t_4)}$$

$$\frac{PR(t_1,t_2,t_3) \; ID(t_1,t_4)}{PR(t_4,t_2,t_3)} \quad \frac{PR(t_1,t_2,t_3) \; ID(t_2,t_4)}{PR(t_1,t_4,t_3)} \quad \frac{PR(t_1,t_2,t_3) \; ID(t_3,t_4)}{PR(t_1,t_2,t_4)}$$

In some arithmetical calculi, these rules are replaced by a more general rule: $\frac{\varphi(t_1) \quad ID(t_1,t_2)}{\varphi(t_2)}$. However, this general rule is problematic (see 2.3.1).

Fourth, there are rules which ensure that SC stands for the graph of a one-to-one function on the set of numbers for which there is a unique number which is not in its range:

$$\frac{SC(t_1,t_2) \quad SC(t_3,t_2)}{ID(t_1,t_3)} \qquad\qquad \exists a(Na \wedge \forall b \neg SC(b,a))$$

$$\frac{Nt_1 \quad \forall b \neg SC(b,t_1) \quad Nt_2 \quad \forall b \neg SC(b,t_2)}{ID(t_1,t_2)}$$

The open sentence $Nx \wedge \forall b \neg SC(b,x)$ expresses the property of being an initial natural number. The second rule states that there is at least one initial number, and the third rule states that there is at most one initial number.

Fifth, there are 'recursive' rules for SM and PR (by repeated applications of these rules one can evaluate arbitrary sums and products):

$$\frac{Nt_1 \quad Nt_2 \quad \forall b \neg SC(b,t_2)}{SM(t_1,t_2,t_1)} \quad \frac{Nt_1 \quad Nt_2 \quad \forall b \neg SC(b,t_2) \quad PR(t_1,t_2,t_3)}{\forall b \neg SC(b,t_3)}$$

$$\frac{SM(t_1,t_2,t_3) \, SC(t_2,t_4) \, SM(t_1,t_4,t_5)}{SC(t_3,t_5)} \quad \frac{SM(t_1,t_2,t_3) \, SC(t_2,t_4) \, SC(t_3,t_5)}{SM(t_1,t_4,t_5)}$$

$$\frac{PR(t_1,t_2,t_3) \, SC(t_2,t_4) \, PR(t_1,t_4,t_5)}{SM(t_3,t_1,t_5)} \quad \frac{PR(t_1,t_2,t_3) \, SC(t_2,t_4) \, SM(t_3,t_1,t_5)}{PR(t_1,t_4,t_5)}$$

Finally, there is the rule of induction. It says that if the initial number satisfies an open sentence φ and if the successor of an arbitrary number that satisfies φ also satisfies φ, then every number satisfies φ:

$$\frac{\begin{array}{cc} [Nx, \forall b \neg SC(b,x)] & [\varphi(y), SC(y,z)] \\ \vdots & \vdots \\ \varphi(x) \qquad Nt \qquad \varphi(z) \end{array}}{\varphi(t)}$$

This completes the list of arithmetical rules of $\mathcal{R}^{Ar}$.

In the following proof, a simple consequence of the rule of induction is employed, namely the following rule of double induction:

$$\frac{[Nx, \forall b \neg SC(b,x), Ny] \quad [Nx, Ny, \forall b \neg SC(b,y)] \quad [\varphi(x,y), SC(x,z), SC(y,v)]}{\varphi(t_1, t_2)}$$
$$\begin{array}{ccccc} \vdots & & \vdots & & \vdots \\ \varphi(x,y) & Nt_1 & \varphi(x,y) & Nt_2 & \varphi(z,v) \end{array}$$

According to this rule, if the initial number is φ-related to every number, if every number is φ-related to the initial number, and if the successors of numbers that are φ-related are also φ-related, then all numbers are φ-related.[2]

I will now present the three rules which govern the binary predicate symbol π. These rules correspond to a recursive definition of π:

$$\frac{N_1 t \quad \forall b \neg SC_1(b,t) \quad N_2 t^* \quad \forall b \neg SC_2(b,t^*)}{\pi(t,t^*)} \ \pi_1$$

$$\frac{\pi(t_1, t_1^*) \quad SC_1(t_1, t_2) \quad SC_2(t_1^*, t_2^*)}{\pi(t_2, t_2^*)} \ \pi_2$$

$$\frac{[N_1 x, \forall b \neg SC_1(b,x), N_2 x^*, \forall b \neg SC_2(b,x^*)][\varphi(x,x^*), SC_1(x,y), SC_2(x^*,y^*)]}{\varphi(t,t^*)}$$
$$\begin{array}{ccc} \vdots & & \vdots \\ \varphi(x,x^*) & \pi(t,t^*) & \varphi(y,y^*) \end{array} \ \pi_3$$

The first two rules introduce two possible grounds for inferring a sentence of the form $\pi(t_1, t_1^*)$; and the third rule states that these are the only grounds for inferring a sentence of this form.

In the following proof, I will make use of some derivable rules which govern π:

$$\frac{\pi(t,t^*)}{N_1 t} \ L_1 \qquad \frac{\pi(t,t^*)}{N_2 t^*} \ L_2 \qquad \frac{N_1 t}{\exists a \pi(t,a)} \ L_3 \qquad \frac{N_2 t^*}{\exists a \pi(a,t^*)} \ L_4$$

$$\frac{\pi(t,t^*) \quad \forall b \neg SC_1(b,t)}{\forall b \neg SC_2(b,t^*)} \ L_5 \qquad \frac{\pi(t,t^*) \quad \forall b \neg SC_2(b,t^*)}{\forall b \neg SC_1(b,t)} \ L_6$$

$$\frac{\pi(t_1, t_1^*) \quad \pi(t_2, t_2^*) \quad SC_1(t_1, t_2)}{SC_2(t_1^*, t_2^*)} \ L_7 \qquad \frac{\pi(t_1, t_1^*) \quad \pi(t_2, t_2^*) \quad SC_2(t_1^*, t_2^*)}{SC_1(t_1, t_2)} \ L_8$$

The rules L_1 and L_2 state that π applies to a pair of objects (a,b) only if a satisfies N_1 and b satisfies N_2.[3] The rule L_3 states that every object that satisfies N_1 is π-related to something. The rule L_4 states that for every object that

---

[2]   Here, I will not show that this rule is derivable. For a proof of the derivability of an analogous rule (in an analogous setting), see Troelstra & van Dalen 1988: 124.

[3]   The derivability of these rules follows from π_3 with $\varphi(t_1, t_1^*) := (N_1 t_1 \wedge N_2 t_1^*)$.

satisfies N_2 there is something which is π-related to it.[4] The remaining rules L_5, L_6, L_7, L_8 state that π is congruent with respect to SC_1 and SC_2.[5]

I will now indicate the proof of T^{Ar}. Suppose that S_1 is an open sentence of $\mathcal{L}_+^{Ar}$ which contains precisely the (distinct) parameters $x_1, \ldots, x_n$, and let S_2 be the sentence that results form the translation of S_1 by systematically replacing the parameters $x_1, \ldots, x_n$ with new (distinct) parameters $x_1^*, \ldots, x_n^*$. Now, the idea is to show, by induction on the complexity of S_1, that the instances of the following schema are $\mathcal{R}_*(\mathcal{L}_*)$-derivable:

$${}^*\mathrm{T}^{Ar} \quad (\pi(x_1, x_1^*) \wedge \ldots \wedge \pi(x_n, x_n^*)) \to (S_1 \leftrightarrow S_2).$$

In particular, one then has shown that the instances of the schema T^{Ar} are $\mathcal{R}_*(\mathcal{L}_*)$-derivable. (The instances of T^{Ar} are those of ${}^*\mathrm{T}^{Ar}$ that do not contain parameters.)

Proof. The schema ${}^*\mathrm{T}^{Ar}$ is proven by induction on the complexity of S_1. For the different types of atomic open sentences, derivations of the following conditionals have to be given:

(1) $\pi(x, x^*) \to (N_1 x \leftrightarrow N_2 x^*)$,

(2) $(\pi(x, x^*) \wedge \pi(y, y^*)) \to (SC_1(x, y) \leftrightarrow SC_2(x^*, y^*))$,

(3) $(\pi(x, x^*) \wedge \pi(y, y^*)) \to (ID_1(x, y) \leftrightarrow ID_2(x^*, y^*))$,

(4) $(\pi(x, x^*) \wedge \pi(y, y^*) \wedge \pi(z, z^*)) \to (SM_1(x, y, z) \leftrightarrow SM_2(x^*, y^*, z^*))$,

(5) $(\pi(x, x^*) \wedge \pi(y, y^*) \wedge \pi(z, z^*)) \to (PR_1(x, y, z) \leftrightarrow PR_2(x^*, y^*, z^*))$.

The conditional (1) is a trivial consequence of L_1 and L_2, and the conditional (2) is a trivial consequence of L_7 and L_8. Here, I will indicate how the derivations of (3) and (4) can be found. (The derivation of (5) is similar to that of (4).)

A derivation of (3) is easily obtained from derivations of the following two open sentences:

$$\varphi(x, y) \ =: \ \forall a \forall b \left((\pi(x, a) \wedge \pi(y, b) \wedge ID_1(x, y)) \to ID_2(a, b)\right),$$

$$\psi(x^*, y^*) \ =: \ \forall a \forall b \left((\pi(a, x^*) \wedge \pi(b, y^*) \wedge ID_2(x^*, y^*)) \to ID_1(a, b)\right).$$

I will present a derivation of $\varphi(x,y)$. (A derivation of $\psi(x^*, y^*)$ can be build by similar means.) Its final application is an application of the rule of double induction (for N_1) with respect to $\varphi(x,y)$. Therefore, one has to show the existence of three derivations:

(i)   a derivation of $\varphi(x,y)$ from $N_1 x$, $N_1 y$, $\forall b \neg SC_1(b,x)$;

(ii) a derivation of $\varphi(x,y)$ from $N_1 x$, $N_1 y$, $\forall b \neg SC_1(b,y)$;

(iii) a derivation of $\varphi(z,v)$ from $\varphi(x,y)$, $SC_1(x,z)$, $SC_1(y,v)$.

I will present derivations of (i) and (iii). (A derivation of (ii) is similar to one of (i).)

As for (i), let S be the open sentence $\pi(x,x^*) \wedge \pi(y,y^*) \wedge ID_1(x,y)$, and suppose that Θ_1 is a derivation of $\forall b \neg SC_2(b, y^*)$ for the assumptions $\forall b \neg SC_1(b,x)$ and $S^{(1)}$. Then, the desired derivation can be obtained by pasting Θ_1 into the following expression:[6]

$$
\cfrac{
\cfrac{
\cfrac{\cfrac{S^{(1)}}{\pi(x,x^*)}}{N_2 x^*} \quad
\cfrac{\cfrac{S^{(1)}}{\forall b \neg SC_1(b,x) \quad \pi(x,x^*)}}{\forall b \neg SC_2(b, x^*)} \quad
\cfrac{\cfrac{S^{(1)}}{\pi(y,y^*)}}{N_2 y^*} \quad
\cfrac{\Theta_1}{\forall b \neg SC_2(b, y^*)}
}{
\cfrac{ID_2(x^*, y^*)}{\cfrac{S \to ID_2(x^*, y^*)^{(1)}}{\varphi(x,y)}}
}
}{}
$$

The following expression is the derivation Θ_1:

$$
\cfrac{
\cfrac{
\cfrac{
\cfrac{\cfrac{S^{(1)}}{ID_1(x,y)} \quad SC_1(z,y)^{(2)}}{SC_1(z,x)} \quad
\cfrac{\forall b \neg SC_1(b,x)}{\neg SC_1(z,x)}
}{
\cfrac{\neg SC_1(z,y)^{(2)}}{\forall b \neg SC_1(b,y)}
} \quad
\cfrac{S^{(1)}}{\pi(y,y^*)}
}{\forall b \neg SC_2(b, y^*)}
}{}
$$

This completes the derivation of $\varphi(x,y)$ from $\forall b \neg SC_1(b,x)$.

As for (iii), let S be the open sentence $\pi(z,z^*) \wedge \pi(v,v^*) \wedge ID_1(z,v)$, and suppose that Θ_2 is a derivation of $\varphi(z,v)$ from the assumptions $\mathcal{A}$:

$$SC_1(x,z), SC_1(y,v), \varphi(x,y), \pi(x,x^*)^{(1)}, \pi(y,y^*)^{(2)}.$$

---

[6]   I indicate derivations by making use of a usual simplification. I only display the sentences to the right of $\succ$, I use superscript numerals at topmost sentences to indicate the introduction of assumptions, and I use further superscript numerals to indicate at which point an assumption is discharged. See 5.1 for details.

The desired derivation can be obtained by pasting Θ_2 into the following expression:

$$
\cfrac{\cfrac{SC_1(x,z)}{N_1 x} \quad \cfrac{\cfrac{SC_1(y,v)}{N_1 y}}{\exists a\pi(y,a))} \quad \Theta_2 \atop \varphi(z,v)}{\cfrac{\exists a\pi(x,a)}{\varphi(z,v)^{(1)}} \quad \varphi(z,v)^{(2)}}
$$

Now, suppose that Θ_3 is a derivation of $SC_2(x^*, y^*)$ from $\mathcal{A}$ and $S^{(3)}$, and suppose that Θ_4 is a derivation of $ID_2(x^*, y^*)$ from $\mathcal{A}$ and $S^{(3)}$. Then, the derivation Θ_2 can be obtained by pasting Θ_3 and Θ_4 into the following expression:

$$
\cfrac{\cfrac{\pi(y,y^*)^{(2)} \quad \cfrac{S^{(3)}}{\pi(v,v^*)} \quad SC_1(y,v) \quad \cfrac{\cfrac{\Theta_3}{SC_2(x^*,z^*)} \quad \cfrac{\Theta_4}{ID_2(x^*,y^*)}}{SC_2(y^*,z^*)}}{SC_2(y^*,v^*)}}{\cfrac{ID_2(z^*,v^*)}{\cfrac{S \rightarrow ID_2(z^*,v^*)^{(3)}}{\varphi(z,v)}}}
$$

The following expression is the derivation Θ_3:

$$
\cfrac{\cfrac{S^{(3)}}{\pi(z,z^*)} \quad SC_1(x,z) \quad \pi(x,x^*)^{(1)}}{SC_2(x^*,z^*)}
$$

If S_α abbreviates $\pi(x,x^*) \wedge \pi(y,y^*) \wedge ID_1(x,y)$, then the following expression is the derivation Θ_4:

$$
\cfrac{\pi(x,x^*)^{(1)} \quad \pi(y,y^*)^{(2)} \quad \cfrac{\cfrac{\cfrac{S^{(3)}}{ID_1(z,v)} \; SC_1(x,z)}{SC_1(x,v)} \quad SC_1(y,v)}{ID_1(x,y)}}{\cfrac{\cfrac{S_\alpha}{ID_2(x^*,y^*)}}{}} \quad \cfrac{\varphi(x,y)}{S_\alpha \rightarrow ID_2(x^*,y^*)}
$$

This completes the derivation of $\varphi(z,v)$ from the assumptions $\varphi(x,y)$, $SC_1(x,z)$, and $SC_1(y,v)$.

I will now turn to a derivation of (4). Such a derivation is easily obtained from derivations of the following two open sentences, $\varphi(y)$ and $\psi(y)$:

$$\forall b^* \forall c \forall c^* (\pi(x,x^*) \rightarrow ((\pi(y,b^*) \wedge \pi(c,c^*) \wedge SM_1(x,y,c)) \rightarrow SM_2(x^*,b^*,c^*))),$$

$$\forall b^* \forall c \forall c^* (\pi(x,x^*) \rightarrow ((\pi(y,b^*) \wedge \pi(c,c^*) \wedge SM_2(x^*,b^*,c^*)) \rightarrow SM_1(x,y,c))).$$

I will present a derivation of $\varphi(y)$. (A derivation of $\psi(y)$ can be build by similar means.) Its finial application is an application of the rule of induction (for N_1) with respect to $\varphi(y)$. Therefore, one has to show the existence of two derivations:

(i) a derivation of $\varphi(y)$ from $N_1 y$, $\forall b \neg SC_1(b, y)$;

(ii) a derivation of $\varphi(z)$ from $\varphi(y)$, $SC_1(y, z)$.

I will present the derivations of (i) and (ii) in turn.

As for (i), let S be the open sentence $\pi(y, y^*) \wedge \pi(z, z^*) \wedge SM_1(x, y, z)$, and let $\mathcal{A}$ be the assumptions $N_1 y$, $\forall b \neg SC_1(b, y)$, $S^{(1)}$, and $\pi(x, x^*)^{(2)}$. Furthermore, suppose that Θ_5 is a derivation of $ID_2(x^*, z^*)$ from $\mathcal{A}$, and suppose that Θ_6 is a derivation of $SM_2(x^*, y^*, z^*)$ from $\mathcal{A}$. Then, the desired derivation can be obtained by pasting Θ_5 and Θ_6 into the following expression:

$$
\cfrac{
\cfrac{
\cfrac{
\cfrac{\cfrac{\Theta_5}{ID_2(x^*, z^*)} \quad \cfrac{\Theta_6}{SM_2(x^*, y^*, x^*)}}{SM_2(x^*, y^*, z^*)}
}{S \to SM_2(x^*, y^*, z^*)^{(1)}}
}{\pi(x, x^*) \to (S \to SM_2(x^*, y^*, z^*))^{(2)}}
}{\varphi(y)}
$$

Now, if Θ_7 is a derivation $ID_2(x^*, z^*)$ from $\pi(x, x^*) \wedge \pi(z, z^*) \wedge ID_1(x, z)$, which is known to exist by the previous part of this proof, then the following expression is the derivation Θ_5:

$$
\cfrac{
\cfrac{
\cfrac{N_1 y \quad \forall b \neg SC_1(b, y) \quad \cfrac{\pi(x, x^*)^{(2)}}{N_1 x}}{SM_1(x, y, x)}
}{ID_1(x, z)} \quad
\cfrac{\cfrac{S^{(1)}}{SM_1(x, y, z)} \quad \pi(x, x^*)^{(2)} \quad \cfrac{S^{(1)}}{\pi(z, z^*)}}{}
}{\pi(x, x^*) \wedge \pi(z, z^*) \wedge ID_1(x, z)}
$$

$$
\vdots \\ \Theta_7 \\ \vdots
$$

$$
ID_2(x^*, z^*)
$$

The following expression is the derivation Θ_6:

$$
\cfrac{
\cfrac{\pi(x, x^*)^{(2)}}{N_2 x^*} \quad \cfrac{\cfrac{S^{(1)}}{\pi(y, y^*)}}{N_2 y^*} \quad \cfrac{\forall b \neg SC_1(b, y) \quad \cfrac{S^{(1)}}{\pi(y, y^*)}}{\forall b \neg SC_2(b, y^*)}
}{SM_2(x^*, y^*, x^*)}
$$

This completes the derivation of $\varphi(y)$ from the assumptions $N_1 y$ and $\forall b \neg SC_1(b, y)$.

As for (ii), let S be the sentence $\pi(z, z^*) \wedge \pi(v, v^*) \wedge SM_1(x, z, v)$, let S_α be the sentence $S \to SM_2(x^*, z^*, v^*)$, and suppose that Θ_8 is a derivation of $SM_2(x^*, z^*, v^*)$ from the assumptions $\mathcal{A}$:

$$
\varphi(y), SC_1(y, z), \pi(x, x^*)^{(1)}, SM_1(x, y, w)^{(2)}, \pi(w, w^*)^{(3)}, \pi(y, y^*)^{(4)}, S^{(5)}.
$$

Then, the desired derivation can be obtained by pasting Θ_8 into the following expression:

$$
\cfrac{
\cfrac{
\cfrac{\pi(x,x^*)^{(1)} \quad \cfrac{SC_1(y,z)}{N_1 y}}{\exists a SM_1(x,y,a)}
\quad
\cfrac{SM_1(x,y,w)^{(2)} \quad \cfrac{\cfrac{SC_1(y,z)}{N_1 y}}{\exists a\pi(y,a)}}{\exists a\pi(w,a)}
\quad
\cfrac{\cfrac{SC_1(y,z)}{N_1 y}}{\exists a\pi(y,a)} \quad \cfrac{\Theta_8}{\cfrac{SM_2(x^*,z^*,v^*)}{S_\alpha^{(5)}}}}{S_\alpha^{(4)}}
}{
\cfrac{\cfrac{S_\alpha^{(3)}}{S_\alpha^{(2)}}}{\cfrac{\pi(x,x^*)\to S_\alpha^{(1)}}{\varphi(z)}}
}
$$

Now, suppose that Θ_9 is a derivation of $SM_2(x^*,y^*,w^*)$ from $\mathcal{A}$, and suppose that Θ_{10} is a derivation of $SC_2(w^*,v^*)$ from $\mathcal{A}$. Then, the derivation Θ_8 is obtained by pasting Θ_9 and Θ_{10} into the following expression:

$$
\cfrac{
\cfrac{\Theta_9}{SM_2(x^*,y^*,w^*)}
\quad
\cfrac{\pi(y,y^*)^{(4)} \quad \cfrac{\cfrac{S^{(5)}}{\pi(z,z^*)} \quad SC_1(y,z)}{SC_2(y^*,z^*)}}{}
\quad
\cfrac{\Theta_{10}}{SC_2(w^*,v^*)}
}{SM_2(x^*,z^*,v^*)}
$$

Now, let S_β be the sentence $\pi(y,y^*)\wedge\pi(z,z^*)\wedge SM_1(x,y,w)$, and let S_γ be the sentence $S_\beta\to SM_2(x^*,y^*,w^*)$. Then, Θ_9 is the following derivation:

$$
\cfrac{
\cfrac{\pi(y,y^*)^{(4)} \quad \cfrac{S^{(5)}}{\pi(z,z^*)} \quad SM_1(x,y,w)^{(2)}}{S_\beta}
\quad
\cfrac{\pi(x,x^*)^{(1)} \quad \cfrac{\varphi(y)}{\pi(x,x^*)\to S_\gamma}}{S_\gamma}
}{SM_2(x^*,y^*,w^*)}
$$

Finally, the following expression is the derivation Θ_{10}:

$$
\cfrac{
\cfrac{SM_1(x,y,w)^{(2)} \quad \cfrac{S^{(5)}}{SM_1(x,z,v)} \quad SC_1(y,z)}{SC_1(w,v)}
\quad
\cfrac{\pi(w,w^*)^{(3)} \quad \cfrac{S^{(5)}}{\pi(v,v^*)}}{}
}{SC_2(w^*,v^*)}
$$

This completes the derivation of $\varphi(z)$ from the assumptions $\varphi(y)$ and $SC_1(y,z)$.

Thus, if S_1 is atomic, then the corresponding instance of $^*T^{Ar}$ is $\mathcal{R}_*(\mathcal{L}_*)$-derivable. It remains to show that an instance of $^*T^{Ar}$ which corresponds to a complex sentence S_1 is $\mathcal{R}_*(\mathcal{L}_*)$-derivable given that those instances are $\mathcal{R}_*(\mathcal{L}_*)$-derivable which correspond to simpler sentences. More precisely, suppose that the instances which correspond to S_1^α, S_1^β, and S_1^γ are $\mathcal{R}_*(\mathcal{L}_*)$-derivable. It then has to be shown that the following more complex instances of $^*T^{Ar}$ are $\mathcal{R}_*(\mathcal{L}_*)$-derivable:

(6) $\pi \to (\neg S_1^\alpha \leftrightarrow \neg S_2^\alpha)$,

(7) $\pi \to ((S_1^\alpha \vee S_1^\beta) \leftrightarrow (S_2^\alpha \vee S_2^\beta))$,

(8) $\pi \to ((S_1^\alpha \wedge S_1^\beta) \leftrightarrow (S_2^\alpha \wedge S_2^\beta))$,

(9) $\pi \to ((S_1^\alpha \to S_1^\beta) \leftrightarrow (S_2^\alpha \to S_2^\beta))$,

(10) $\pi \to (\forall a(N_1 a \to S_1^\gamma(a)) \leftrightarrow \forall a(N_2 a \to S_2^\gamma(a)))$,

(11) $\pi \to (\exists a(N_1 a \wedge S_1^\gamma(a)) \leftrightarrow \exists a(N_2 a \wedge S_2^\gamma(a)))$,

where π abbreviates $\pi(x_1, x_1^*) \wedge \ldots \wedge \pi(x_n, x_n^*)$.

Now, the derivations of (6)-(9) are trivially obtainable from derivations of the instances that correspond to S_1^α and S_1^β. Furthermore, the derivations of (10) and (11) are similar, and so I will confine myself to show how a derivation of (10) can be obtained from a derivation Θ_{11} of the instance that corresponds to S_1^γ:

$$(\pi \wedge \pi(x, x^*)) \to (S_1^\gamma(x) \leftrightarrow S_2^\gamma(x^*)),$$

where $x, x_1, \ldots, x_n$ are the parameters of $S_1^\gamma(x)$, and where $x^*, x_1^*, \ldots, x_n^*$ are the parameters of $S_2^\gamma(x^*)$.

To this end, suppose that Θ_{12} is a derivation of $S_1^\gamma(x)$ from the assumptions $(\forall a(N_1 a \to S_1^\gamma(a)))^{(2)}$ and $\pi(x, x^*)^{(4)}$, and suppose that Θ_{13} is a derivation of $S_1^\gamma(x) \to S_2^\gamma(x^*)$ from the assumptions $\pi^{(1)}$ and $\pi(x, x^*)^{(4)}$. Then, the desired derivation can be obtained by pasting Θ_{12} and Θ_{13} into the following expression:

$$\cfrac{\cfrac{N_2 x^{*(3)}}{\exists a \pi(a, x^*)} \quad \cfrac{\cfrac{\Theta_{12}}{S_1^\gamma(x)} \quad \cfrac{\Theta_{13}}{S_1^\gamma(x) \to S_2^\gamma(x^*)}}{\cfrac{S_2^\gamma(x^*)}{\cfrac{S_2^\gamma(x^*)^{(4)}}{\cfrac{N_2 x^* \to S_2^\gamma(x^*)^{(3)}}{\forall a(N_2 a \to S_2^\gamma(a))}}}}}{\cfrac{\forall a(N_1 a \to S_1^\gamma(a)) \to \forall a(N_2 a \to S_2^\gamma(a))^{(2)}}{\pi \to (\forall a(N_1 a \to S_1^\gamma(a)) \to \forall a(N_2 a \to S_2^\gamma(a)))^{(1)}}}$$

The following expression is the derivation Θ_{12}:

$$\cfrac{\cfrac{\pi(x, x^*)^{(4)}}{N_1 x} \quad \cfrac{(\forall a(N_1(a) \to S_1^\gamma(a)))^{(2)}}{N_1 x \to S_1^\gamma(x)}}{S_1^\gamma(x)}$$

The following expressions is the derivation Θ_{13}:

$$\cfrac{\cfrac{\cfrac{\Theta_{11}}{(\pi \wedge \pi(x, x^*)) \to (S_1^\gamma(x) \leftrightarrow S_2^\gamma(x^*))} \quad \cfrac{\pi^{(1)} \quad \pi(x, x^*)^{(4)}}{\pi \wedge \pi(x, x^*)}}{S_1^\gamma(x) \leftrightarrow S_2^\gamma(x^*)}}{S_1^\gamma(x) \to S_2^\gamma(x^*)}$$

This completes the derivation of (10). $\qquad\qquad\qquad\qquad\qquad\qquad\quad$ $\square$

A.2 Set-Theoretic Uniqueness

In this appendix, I will first present the set-theoretic calculus $\mathcal{R}^{Cl}$. Then, I will indicate how the theorem T^{Cl} can be proven.

Apart from the rules of the arithmetical calculus $\mathcal{R}^{Ar}$, the calculus $\mathcal{R}^{Cl}$ contains the following set-theoretic rules. First, there are rules which express that $\in$ applies to pairs of numbers and sets of numbers and that $\in$ respects identity of the natural numbers:

$$\frac{t_1 \in t_2}{Nt_1} \qquad \frac{t_1 \in t_2}{Ct_2} \qquad \frac{t_1 \in t_2 \quad ID(t_1,t_3)}{t_3 \in t_2}$$

(One may note that the third of these rules is not needed for the following proof.) Second, there is the following comprehension rule:

$$\overline{\exists a(Ca \wedge \forall b(b \in a \leftrightarrow Nb \wedge \varphi(a)))}$$

It states that there is a set of numbers which contains precisely those numbers that satisfy φ. This completes the list of set-theoretic rules of $\mathcal{R}^{Cl}$.

I will now present the proof of T^{Cl}. I will use the following abbreviation:

$$\tau(t,t^*) := C_1 t \wedge C_2 t^* \wedge \forall a(a \in_1 t \leftrightarrow a \in_2 t^*).$$

Suppose that S_1 is an open sentence of $^1\mathcal{L}_+^{Cl}$ which contains precisely the (distinct) parameters $x_1, \ldots, x_n$ in set-theoretic positions, and let S_2 be the sentence that results from the translation of S_1 by systematically replacing the parameters $x_1, \ldots, x_n$ with new (distinct) parameters $x_1^*, \ldots, x_n^*$. Then, the idea is to show, by induction on the complexity of S_1, that the following conditional is $\mathcal{R}_*(\mathcal{L}_*)$-derivable:

$$^*\mathrm{T}^{Cl} \quad (\tau(x_1,x_1^*) \wedge \ldots \wedge \tau(x_n,x_n^*)) \to (S_1 \leftrightarrow S_2).$$

In particular, one then has shown that the instances of the schema T^{Cl} are $\mathcal{R}_*(\mathcal{L}_*)$-derivable. (The instances of T^{Cl} are those of $^*\mathrm{T}^{Cl}$ that do not contain parameters.)

Proof. The schema $^*\mathrm{T}^{Cl}$ is proven by induction on the complexity of S_1. First, there are the following types of arithmetical atomic open sentences:

$$Nx, ID(x,y), SC(x,y), SM(x,y,z), PR(x,y,z).$$

For each of these one only needs to find a derivation of a sentence of the form $S \leftrightarrow S$ (because they do not contain set-theoretic predicates).

Second, there are two types of set-theoretic atomic open sentences: $C_1 x$ and $z \in_1 x$. Here, one has to find derivations of the following open sentences:

$$\tau(x,x^*) \to (C_1 x \leftrightarrow C_2 x^*),$$

$$\tau(x,x^*) \to (z \in_1 x \leftrightarrow z \in_2 x^*).$$

The existence of these derivations is an immediate consequence of the definition of τ.

Thus, if S_1 is atomic, then the corresponding instance of $^*\mathrm{T}^{Cl}$ is $\mathcal{R}_*(\mathcal{L}_*)$-derivable. It remains to show that an instance of $^*\mathrm{T}^{Cl}$ which corresponds to a complex sentence S_1 is $\mathcal{R}_*(\mathcal{L}_*)$-derivable given that those instances are $\mathcal{R}_*(\mathcal{L}_*)$-derivable which correspond to simpler sentences.

On the one hand, there are complex open sentences of the following types:

$$\neg S_1^\alpha, S_1^\alpha \vee S_1^\beta, S_1^\alpha \wedge S_1^\beta, S_1^\alpha \to S_1^\beta, \forall a(Na \to S_1^\gamma(a)), \exists a(Na \wedge S_1^\gamma(a)).$$

For each of these sentences it is easy to obtain a derivation of the corresponding instance of $^*\mathrm{T}^{Cl}$ from derivations of the instances of $^*\mathrm{T}^{Cl}$ that correspond to S_1^α, S_1^β, and S_1^γ.

On the other hand, there are complex open sentences of the following two types:

$$\forall a(C_1 a \to S_1^\gamma(a)), \quad \exists a(C_1 a \wedge S_1^\gamma(a)).$$

These cases are very similar, and I will confine myself to deal with the former one. If τ abbreviates $\tau(x_1, x_1^*) \wedge \ldots \wedge \tau(x_n, x_n^*)$, then one has to present a derivation of the following open sentence:

$(*)$ $\tau \to (\forall a(C_1 a \to S_1^\gamma(a)) \leftrightarrow \forall a(C_2 a \to S_2^\gamma(a)))$.

To this end, one may inductively assume that there is a derivation, Θ_1, of the open sentence: $(\tau \wedge \tau(x, x^*)) \to (S_1^\gamma(x) \leftrightarrow S_2^\gamma(x^*))$. Now, a derivation of $(*)$ is easily obtained from derivations of the following two open sentences:

$$\varphi := \tau \to (\forall a(C_1 a \to S_1^\gamma(a)) \to \forall a(C_2 a \to S_2^\gamma(a))),$$

$$\psi := \tau \to (\forall a(C_2 a \to S_2^\gamma(a)) \to \forall a(C_1 a \to S_1^\gamma(a))).$$

I will present a derivation of φ. (A derivation of ψ can be obtained by switching the indices of the set-theoretic predicates.)

Suppose that Θ_2 is a derivation of $S_2^\gamma(x^*)$ from the assumptions $\mathcal{A}$:

$$\tau^{(1)}, \forall a(C_1 a \to S_1^\gamma(a))^{(2)}, C_1 x \wedge \forall b(b \in_1 x \leftrightarrow (Nb \wedge b \in_2 x^*))^{(3)}, C_2 x^{*(4)}.$$

Then, the derivation of $(*)$ can be obtained by pasting Θ_2 into the following expression:

$$\cfrac{\cfrac{\exists a(C_1 a \wedge (\forall b(b \in_1 a \leftrightarrow (Nb \wedge b \in_2 x^*)))) \quad \cfrac{\Theta_2}{S_2^\gamma(x^*)}}{\cfrac{S_2^\gamma(x^*)^{(3)}}{\cfrac{C_2 x^* \to S_2^\gamma(x^*)^{(4)}}{\cfrac{\forall a(C_2 a \to S_2^\gamma(a))}{\cfrac{\forall a(C_1 a \to S_1^\gamma(a)) \to \forall a(C_2 a \to S_2^\gamma(a))^{(2)}}{\tau \to (\forall a(C_1 a \to S_1^\gamma(a)) \to \forall a(C_2 a \to S_2^\gamma(a)))^{(1)}}}}}}}$$

Suppose that Θ_3 is a derivation of $S_1^\gamma(x) \to S_2^\gamma(x^*)$ from $\mathcal{A}$. Then, the derivation Θ_2 can be obtained by pasting Θ_3 into the following expression:

$$\cfrac{\cfrac{\cfrac{\forall a(C_1 a \to S_1^\gamma(a))^{(2)}}{C_1 x \to S_1^\gamma(x)} \quad \cfrac{C_1 x \wedge \forall b(b \in_1 x \leftrightarrow (Nb \wedge b \in_2 x^*))^{(3)}}{C_1 x}}{S_1^\gamma(x)} \quad \cfrac{\Theta_3}{S_1^\gamma(x) \to S_2^\gamma(x^*)}}{S_2^\gamma(x^*)}$$

Now, suppose that Θ_4 is a derivation of $\tau(x, x^*)$ from $\mathcal{A}$. Then, the derivation Θ_3 can be obtained by pasting Θ_1 and Θ_4 into the following expression:

$$\cfrac{\cfrac{\cfrac{\Theta_1}{(\tau \wedge \tau(x, x^*)) \to (S_1^\gamma(x) \leftrightarrow S_2^\gamma(x^*))} \quad \cfrac{\tau^{(1)} \quad \cfrac{\Theta_4}{\tau(x, x^*)}}{\tau \wedge \tau(x, x^*)}}{S_1^\gamma(x) \leftrightarrow S_2^\gamma(x^*)}}{S_1^\gamma(x) \to S_2^\gamma(x^*)}$$

By using the rule $\frac{t_1 \in_2 t_2}{Nt_1}$, one can easily find the derivation Θ_4 from $\mathcal{A}$. This then completes the derivation of $(*)$. $\qquad\square$

A.3   Function-Theoretic Uniqueness

In this appendix, I will first present the function-theoretic calculus $\mathcal{R}^{Seq}$. Then, I will indicate how the theorem T^{Seq} can be proven.

Apart from the rules of the arithmetical calculus $\mathcal{R}^{Ar}$, the calculus $\mathcal{R}^{Seq}$ contains the following function-theoretic rules. First, there are rules which express that AP applies to triples of sequences of numbers, numbers, and numbers, and that AP respects identity of the natural numbers:

$$\frac{AP(t_1, t_2, t_3)}{SQ\, t_1} \quad \frac{AP(t_1, t_2, t_3)}{Nt_2} \quad \frac{AP(t_1, t_2, t_3)}{Nt_3}$$

$$\frac{AP(t_1, t_2, t_3) \quad ID(t_2, t_4)}{AP(t_1, t_4, t_3)} \quad \frac{AP(t_1, t_2, t_3) \quad ID(t_3, t_4)}{AP(t_1, t_2, t_4)}$$

(As in the set-theoretic case, one may note that the forth and fifth of these rules are not needed for the following proof.) Second, there are two rules which have no counterparts in the set-theoretic case:

$$\frac{AP(t_1, t_2, t_3) \quad AP(t_1, t_2, t_4)}{ID(t_3, t_4)} \quad \frac{SQ\, t_1 \quad Nt_2}{\exists a\, AP(t_1, t_2, a)}$$

These rules express that AP is functional and that a sequence maps every number to some number. Third, there is the following comprehension rule:

$$\frac{\cfrac{[\varphi(x, y)]}{\vdots} \quad \cfrac{[\varphi(x, y)]}{\vdots} \quad \cfrac{[\varphi(x, y), \varphi(x, z)]}{\vdots} \quad \cfrac{[Nx]}{\vdots}}{\exists a(SQ\, a \wedge \forall b \forall c(AP(a, b, c) \leftrightarrow \varphi(b, c)))}$$

with premises Nx, Ny, $ID(y, z)$, $\exists a\varphi(x, a)$.

It states that if φ applies only to pairs of numbers, is functional, and relates every number to some number, then there is a sequence that maps a number b to a number c iff the pair (b, c) satisfies φ. This completes the list of function-theoretic rules of $\mathcal{R}^{Seq}$.

I will now present the proof of T^{Seq}. I will use the following abbreviation:

$$\sigma(t, t^*) := SQ_1 t \wedge SQ_2 t^* \wedge \forall a \forall b (AP(t, a, b) \leftrightarrow AP(t^*, a, b)).$$

Suppose that S_1 is an open sentence of $^1\mathcal{L}_+^{Seq}$ which contains precisely the (distinct) parameters $x_1, \ldots, x_n$ in function-theoretic positions, and let S_2 be the sentence that results from the translation of S_1 by systematically replacing the parameters $x_1, \ldots, x_n$ with new (distinct) parameters $x_1^*, \ldots, x_n^*$. Then, the idea is to show, by induction on the complexity of S_1, that the following conditional is $\mathcal{R}_*(\mathcal{L}_*)$-derivable:

$^*\mathbf{T}^{Seq}$ $\quad (\sigma(x_1, x_1^*) \wedge \ldots \wedge \sigma(x_n, x_n^*)) \to (S_1 \leftrightarrow S_2).$

In particular, one then has shown that the instances of the schema T^{Seq} are $\mathcal{R}_*(\mathcal{L}_*)$-derivable. (The instances of T^{Seq} are those of $^*\mathrm{T}^{Seq}$ that do not contain parameters.)

Proof. The schema $^*\mathrm{T}^{Seq}$ is proven by induction on the complexity of S_1. The proof is very similar to the proof of $^*\mathrm{T}^{Cl}$. There are only two differences. First, there are now the following two types of function-theoretic atomic open sentences: $SQ_1 x$ and $AP_1(x, y, z)$. Here, one has to find derivations of the following open sentences:

$$\sigma(x, x^*) \to (SQ_1 x \leftrightarrow SQ_2 x^*),$$

$$\sigma(x, x^*) \to (AP_1(x, y, z) \leftrightarrow AP_2(x^*, y, z)).$$

The existence of these derivations directly follows from the definition of σ.
Second, there are also two new types of complex open sentences:

$$\forall a(SQ_1 a \to S_1^\gamma(a)), \quad \exists a(SQ_1 a \wedge S_1^\gamma(a)).$$

If σ abbreviates $\sigma(x_1, x_1^*) \wedge \ldots \wedge \sigma(x_n, x_n^*)$, then one has to present derivations of the following open sentences:

($\sharp_1$) $\sigma \to (\forall a(SQ_1 a \to S_1^\gamma(a)) \leftrightarrow \forall a(SQ_2 a \to S_2^\gamma(a)))$,

($\sharp_2$) $\sigma \to (\exists a(SQ_1 a \wedge S_1^\gamma(a)) \leftrightarrow \exists a(SQ_2 a \to S_2^\gamma(a)))$,

where one may inductively assume that there is a derivation of the open sentence $(\sigma \wedge \sigma(x, x^*)) \to (S_1^\gamma(x) \leftrightarrow S_2^\gamma(x^*))$.

However, derivations of ($\sharp_1$) and ($\sharp_2$) can be build in analogy to the corresponding derivations from the set-theoretic case. In sum, the proof of $^*\mathrm{T}^{Seq}$ is a trivial variant of the proof of $^*\mathrm{T}^{Cl}$. $\qquad\square$

Bibliography

Aberdein, Andrew and Read, Stephen 2009: "The Philosophy of Alternative Logics", in L. Haaparanta (ed.) 2009: *The Development of Modern Logic*, Oxford: Oxford University Press, 613-723.

Abney, Steven 1987: *The English Noun Phrase in its Sentential Aspect*, Ph.D. Dissertation, MIT.

Aczel, Peter 1978: "The Type Theoretic Interpretation of Constructive Set Theory", in A. Macintyre et al. (eds.) 1978: *Logic Colloquium '77*, Amsterdam and New York: North-Holland, 55-66.

Aczel, Peter and Rathjen, Michael 2001: "Notes on Constructive Set Theory", Institut Mittag-Leffler, Preprint 40.

Anderson, Alan Ross and Belnap, Nuel D. 1975: *Entailment: The Logic of Relevance and Neccessity, Vol. I*, Princeton: Princeton University Press.

Appiah, Kwame Anthony 1985: "Verification and the Manifestation of Meaning", *Proceedings of the Aristotelian Society (Suppl. Vol.)* 59, 17-31.

Appiah, Kwame Anthony 1986: *For Truth in Semantics*, Oxford: Basil Blackwell.

van Atten, Mark and van Dalen, Dirk 2002: "Arguments for the Continuity Principle", *The Bulletin of Symbolic Logic* 8 (3), 329-47.

van Atten, Mark 2011: "Luitzen Egbertus Jan Brouwer", in E. N. Zalta (ed.) 2011: *The Stanford Encyclopedia of Philosophy* (Summer 2011 Edition), URL = <http://plato.stanford.edu/archives/sum2011/entries/brouwer/>.

van Atten, Mark 2014: "The Development of Intuitionistic Logic", in E. N. Zalta (ed.) 2014: *The Stanford Encyclopedia of Philosophy* (Spring 2014 Edition), URL = <http://plato.stanford.edu/archives/spr2014/entries/intuitionistic-logic-development/>.

Auxier, Randalle E. and Hahn, Lewis Edwin (eds.) 2007: *The Philosophy of Michael Dummett*, Chicago and LaSalle: Open Court.

Beeson, Michael J. 1985: *Foundations of Constructive Mathematics: Metamathematical Studies*, Berlin: Springer.

Bencivenga, Ermanno 2002: "Free Logics", in Gabbay & Guenthner 2002: 147-96.

Bendall, Kent 1978: "Natural Deduction, Separation, and the Meaning of Logical Operators", *Journal of Philosophical Logic* 7 (1), 245-76.

Boeckx, Cedric 2010: *Language in Cognition. Uncovering Mental Structures and the Rules Behind Them*, Oxford: Wiley-Blackwell.

Boolos, George 1984: "To Be is to Be a Value of a Variable (or to Be Some Values of Some Variables)", *The Journal of Philosophy*, 81 (8), 430-49.

Boolos, George 1985: "Nominalist Platonism", *The Philosophical Review*, 94 (3), 327-44.

Boričić, Branislav R. 1985: "On Sequence-Conclusion Natural Deduction Systems", *Journal of Philosophical Logic* 14 (4), 359-77.

Boulter, Stephen 2001: "Whose Challenge? Which Semantics?", *Synthese*, 126 (1-2), 325-37.

Brouwer, Luitzen Egbertus Jan 1907: "Over de grondslagen der wiskunde", English translation in Heyting 1975: 11-101.

Brouwer, Luitzen Egbertus Jan 1908: "De onbetrouwbaarheid der logische principes", English translation in Heyting 1975: 107-11.

Brouwer, Luitzen Egbertus Jan 1918: "Begründung der Mengenlehre Unabhängig vom Logischen Satz vom Ausgeschlossenen Dritten", in Heyting 1975: 150-90.

Brouwer, Luitzen Egbertus Jan 1947: "Richtlijnen der intuïtionistische wiskunde", English translation in Heyting 1975: 477.

Brouwer, Luitzen Egbertus Jan 1948: "Consciousness, Philosophy, and Mathematics", in Heyting 1975: 480-94.

Brouwer, Luitzen Egbertus Jan 1952: "Historical Background, Principles and Methods of Intuitionism", in Heyting 1975: 508-15.

Brouwer, Luitzen Egbertus Jan 1955: "The Effect of Intuitionism on Classical Algebra of Logic", in Heyting 1975: 551-4.

Burgess, John P. 1984: "Dummett's Case for Intuitionism", *History and Philosophy of Logic* 5 (2), 177-94.

Burgess, John P. 2005: "No Requirement of Relevance", in Shapiro 2005: 727-50.

Byrne, Darragh 2005: "Compositionality and the Manifestation Challenge", *Synthese* 144 (1), 101-36.

Campell, John 1982: "Knowledge and Understanding", *The Philosophical Quarterly* 32 (126), 17-34.

Carnap, Rudolf 1934: *Logische Syntax der Sprache*, English translation 1937: *The Logical Syntax of Language*, London: Routledge.

Carnap, Rudolf 1963: "Intellectual Autobiography", in P. A. Schlipp (ed.) 1963: *The Philosophy of Rudolf Carnap*, LaSalle: Open Court, 1-84.

Chomsky, Noam 1986: *Knowledge of Language. Its Nature, Origin, and Use*, New York: Praeger.

van Dalen, Dirk 1999: *Mystic, Geometer, and Intuitionist. The Life of L.E.J. Brouwer. Vol. 1. The Dawning Revolution*, Oxford: Clarendon Press.

van Dalen, Dirk 2005: *Mystic, Geometer, and Intuitionist. The Life of L.E.J. Brouwer. Vol. 2. Hope and Disillusion*, Oxford: Clarendon Press.

Davidson, Donald 1967a: "Truth and Meaning", in Davidson 2001: 17-36.

Davidson, Donald 1967b: "The Logical Form of Action Sentences", in D. Davidson 1980: *Essays on Actions and Events*, Oxford: Clarendon Press, 105-22.

Davidson, Donald 1973: "Radical Interpretation", in Davidson 2001: 125-39.

Davidson, Donald 1976: "Reply to Foster", in Davidson 2001: 171-9.

Davidson, Donald 2001: *Inquiries into Truth and Interpretation*, 2nd edition, Oxford: Clarendon Press.

Davies, Martin 1981: *Meaning, Quantification, Necessity: Themes in Philosophical Logic*, London: Routledge and Kegan Paul.

Detlefsen, Michael 1990: "Brouwerian Intuitionism", *Mind* 99 (396), 501-34.

Devitt, Michael 1983: "Dummett's Anti-Realism", *The Journal of Philosophy* 80 (2), 73-99.

Diaconescu, Radu 1975: "Axiom of Choice and Complementation", *Proceedings of the American Mathematical Society* 51, 176-8.

Dummett, Michael 1959: "Truth", in Dummett 1978a: 1-19.

Dummett, Michael 1963: "Realism", in Dummett 1978a: 145-65.

Dummett, Michael 1973a: *Frege. Philosophy of Language*, 2nd edition 1981, Cambridge MA: Havard University Press.

Dummett, Michael 1973b: "The Philosophical Basis of Intuitionistic Logic", in Dummett 1978a: 215-47.

Dummett, Michael 1975: "What is a Theory of Meaning? (I)", in Dummett 1993: 1-33.

Dummett, Michael 1976: "What is a Theory of Meaning? (II)", in Dummett 1993: 34-93.

Dummett, Michael 1977: *Elements of Intuitionism*, 2nd edition 2000, Oxford: Clarendon Press.

Dummett, Michael 1978a: *Truth and Other Enigmas*, Cambridge MA: Havard University Press.

Dummett, Michael 1978b: "What do I Know when I Know a Language?", in Dummett 1993: 94-105.

Dummett, Michael 1983: "Language and Truth", in Dummett 1993: 117-65.

Dummett, Michael 1991: *The Logical Basis of Metaphysics*, Cambridge MA: Harvard University Press.

Dummett, Michael 1993: *The Seas of Language*, Oxford: Clarendon Press.

Dummett, Michael 1994: "Reply to Prawitz", in McGuinness & Oliveri 1994: 292-8.

Dummett, Michael 2007: "Reply to Richard G. Heck, Jr.", in Auxier & Hahn 2007: 558-65.

Dunn, Michael J. and Restall, Greg 2002: "Relevance Logic", in D. Gabbay and F. Guenthner (eds.): *Handbook of Philosophical Logic. Volume 6*, Dordrecht: Kluwer, 1-128.

Edgington, Dorothy 1985: "Verification and the Manifestation of Meaning", *Proceedings of the Aristotelian Society (Suppl. Vol.)* 59, 33-52.

Edgington, Dorothy 1995: "On Conditionals", *Mind*, 104 (414), 235-329.

Evans, Gareth and McDowell, John Henry (eds.) 1976: *Truth and Meaning*, Oxford: Oxford University Press.

Evans, Gareth 1982: *The Varieties of Reference*, Oxford: Clarendon Press.

Feferman, Solomon 2012: "And so on ...: Reasoning with Infinite Diagrams", *Synthese* 186 (1), 371-86.

Field, Hartry 2008: *Saving Truth from Paradox*, Oxford: Oxford University Press.

Field, Hartry 2009a: "Pluralism in Logic", *The Review of Symbolic Logic* 2 (2), 342-59.

Field, Hartry Field 2009b: "Epistemology Without Metaphysics", *Philosophical Studies* 143 (2), 249-90.

Field, Hartry Field 2009c: "What is the Normative Role of Logic?", *Proceedings of the Aristotelian Society* 83 (1), 251-68.

Fine, Kit 2007: *Semantic Relationism*, Oxford: Wiley-Blackwell.

Fine, Kit 2010: "Semantic Necessity", in B. Hale and A. Hoffmann (eds.) 2001: *Modality. Metaphysics, Logic, and Epistemology*, Oxford: Oxford University Press, 65-80.

Fine, Kit 2012: "Guide to Ground", in F. Correia and B. Schnieder (eds.) 2012: *Metaphysical Grounding*, Cambridge: Cambridge University Press, 37-80.

Fitch, Frederic B. 1963: "A Logical Analysis of Some Value Concepts", *The Journal of Symbolic Logic* 28 (2), 135-42; repr. in Salerno 2009: 21-28.

Foster, John A. 1976: "Meaning and Truth Theory", in Evans & McDowell 1976: 1-32.

Francez, Nissim and Dyckhoff, Roy 2012: "A Note on Harmony", *Journal of Philosophical Logic* 41 (3), 613-28.

Franchella, Miriam 1995: "L.E.J. Brouwer: Toward Intuitionistic Logic", *Historia Mathematica* 22 (3), 304-22.

Frege, Gottlob 1879: *Begriffsschrift. Eine der arithmetischen nachgebildete Formelsprache des reinen Denkens*, Halle: Verlag von Louis Nebert.

Frege, Gottlob 1881-3a: "Dialog mit Pünjer über Existenz", in H. Hermes et al. (eds.) 1983: *Frege. Nachgelassene Schriften*, Hamburg: Meiner, 53-67.

Frege, Gottlob 1881-3b: "Kernsätze zur Logik", in H. Hermes et al. (eds.) 1983: *Frege. Nachgelassene Schriften*, Hamburg: Meiner, 189-90.

Frege, Gottlob 1893: *Grundgesetze der Arithmetik. Bd. 1*, reprinted 1962, Hildesheim: Olms.

Frege, Gottlob 1918: "Der Gedanke. Eine Logische Untersuchung", *Beiträge zur Philosophie des Deutschen Idealismus* 2 (1918-1919), 58-77.

Gabbay, Dov M. and Guenthner, F. (eds.) 2002: *Handbook of Philosophical Logic. 2nd Edition. Volume 5*, Dordrecht: Kluwer.

Gentzen, Gerhard 1934: "Untersuchungen über das logische Schließen. I", *Mathematische Zeitschrift* 39 (2), 176-210.

George, Alexander 1984: "On Devitt on Dummett", *The Journal of Philosophy* 81 (9), 516-27.

Gödel, Kurt 1953: "Is Mathematics Syntax of Language?", in S. Feferman et al. (eds.) 1995: *Kurt Gödel Collected Works. Vol. III*, Oxford: Oxford University Press, 334-64.

Gödel, Kurt 1958: "Über eine bisher noch nicht benützte Erweiterung des finiten Standpunktes", *Dialectica* 12 (3-4), 280-7.

Goodman, Nicolas D. and Myhill, John R. 1978: "Choice Implies Excluded Middle", *Zeitschrift für mathematische Logik und Grundlagen der Mathematik* 24 (5), 461.

Harris, John H. 1982: "What's so Logical about the 'Logical' Axioms", *Studia Logica* 41 (2-3), 159-71.

Hart, W. D. and McGinn, Colin 1976: "Knowledge and Necessity", *Journal of Philosophical Logic* 5 (2), 205-8.

Haverkamp, Nick 2011: "Nothing But Objects", *Grazer Philosophische Studien* 82, 209-37.

Heck, Richard G., Jr. 2007: "Use and Meaning", in Auxier & Hahn 2007: 531-57.

Heim, Irene and Kratzer, Angelika 1998: *Semantics in Generative Grammar*, Oxford: Blackwell.

Heyting, Arend 1930a: "Die formalen Regeln der intuitionistischen Mathematik II", *Sitzungsberichte der Preussischen Akademie der Wissenschaften*, 158-69.

Heyting, Arend 1930b: "Sur la logique intuitionniste", *Académie Royale de Belgique, Bulletin de la Classe des Sciences* 16, 957-63.

Heyting, Arend 1931: "Die intuitionistische Grundlegung der Mathematik", *Erkenntnis* 2 (1), 106-15.

Heyting, Arend 1934: *Mathematische Grundlagenforschung, Intuitionismus, Beweistheorie*, Berlin: Springer.

Heyting, Arend 1956: *Intuitionism. An Introduction*, 3rd edition 1971, Amsterdam: North-Holland.

Heyting, Arend (ed.) 1975: *L.E.J. Brouwer. Collected Works. Vol. 1*, Amsterdam: North-Holland.

Higginbotham, James 1992: "Truth and Understanding", *Philosophical Studies* 65 (1-2), 3-16.

Hinzen, Wolfram 2006: *Mind Design and Minimal Syntax*, Oxford: Oxford University Press.

Hodes, Harold 2004: "On the Sense and Reference of a Logical Constant", *The Philosophical Quarterly* 54 (214), 134-65.

Hoeltje, Miguel 2012: *Wahrheit, Bedeutung und Form*, Paderborn: Mentis.

Humberstone, Lloyd 2000: "The Revival of Rejective Negation", *Journal of Philosophical Logic* 29 (4), 331-81.

Humberstone, Lloyd 2011: *The Connectives*, Cambridge MA: MIT Press.

Kanamori, Akihiro 2004: "Zermelo and Set Theory", *Bulletin of Symbolic Logic* 10 (4), 487-553.

Kleene, Stephen Cole 1945: "On the Interpretation of Intuitionistic Number Theory", *Journal of Symbolic Logic* 10 (4), 109-24.

Kleene, Stephen Cole 1952: *Introduction to Metamathematics*, New York: van Nostrand.

Kölbel, Max 2001: "Two Dogmas of Davidsonian Semantics", *The Journal of Philosophy* 98 (12), 613-35.

Kolmogorov, Andrey N. 1932: "Zur Deutung der intuitionistischen Logik", *Mathematische Zeitschrift* 35, 58-65.

Kneale, William C. 1956: "The Province of Logic", in H. D. Lewis (ed.) 1956: *Contemporary British Philosophy: Third Series*, London: Allen and Unwin, 237-61.

Kratzer, Angelika 1986: "Conditionals", *Chicago Linguistics Society*, 22 (2), 1-15.

Kratzer, Angelika 1991: "Conditionals", in A. von Stechow and D. Wunderlich (eds.) 1991: *Semantik: Ein internationales Handbuch zeitgenössischer Forschung*, Berlin: de Gruyter, 651-6.

Kreisel, Georg 1962: "Foundations of Intuitionistic Logic", in E. Nagel et al. (eds.) 1962: *Logic, Methodology, and Philosophy of Science*, Stanford: Stanford University Press, 198-210.

Kripke, Saul 1972: *Naming and Necessity*, with a new preface (1980), Oxford: Blackwell.

Kripke, Saul 1982: *Wittgenstein on Rules and Private Language*, Cambridge MA: Harvard University Press.

Künne, Wolfgang 1997: "First Person Propositions", in W. Künne et al. (eds.) 1997: *Direct Reference, Indexicality, and Propositional Attitudes*, Stanford: CSLI Publications, 49-68.

Künne, Wolfgang 2003: *Conceptions of Truth*, Oxford: Oxford University Press.

Künne, Wolfgang 2008: "Replies to Commentators", *Dialectica* 62 (3), 385-401.

von Kutschera, Franz 1962: "Zum Deduktionsbegriff der klassischen Prädikatenlogik erster Stufe", in W. Britzelmayr et al. (eds.) 1962: *Logik und Logikkalkül*, Freiburg: Alber, 211-36.

Lambert, Karel 2001: "Free Logics", in L. Goble (ed.) 2001: *The Blackwell Guide to Philosophical Logic*, Oxford: Blackwell, 258-79.

Larson, Richard and Segal, Gabriel 1995: *Knowledge of Meaning*, Cambridge MA: MIT Press.

Lehmann, Scott 2002: "More Free Logic", in Gabbay & Guenthner 2002: 197-259.

Leitgeb, Hannes 2009: "Formal and Informal Provability", in O. Bueno and Ø. Linnebo (eds.) 2009: *New Waves in Philosophy of Mathematics*, London: Palgrave Macmillan, 263-99.

Lepore, Ernest and Loewer, Barry M. 1989: "Dual Aspect Semantics", in S. Silvers (ed.) 1989: *Rerepresentation. Readings in the Philosophy of Mental Representation*, Dordrecht: Kluwer, 161-88.

Lepore, Ernest and Ludwig, Kirk 2007: *Donald Davidson's Truth-Theoretic Semantics*, Oxford: Oxford University Press.

Lewis, David 1970: "General Semantics", *Synthese* 22 (1-2), 18-67.

Lewis, David 1975: "Adverbs of Quantification", in D. Lewis 1998: *Papers in Philosophical Logic*, Cambridge: University Press, 5-20.

Lievers, Menno 1998: "Two Versions of the Manifestation Argument", *Synthese* 115 (2), 199-227.

Ludwig, Kirk 2002: "What is the Role of a Truth Theory in a Meaning Theory", in J. K. Campbell et al. (eds.) 2002: *Meaning and Truth: Investigations in Philosophical Semantics*, New York: Seven Bridges Press, 142-63.

Mares, Edwin D. 2004: *Relevant Logic. A Philosophical Interpretation*, Cambridge: University Press.

Martin, Richard Milton 1958: *Truth and Denotation: A Study in Semantical Theory*, Chicago: University of Chicago Press.

McCarty, David Charles 1987: "Variations on a Thesis: Intuitionism and Computability", *Notre Dame Journal of Formal Logic* 28 (4), 536-80.

McCarty, David Charles 2005: "Intuitionism in Mathematics", in Shapiro 2005, 356-87.

McCarty, David Charles 2008a: "Intuitionism and Logical Syntax", *Philosophia Mathematica* 16 (1), 56-77.

McCarty, David Charles 2008b: "The New Intuitionism", in M. van Atten et al. (eds.) 2008: *One Hundred Years of Intuitionism (1907-2007)*, Berlin: Birkhäuser, 37-49.

McCarty, David Charles 2009: "Constructivism in Mathematics", in A. D. Irvine (ed.) 2009: *Philosophy of Mathematics. Handbook of The Philosophy of Science*, Amsterdam: Elsevier, 311-44.

McDowell, John Henry 1976: "Truth-conditions, Bivalence and Verificationism", in Evans & McDowell 1976: 42-66.

McDowell, John Henry 1981: "Anti-Realism and the Epistemology of Understanding", in H. Parret and J. Bouveresse (eds.) 1981: *Meaning and Understanding*, Berlin and New York: de Gruyter, 225-48.

McGinn, Colin 1980: "Truth and Use", in M. Platts (ed.) 1980: *Reference, Truth and Reality*, London: Routledge and Kegan Paul, 19-40.

McGuinness, Brian and Oliveri, Gianluigi (eds.) 1994: *The Philosophy of Michael Dummett*, Dordrecht: Kluwer.

Miller, Alexander 2002: "What is the Manifestation Argument?", *The Pacific Philosophical Quarterly* 83 (4), 352-83.

Milne, Peter 1994: "Classical Harmony: Rules of Inference and the Meaning of the Logical Constants", *Synthese* 100 (1), 49-94.

Milne, Peter 2002: "Harmony, Purity, Simplicity and a 'Seemingly Magical Fact' ", *The Monist* 85 (4), 498-535.

Montague, Richard 1970a: "English as a Formal Language", in B. Visentini et al. (eds.): *Linguaggi nella Societa et nella Technica*, Milan: Edizioni di Communita, 188-221; repr. in Thomason 1974.

Montague, Richard 1970b: "Universal Grammar", *Theoria* 36, 373-98; repr. in Thomason 1974.

Murzi, Julien 2012: "Manifestability and Epistemic Truth", *Topoi* 31 (1), 17-26.

Napoli, Ernesto 2006: "Negation", *Grazer Philosophische Studien* 72 (1), 233-52.

Pagin, Peter 1998: "Bivalence: Meaning Theory vs Metaphysics", *Theoria* 64 (2-3), 157-86.

Pagin, Peter 2008: "What is Communicative Success?", *The Canadian Journal of Philosophy* 38 (1), 85-115.

Pagin, Peter 2009a: "Intuitionism and the Anti-Justification of Bivalence", in S. Lindström et. al. (eds.) 2009: *Logicism, Intuitionism, and Formalism. What has Become of Them?*, Amsterdam: Springer, 221-36.

Pagin, Peter 2009b: "Compositionality, Understanding, and Proofs", *Mind* 118 (471), 713-37.

Pagin, Peter and Westerståhl, Dag 2010a: "Compositionality I: Definitions and Variants", *Philosophy Compass* 5 (3), 250-64.

Pagin, Peter and Westerståhl, Dag 2010b: "Compositionality II: Arguments and Problems", *Philosophy Compass* 5 (3), 265-82.

Peacocke, Christopher 1987: "Understanding the Logical Constants: A Realist's Account", *Proceedings of the British Academy* 73, 153-200.

Pettit, Dean 2002: "Why Knowledge is Unnecessary for Understanding Language", *Mind* 111 (443), 519-50.

Popper, Karl Raimund 1947: "New Foundations for Logic", *Mind*, 56, 193-235.

Popper, Karl Raimund 1948: "On the Theory of Deductions. II. The Definitions of Classical and Intuitionistic Negation", *Indagationes Mathematicae* 10, 111-20.

Posy, Carl J. 2007: "Free Logics", in Dov M. Gabbay and J. Woods (eds.) 2007: *Handbook of the History of Logic. Vol. 8*, Amsterdam: Elsevier, 633-80.

Prawitz, Dag 1965: *Natural Deduction: A Proof-Theoretical Study*, Stockholm: Almqvist and Wiksell.

Prawitz, Dag 1973: "Towards a Foundation of a General Proof Theory", in P. Suppes et al. (eds.) 1973: *Logic, Methodology, and Philosophy of Science IV*, Amsterdam: North-Holland, 225-50.

Prawitz, Dag 1977: "Meaning and Proof: On the Conflict Between Classical and Intuitionistic Logic", *Theoria* 43 (1), 2-40.

Prawitz, Dag 1994: "Meaning Theory and Anti-Realism", in McGuinness & Oliveri 1994: 79-89.

Prawitz, Dag 1998: "Comments on the Papers", *Theoria* 64 (2-3), 283-337.

Prawitz, Dag 2009: "Inference and Knowledge", in M. Pelis (ed.) 2009: *The Logica Yearbook 2008*, London: College Publications, 175-92.

Prawitz, Dag 2012a: "The Epistemic Significance of Valid Inference", *Synthese* 187 (3), 887-898.

Prawitz, Dag 2012b: "Truth as an Epistemic Norm", *Topoi* 31 (1), 9-16.

Prawitz, Dag (unpublished): "Truth and Proof: Ontological or Epistemic Concepts?", unpublished draft.

Putnam, Hilary 1975: "The Meaning of 'Meaning' ", *Minnesota Studies in the Philosophy of Science* **7**, 131-93.

Quine, Willard van Orman 1940: *Mathematical Logic*, Harvard: Harvard University Press.

Quine, Willard van Orman 1970: *The Philosophy of Logic*, Cambridge MA: Harvard University Press.

Ramsey, Frank Plumpton 1927: "Facts and Propositions", *Proceedings of the Aristotelian Society (Suppl. Vol.)* 7 (1), 153-70.

Rayo, Agustin and Yablo, Stephen 2001: "Nominalism Through de-Nominalization", *Noûs* **35** (1), 74-92.

Read, Stephen 1988: *Relevant Logic: A Philosophical Examination of Inference*, Oxford: Blackwell.

Read, Stephen 2000: "Harmony and Autonomy in Classical Logic", *Journal of Philosophical Logic* 29 (2), 123-54.

Read, Stephen 2010: "General-Elimination Harmony and the Meaning of the Logical Constants", *Journal of Philosophical Logic* 39 (5), 557-76.

Resnik, Michael D. 1985: "Logic: Normative or Descriptive? The Ethics of Belief or a Branch of Psychology?", *Philosophy of Science* 52 (2), 221-238.

Restall, Greg 2005: "Multiple Conclusions", in P. Hajek et al. (eds.) 2005: *Logic, Methodology, and Philosophy of Science: Proceedings of the Twelfth International Congress*, London: King's College Publications, 189-205.

Routley, Richard, Plumwood, Val, Meyer, Robert K., and Brady, Ross T. 1982: *Relevant Logics and Their Rivals: Part 1. The Basic Philosophical and Semantical Theory*, Atascadero: Ridgeview.

Rumfitt, Ian 2000: " 'Yes' and 'No' ", *Mind* 109 (436), 781-823.

Rumfitt, Ian 2007: "Asserting and Excluding: Steps Towards an Anti-Realist Account of Classical Consequence", in Auxier and Hahn 2007, 639-93.

Rumfitt, Ian 2008: "Knowledge by Deduction", *Grazer Philosophische Studien* 77 (1), 61-84.

Russell, Bertrand and Whitehead, Alfred North 1910: *Principia Mathematica. Volume 1*, Cambridge: Cambridge University Press.

Sainsbury, Mark 2002: *Departing from Frege. Essays in the Philosophy of Language*, London: Routledge.

Sainsbury, Mark 2005: *Reference without Referents*, Oxford: Oxford University Press.

Salerno, Joseph 2009: "Knowability Noir: 1945-1963", in J. Salerno (ed.) 2009: *New Essays on the Knowability Paradox*, Oxford: Oxford University Press, 29-48.

Ščedrov, Andre 1985: "Intuitionistic Set Theory", in L. A. Garrubgtib et al. (eds.) 1985: *Harvey Friedman's Research on the Foundations of Mathematics*, Amsterdam: Elsevier, 257-84.

Schechter, Joshua 2011: "Juxtaposition: A New Way to Combine Logics", *The Review of Symbolic Logic* 4 (4), 560-606.

Schiffer, Stephen 2003: *The Things We Mean*, Oxford: Oxford University Press.

Schroeder-Heister, Peter 1984: "A Natural Extension of Natural Deduction", *The Journal of Symbolic Logic* 49 (4), 1284-1300.

Shapiro, Stewart (ed.) 2005: *The Oxford Handbook of Philosophy of Mathematics and Logic*, Oxford: Oxford University Press.

Shieh, Sanford 1998: "On the Conceptual Foundations of Anti-Realism", *Synthese* 115 (1), 33-70.

Shoesmith, D. J. and Smiley, Timothy John 1978: *Multiple-Conclusion Logic*, Cambridge: Cambridge University Press.

Slater, Barry Hartley 2008: "Harmonising Natural Deduction", *Synthese* 163 (2), 187-98.

Smiley, Timothy John 1996: "Rejection", *Analysis* 56 (1), 1-9.

Soames, Scott 2008: "Truth and Meaning: in Perspective", *Midwest Studies in Philosophy* 32 (1), 1-19.

Steinberger, Florian 2011: "Why Conclusions Should Remain Single", *Journal of Philosophical Logic* 40 (3), 333-55.

Strawson, Peter Frederick 1974: "On Understanding the Structure of One's Language", in P. F. Strawson 1974: *Freedom and Resentment and Other Essays*, London: Methuen, 198-207.

Sundholm, Göran 1983: "Constructions, Proofs and the Meaning of Logical Constants", *Journal of Philosophical Logic* 12 (2), 151-72.

Sundholm, Göran 1994: "Existence, Proof and Truth-Making: A Perspective on the Intuitionistic Conception of Truth", *Topoi* 13 (2), 117-26.

Sundholm, Göran 1997: "Implicit Epistemic Aspects of Constructive Logic", *Journal of Logic, Language, and Information* 6 (2), 191-212.

Sundholm, Göran 1998: "Proofs as Acts and Proofs as Objects: Some Questions for Dag Prawitz", *Theoria* 64 (2-3), 187-216.

Sundholm, Göran 2004: "Antirealism and the Roles of Truth", in I. Niiniluoto et al. (eds.) 2004: *Handbook of Epsitemology*, Dordrecht: Kluwer, 437-66.

Sundholm, Göran 2008: "Summa de Veritate Hamburgensis: Truth According to Wolfgang Künne, *Dialectica* 62 (3), 359-71.

de Swart, Harrie C. M. 1992: "Spreads or Choice Sequences?", *History and Philosophy of Logic* 13 (2), 203-13.

Szabo, M. E. (ed.) 1969: *The Collected Papers of Gerhard Gentzen*, Amsterdam: North-Holland.

Tarski, Alfred 1931: "The Concept of Truth in Formalized Languages", English translation in Tarski 1956: 52-267.

Tarski, Alfred 1936: "On the Concept of Logical Consequence", in Tarski 1956: 407-20.

Tarski, Alfred 1956: *Logic, Semantics, Metamathematics*, Oxford: Clarendon Press.

Tennant, Neil 1987: *Anti-Realism and Logic*, Oxford: Oxford University Press.

Tennant, Neil 1997: *The Taming of the True*, Oxford: Oxford University Press.

Tennant, Neil 1999: "Negation, Absurdity and Contrariness", in D. Gabbay and H. Wansing (eds.) 1999: *What is Negation?*, Dordrecht: Kluwer, 199-222.

Thomason, Richmond H. (ed.) 1974: *Formal Philosophy: Selected Papers of Richard Montague*, New Haven: Yale University Press.

Troelstra, Anna Sjerp 1973: "Notes on Intuitionistic Second-Order Arithmetic", in A. Mathias and H. Rogers (eds.) 1973: *Cambridge Summer School in Mathematical Logic*, Berlin: Springer, 171-205.

Troelstra, Anna Sjerp 1977: "Aspects of Constructive Mathematics", in J. Barwise (ed.) 1977: *Handbook of Mathematical Logic*, Amsterdam: North-Holland, 973-1052.

Troelstra, Anna Sjerp and van Dalen, Dirk 1988: *Constructivism in Mathematics. An Introduction. Vol. 1*, Amsterdam: Elsevier.

Troelstra, Anna Sjerp and Schwichtenberg, Helmut 2000: *Basic Proof Theory*, 2nd edition, Cambridge: Cambridge University Press.

Weir, Alan 1985: "Rejoinder to Tennant", *Analysis* 45 (2), 68-72.

Wiggins, David 1997: "Meaning and Truth Conditions: From Frege's Grand Design to Davidson's", in B. Hale and C. Wright (eds.) 1997: *A Companion to the Philosophy of Language*, Oxford: Blackwell, 3-28.

Williamson, Timothy 1982: "Intuitionism Disproved?", *Analysis* 42 (4), 203-7.

Williamson, Timothy 1994: *Vagueness*, London: Routledge.

Williamson, Timothy 2000: *Knowledge and Its Limits*, Oxford: Oxford University Press.

Williamson, Timothy 2006: "Conceptual Truth", *Aristotelian Society (Suppl. Vol.)* 80 (1), 1-41.

Williamson, Timothy 2007a: *The Philosophy of Philosophy*, Oxford: Wiley-Blackwell.

Williamson, Timothy 2007b: "Absolute Identity and Absolute Generality", in G. Uzquiano and A. Rayo (eds.) 2007: *Absolute Generality*, Oxford: Oxford University Press, 369-89.

Wright, Crispin 1980: "Realism, Truth-Value Links, Other Minds and the Past", in Wright 1993: 85-106.

Wright, Crispin 1993: *Realism, Meaning and Truth*, 2nd edition, Oxford: Basil Blackwell.

Index